IDENTITIES

FUNDAMENTAL IDENTITIES

1. $\sec\theta = \dfrac{1}{\cos\theta}$ **2.** $\csc\theta = \dfrac{1}{\sin\theta}$

3. $\cot\theta = \dfrac{1}{\tan\theta}$ **4.** $\tan\theta = \dfrac{\sin\theta}{\cos\theta}$

5. $\cot\theta = \dfrac{\cos\theta}{\sin\theta}$ **6.** $\cos^2\theta + \sin^2\theta = 1$

7. $1 + \tan^2\theta = \sec^2\theta$

8. $1 + \cot^2\theta = \csc^2\theta$

COFUNCTION IDENTITIES

9. $\cos\left(\dfrac{\pi}{2} - \theta\right) = \sin\theta$

10. $\cos\left(\dfrac{\pi}{2} - \theta\right) = \sin\theta$

11. $\cos\left(\dfrac{\pi}{2} - \theta\right) = \sin\theta$

OPPOSITE–ANGLE IDENTITIES

12. $\cos(-\theta) = \cos\theta$

13. $\sin(-\theta) = -\sin\theta$

14. $\tan(-\theta) = -\tan\theta$

ADDITION LAWS

15. $\cos(\alpha + \beta) = \cos\alpha\cos\beta - \sin\alpha\sin\beta$

16. $\cos(\alpha - \beta) = \cos\alpha\cos\beta + \sin\alpha\sin\beta$

17. $\sin(\alpha + \beta) = \sin\alpha\cos\beta + \cos\alpha\sin\beta$

18. $\sin(\alpha - \beta) = \sin\alpha\cos\beta - \cos\alpha\sin\beta$

19. $\tan(\alpha + \beta) = \dfrac{\tan\alpha + \tan\beta}{1 - \tan\alpha\tan\beta}$

20. $\tan(\alpha - \beta) = \dfrac{\tan\alpha - \tan\beta}{1 + \tan\alpha\tan\beta}$

DOUBLE–ANGLE IDENTITIES

21. $\cos 2\theta = \cos^2\theta - \sin^2\theta$
$= 2\cos^2\theta - 1$
$= 1 - 2\sin^2\theta$

22. $\sin 2\theta = 2\sin\theta\cos\theta$

23. $\tan 2\theta = \dfrac{2\tan\theta}{1 - \tan^2\theta}$

HALF-ANGLE IDENITIES

24. $\cos\dfrac{1}{2}\theta = \pm\sqrt{\dfrac{1 + \cos\theta}{2}}$

25. $\sin\dfrac{1}{2}\theta = \pm\sqrt{\dfrac{1 - \cos\theta}{2}}$

26. $\tan\dfrac{1}{2}\theta = \dfrac{1 - \cos\theta}{\sin\theta} = \dfrac{\sin\theta}{1 + \cos\theta}$

PRODUCT–TO–SUM IDENTITIES

27. $2\cos\alpha\cos\beta = \cos(\alpha - \beta) + \cos(\alpha + \beta)$

28. $2\sin\alpha\sin\beta = \cos(\alpha - \beta) - \cos(\alpha + \beta)$

28. $2\sin\alpha\cos\beta = \sin(\alpha + \beta) + \sin(\alpha - \beta)$

30. $2\cos\alpha\sin\beta = \sin(\alpha + \beta) - \sin(\alpha - \beta)$

SUM–TO–PRODUCT IDENTITIES

31. $\cos\alpha + \cos\beta = 2\cos\left(\dfrac{\alpha + \beta}{2}\right)\cos\left(\dfrac{\alpha - \beta}{2}\right)$

32. $\cos\alpha - \cos\beta = -2\sin\left(\dfrac{\alpha + \beta}{2}\right)\sin\left(\dfrac{\alpha - \beta}{2}\right)$

33. $\sin\alpha + \sin\beta = 2\sin\left(\dfrac{\alpha + \beta}{2}\right)\cos\left(\dfrac{\alpha - \beta}{2}\right)$

34. $\sin\alpha - \sin\beta = 2\sin\left(\dfrac{\alpha - \beta}{2}\right)\cos\left(\dfrac{\alpha + \beta}{2}\right)$

INVERSE IDENTITIES

35. $\operatorname{Arccot} x = \begin{cases} \operatorname{Arctan}(1/x) & \text{if } x > 0 \\ \pi/2 & \text{if } x = 0 \\ \pi + \operatorname{Arctan}(1/x) & \text{if } x < 0 \end{cases}$

36. $\operatorname{Arcsec} x = \operatorname{Arccos}(1/x)$ if $x \geqslant 1$ or $x \leqslant -1$

37. $\operatorname{Arccsc} x = \operatorname{Arcsin}(1/x)$ if $x \geqslant 1$ or $x \leqslant -1$

REDUCTION IDENTITY

38. $a\sin\theta + b\cos\theta = \sqrt{a^2 + b^2}\,\sin(\theta + \alpha)$,
where α is chosen so that

$$\cos\alpha = \dfrac{a}{\sqrt{a^2 + b^2}} \quad \text{and} \quad \sin\alpha = \dfrac{b}{\sqrt{a^2 + b^2}}$$

TRIGONOMETRY
FOR COLLEGE STUDENTS

SIXTH EDITION

Karl Smith has gained a reputation as a master teacher and guest lecturer. He has been a department chairperson and is past president of the American Mathematical Association of Two-Year Colleges. He has served on numerous boards, including the Conference Board of Mathematical Sciences (which serves as a liaison among the National Academy of Sciences, the federal government, and the mathematical community in the United States), the Council of Scientific Society Presidents, the Board of the California Mathematics Council for Community Colleges, and was invited to participate in the National Summit of Mathematics Education. He received his B.A. and M.A. from the University of California at Los Angeles, and his Ph.D. from Southeastern University. In his spare time he enjoys running and swimming.

The Precalculus Series
by Karl J. Smith

Precalculus with Graphing and Problem Solving
Fifth Edition

Trigonometry for College Students
Sixth Edition

College Algebra

The Smith Business Series

Finite Mathematics
Third Edition

Calculus with Applications
Second Edition

College Mathematics
Second Edition

The Brooks/Cole One-Unit Series
by Karl J. Smith

Analytic Geometry
Second Edition

Symbolic Logic
Second Edition

Problem Solving

Geometry

Other Brooks/Cole Titles
by Karl J. Smith

The Nature of Mathematics
Sixth Edition

Mathematics: Its Power and Utility
Third Edition

Essentials of Trigonometry
Second Edition

Beginning Algebra for College Students
Fourth Edition
(with Patrick J. Boyle)

College Algebra
Fourth Edition
(with Patrick J. Boyle)

Primer for College Algebra
(with Patrick J. Boyle)

Study Guide for Algebra
(with Patrick J. Boyle)

TRIGONOMETRY
FOR COLLEGE STUDENTS

SIXTH EDITION

Karl J. Smith, Ph.D

Brooks/Cole Publishing Company
Pacific Grove, California

I⊤P ™ The trademark ITP is used under license.

Brooks/Cole Publishing Company
A Division of Wadsworth, Inc.
© 1994 by Wadsworth, Inc., Belmont, California 94002.

Printed in the United States of America

10   9   8   8   7   6   5   4   3   2   1

Library of Congress Cataloging-in-Publication Data

Smith, Karl J.
Trigonometry for college students / Karl J. Smith.—6th ed.
Includes index.
ISBN 0-534-16788-8
1. Trigonometry.   I. Title.
QA531.S653 1994   93–30497
516.24—dc20   CIP

Publisher ▮ Gary W. Ostedt
Editorial Associate ▮ Carol Ann Benedict
Production Services Manager ▮ Joan Marsh
Production Services ▮ TECH*arts*, Susan Reiland
Production Editor ▮ Stephanie Kuhns
Manuscript Editor ▮ Christine Levesque
Interior Design ▮ Kathi Townes, TECH*arts*
Cover Design ▮ Cheryl Carrington
Cover Photo ▮ TIB/West, Douglas Struthers
Interior Illustration ▮ TECH*arts;* Lori Heckelman; Reese Thornton
Photo Researchers ▮ Stuart Kenter; Mary Kay Hancharick
Typesetting ▮ Weimer Graphics
Indexer ▮ Helen Berliner
Cover Printing ▮ Phoenix Color Corporation
Printing and Binding ▮ R.R. Donnelley & Sons Co.

This book is dedicated to my son, Shannon Joseph Smith

Shannon is a student at the University of California, Davis. He loves people and in addition to his studies is working as an Emergency Medical Technician. His avocation is contemporary Christian music, and he not only performs but writes and composes. I am very proud of him, and dedicate this book to him with love.

PREFACE

What Is Trigonometry?

Trigonometry is usually considered to be the transition course between high school algebra and calculus. It changes the focus from the development of manipulative skills to the understanding of mathematical concepts. Quite simply, six functions, called the *trigonometric functions* are defined in the first chapter, and then the remainder of the course is devoted to investigating and understanding these functions, their graphs, their relationships to one another, and ways in which they can be used in a variety of applications.

About This Book

This book was written for a two- or three-credit college course in trigonometry. The prerequisite is a second-year high school algebra course or equivalent. Even though a course in geometry is also a desirable prerequisite, I have found that more and more students are taking trigonometry without having taken high school geometry; consequently, in addition to a review of geometry in Appendix C, I've included in the text all of the necessary ideas from geometry as they are needed.

To the Student

Each new course begins with a sense of anticipation, and this course is probably no exception. You are enrolling in a course and your instructor has adopted this textbook. What can you expect? You will learn about the *trigonometric functions* and their properties. As you read the description of trigonometry in the first paragraph, you see that the course depends on one important definition. If you have had a right-triangle definition of these functions in a previous course, you will notice that the definition in Chapter 1 is much more general (we will introduce the right-triangle definition in Chapter 5). It is important that you thoroughly understand this definition as it is discussed in Chapter 1.

You will find that the course will be much easier if you *read the textbook prior to discussing it in class*. Do not assume you are finished with the day's work as soon as you have finished the day's homework assignment. Read, study, and *understand* the important ideas. There will be some things you will need to *memorize*. I have tried to make it easy for you to know what you need to do. Definitions and formulas are enclosed in boxes; new terms are shown in boldface and are again listed at the end of each chapter. Each chapter begins with an overview and ends with a list of objectives, so you should know exactly what you need to be able to do. In addition to the answers to the odd-numbered problems at the back of the book, I have included the answers to all the problems in the sample tests and cumulative tests to help you prepare for examinations.

Use your book and make it you own. You probably were shocked by the price of this book, but Brooks/Cole has done everything possible to keep down the cost. Nevertheless, books are expensive, and I hope you see the money you pay for this book as an investment that will pay off not only for this course, but also as a valuable reference book in your future mathematics courses. If you intend to continue your study in mathematics, you will want to make this book part of your professional library so that you will have access to the notation and formulas developed in this course.

To the Instructor

▌ Overview

Trigonometry for College Students, Sixth Edition, defines the trigonometric functions on a unit circle in the third section of the book and then almost immediately generalizes this definition in terms of any point on the terminal side of the angle. You will also notice that radian measure is defined in the second section and then the book emphasizes working with radian measure of angles as real numbers. I have found that most students have had an introduction to degree measure and an early introduction of radian and real numbers builds a solid foundation for more advanced mathematics courses.

If you are familiar with the fifth edition, then you will notice that the primary change in this edition is the inclusion of **Calculator Comments** to give optional advice about graphing calculators. In particular, I have found the Texas Instruments TI-81 and TI-82 calculators particularly easy to use. I have deleted the specific keystrokes using algebraic and RPN logic because the new generation of calculators is using a logic that follows the algebraic notation exactly. This means that there is a great deal more variety in the algorithms used by students taking trigonometry. Therefore, I have changed the calculator descriptions to be generic rather then specific.

Pay particular attention to the changes made in the introduction of the inverse trigonometric functions. I have not written this material to be familiar and comfortable for you, the instructor, but have written it to overcome one of the most difficult stumbling blocks for most students seeing this topic for the first time. I have found that *students perform best* when they are **first** introduced to the concept of equation solving (using a calculator with graphical reinforcement), and **then** are introduced to the concept of an inverse trigonometric function. As an instructor, and in the first several editions of this book, my mathematics background said that the concept (inverse functions) should be presented first, followed by the concrete (solving equations). However, I noticed a great many students leaving the course without a firm foundation on either inverse functions or solving trigonometric equations. After teaching the concrete first (solving equations) and then the abstract (inverse functions), I have found that my students not only perform better on examinations, but also seem to have a much better grasp and understanding of the *concept* of inverse function. Do not judge the order of my presentation on what you have done in the past without first giving the presentation in this text a trial.

Another new feature of this addition is the inclusion of *Discussion Problems*. These are included in every chapter review and have been added to many of the exercise sets. I believe that students should be given the opportunity to write and communicate mathematics. In order to prepare our students for the 21st century, it is not enough that they be able to "solve problems and get answers," but it is also important that their level of understanding be sufficient for them to communicate their knowledge. They also need to be introduced to library research in order to answer questions. The *Applications for Further Study* can be used as group research projects.

If you are not familiar with prior editions of this book, I would like you to look at the following features:

▮ **Chapter 1**   The emphasis is on understanding that we are indeed dealing with *functions* of an angle, which can be measured in degree, radians, or real numbers.

▮ **Chapter 2**   Graphing is introduced early in my text and I give it special emphasis. You will find an innovative and unique method, called *framing*; with this technique you can quickly and efficiently communicate the concepts of amplitude, period, and phase displacement without the usual difficulties of plotting "ugly coordinates." I'm convinced that by using this technique your students will have a better understanding of graphing more complicated equations—with less class time. It also changes the focus of graphing from what it has been in the past (plotting points) to what it needs to be in the 1990s. Today's (and certainly tomorrow's) technology makes it easier to graph the trigonometric functions *if* we know how to appropriately choose the scale and scaling factors. My framing technique emphasizes the choice of the scaling numbers.

▮ **Chapter 3**   One of the core topics for any college trigonometry course is proving identities. You will find a multitude of graded identity problems in this chapter, as well as a summary of all identities listed inside the front cover of the book. Notice that **two** entire sections are devoted to using the eight fundamental identities to prove other identities, as well as one section devoted to recognizing when a given equation is not an identity.

▮ **Chapter 4**   One of the topics that often distinguishes one trigonometry book from another is the treatment of inverse trigonometric functions and solving identities. These topics are often cluttered and I have taken great care in separating the solving of equations with inverse relations from the limitations necessary for the inverse trigonometric functions for the students.

▮ **Chapter 5**   The other core topic for a trigonometric course is solving triangles. I have emphasized concepts rather than mechanics and thus have taken time to help the student formulate which method to choose. I think that the immediate introduction to the law of cosines eliminates the confusion that sometimes arises with the law of sines and its ambiguous case. The law of cosines eliminates the need to even introduce the ambiguous case, since the quadratic formula leads directly to the proper analysis of the problem. In other words, the ambiguous case is a problem in geometry and for the law of sines, but presents no problem when the law of cosines is used. If you prefer to solve right triangles as an introductory topic, you might look at my other trigonometric text, *Essentials of Trigonometry, Second Edition*.

Chapters 6, 7, and 8 are optional.

▌ **Chapter 6** An entire chapter is devoted to the complex numbers and De Moivre's Formula. I have found that students often confuse complex numbers and polar forms, so in this book I've separated these ideas into different chapters.

▌ **Chapter 7** You will find one of the most complete introductions to polar-form curves in this chapter. I was able to devote more time to this topic by separating Chapters 6 and 7 into two chapters.

▌ **Chapter 8** In many schools, trigonometry is the last precalculus course, and in such cases it is desirable to introduce conic sections prior to beginning calculus. I have included this chapter for those schools that find it worthwhile not only to review the conic sections, but also to apply trigonometry by doing rotations of the conics and conics in polar form.

▌ **Problems**

I provide abundant examples and problems for both drill and theory. The problems are divided into A, B, and C categories by order of difficulty, and in matched pairs so one problem has an answer and the other does not. The student is generally required to work at the A and B level, whereas C problems are designed for an honors class or to present a challenge to most students.

In making up assignments you need to have some idea about the problems before you make the assignment, and you do not want any surprises with a difficult problem embedded within a list of simple problems. For this reason, I've taken care in developing a *standard mathematics assignment*. I've written the problem sets so that all are about the same length and, by using a, b, and c parts for particular problems, I have distributed different types of problems and difficulty levels within each problem set in basically the same way. This means, for example, that you could assign the same problems day after day and have problem sets with consistent length and difficulty. For example, when I teach from this book I assign Problems 3–30, multiples of 3, so that I am sure the student will have a uniform assignment each day and will have practice with all of the important concepts. Because some instructors like to make spiral assignments, I have designed the problems so that they are also easy to assign; for example, problems 5–30, multiples of 5 of the current sections, and Problems 7, 14, 21, and 28 of the previous section.

TYPES OF PROBLEMS:

▌ **A and B Problems** Matched pairs; both drill and applied.

▌ **C Problems** Challenging problems.

▌ **Discussion Problems** The ability to describe, explain, or otherwise discuss a concept is an important skill to develop. For this reason discussion problems are included throughout the book.

▌ **Applications for Further Study**   These problems are included to illustrate some of the useful applications at length.

▌ **Historical Questions**   These problems are included to put trigonometry in its historical perspective.

▌ **Calculator and Computer Problems**   Today's technology cannot be ignored, nor can it be forced. It is assumed that all students have inexpensive calculators, but the more advanced graphics calculators and computers are treated as optional.

▌ **Supplementary Materials**

▌ *Answer Book* provides the answers to all the problems in the text.

▌ *EXPTest Question Bank* provides approximately 1,000 electronically accessible test items in IBM format.

▌ *Solutions and Study Guide* provides detailed step-by-step solutions to selected odd-numbered problems, and section-by-section hints and suggestions.

▌ *TEMATH: A Tool for Exploring Mathematics* by Robert Kowalczyk and Adam Haufknecht provides graphing technology for the Macintosh and is available from Brooks/Cole.

▌ *Converge* by John R. Mowbray provides graphing technology for the IBM and is available from JEMware, 567 South King Street, Suite 178, Honolulu, Hawaii, 96813 (phone, 808 / 523-9911).

Acknowledgments

I am grateful to the following persons who reviewed the manuscript for this edition:

Carol Benson, *Illinois State University*
Patricia Crawley, *Wichita State University*
Lynn R. Eisenberg, *Durham Tech Community College*
Patricia J. Ernst, *St. Cloud State University*
Phyllis Faw, *Oklahoma City College*
Charlotte M. Grossbeck, *State University of New York–Cobleskill*
Shirley Howard, *Okaloosa-Walton Community College*
Twila Dunavant Miller, *Blinn College*
Keith Shuert, *Oakland Community College–Royal Oak Campus*
Norwood Taylor, *Sandhills Community College*

I would also like to thank the reviewers of the previous editions:

Donald Adlong, *University of Central Arkansas*
Susan Barker, *Atlantic Community College*
Roger Breen, *Florida Junior College*
P. D. Briggs, *East Central University*
Wayne Britt, *Louisiana State University*
Dawson Carr, *Sandhills Community College*
Louis Chatterly, *Brigham Young University*

Grace De Velbiss, *Sinclair Community College*
Arthur Dull, *Diablo Valley College*
Marjorie Freeman, *University of Houston Downtown College*
Charles R. Friese, *North Dakota State University*
Carolyn Funk, *Thornton Community College*
Gerald Goff, *Oklahoma State University*
Paul W. Haggard, *East Carolina University*
Franz X. Heirgeist, *West Virginia University*
Louis F. Hoelzle, *Bucks County Community College*
Libby Holt, *Florida Community College at Jacksonville*
Stephen Johnson, *Linn-Benton Community College*
Paula Kemp, *Southwest Missouri State University*
Elton Lacey, *Texas A & M University*
Pamela Littleton, *Tarleton State University*
Michael Morgan, *Linn-Benton Community College*
Mustapha Munem, *Macomb County Community College*
Carla B. Oviatt, *Montgomery College*
Charlene Pappin, *Wright College*
William L. Perry, *Texas A & M University*
Beverly Rich, *Illinois State University*
Dick Schulz, *Charles County Community College*
Edwin Schulz, *Texas Tech University*
Thomas Seremet, *Charles County Community College*
Richard Slinkman, *Bemidji State University*
V. Merriline Smith, *California State Polytechnic University*
John Spellmann, *Southwest Texas State University*
Monty Strauss, *Texas Tech University*
Dorothy Sulock, *University of North Carolina*
Donna Szott, *Community College of Allegheny County*
Robert J. Wisner, *New Mexico State University*
Karen Zobrist, *Louisiana State University*

KARL J. SMITH

CONTENTS

Nicholas Copernicus

1 **Trigonometric Functions**   3

1.1   Angles and Degree Measure   4
1.2   Radian Measure of Angles   8
1.3   Trigonometric Functions on a Unit Circle   17
1.4   Trigonometric Functions of Any Angle   27
1.5   Trigonometric Functions of Real Numbers   32
1.6   Summary and Review   38

APPLICATION FOR FURTHER STUDY
   ▌ Angular and Linear Velocity—From Windmills to Pistons   42

Hipparchus

2 **Graphs of Trigonometric Functions**   49

2.1   Graphs of Cosine, Sine, and Tangent Functions   50
2.2   General Cosine, Sine, and Tangent Curves   59
2.3   Graphs of Secant, Cosecant, and Cotangent Functions   69
2.4   Addition to Ordinates   74
2.5   Summary and Review   79

APPLICATION FOR FURTHER STUDY
   ▌ Computer Graphing and Trigonometric Curves   82

CUMULATIVE REVIEW ▌ Chapters 1–2   89

Karl Gauss

3 **Fundamental Trigonometric Identities**   93

3.1   Proving Identities   94
3.2   Proving Identities (Continued)   99
3.3   Disproving Identities   102
3.4   Cofunction and Opposite-Angle Identities   105
3.5   Addition Laws   112
3.6   Double-Angle and Half-Angle Identities   116
3.7   Product and Sum Identities   124
3.8   Summary and Review   127

APPLICATION FOR FURTHER STUDY

 Sunrise, Sunset, and Linear Interpolation   131

4 **Trigonometric Equations and Inverse Functions**   135

4.1 Introduction to Trigonometric Equations   136
4.2 Trigonometric Equations with Multiple Angles   147
4.3 Inverse Trigonometric Functions   152
4.4 A Reduction Identity   161
4.5 Summary and Review   167

APPLICATION FOR FURTHER STUDY

Waves, Harmonic Motion, and Resonance   167

CUMULATIVE REVIEW ▮ Chapters 3–4   175

5 **Solving Triangles**   179

5.1 Right Triangles   180
5.2 Law of Cosines   189
5.3 Law of Sines   197
5.4 Areas and Volumes   205
5.5 Vector Triangles   209
5.6 Vector Operations   216
5.7 Summary and Review   223

APPLICATION FOR FURTHER STUDY

Parking Lot Design   228

6 **Complex Numbers**   231

6.1 The Imaginary Unit and Complex Numbers   232
6.2 Trigonometric Form of Complex Numbers   236
6.3 De Moivre's Formula   241
6.4 Summary and Review   247

APPLICATION FOR FURTHER STUDY

Electric Generator   250

Leonard Euler

Benjamin Banneker

Abraham De Moivre

Charlotte Angas Scott

7 **Polar Curves** 253

7.1 Polar Coordinates 254
7.2 Graphing Polar-Form Equations 259
7.3 Special Polar-Form Equations 263
7.4 Parametric Equations 269
7.5 Intersection of Polar-Form Curves 27
7.6 Summary and Review 282

APPLICATION FOR FURTHER STUDY
▌ Communication of Bees 285

CUMULATIVE REVIEW ▌ Chapters 5–7 287

8 **Conic Sections** 291

8.1 Parabolas 292
8.2 Ellipses 298
8.3 Hyperbolas 306
8.4 General Conic Sections 312
8.5 Rotations 317
8.6 Conic Sections in Polar Form 324
8.7 Summary and Review 330

APPLICATION FOR FURTHER STUDY
▌ Planetary Orbits 333

Appendixes 338
A Calculators 338
B Accuracy and Rounding 342
C Review of Geometry 346
D Selected Answers 350

Index 379

René Descartes

TRIGONOMETRY
FOR COLLEGE STUDENTS

SIXTH EDITION

Nicholas Copernicus
(1473–1543)

■ Mathematics is a type of thought which seems ingrained in the human mind, which manifests itself to some extent with even the primitive races, and which is developed to a high degree with the growth of civilization .... A type of thought, a body of results, so essentially characteristic of the human mind, so little influenced by environment, so uniformly present in every civilization, is one of which no well-informed mind today can be ignorant.

J. W. A. YOUNG
The Teaching of Mathematics
(London, 1907), p.14

The origins of trigonometry are obscure, but we do know that it began more than 2,000 years ago with the Mesopotamian, Babylonian, and Egyptian civilizations. The word *trigonometry* comes from the words "triangle" (*trigon*) and "measurement" (*metry*). Some of the early names associated with trigonometry are Ptolemy, known as Claudius Ptolemy, Hipparchus, and Menelaus. Their goal was to build a quantitative astronomy that could be used to predict the paths and positions of the heavenly bodies, and to aid in telling time, calendar reckoning, navigating, and studying geography.

Nicholas Copernicus is probably best known as the astronomer who revolutionized the world with his heliocentric theory of the universe, but in his book *De revolutionibus orbium coelestium* he also developed a substantial amount of

trigonometry. This book was published in the year of his death; as a matter of fact, the first copy off the press was rushed to him as he lay on his deathbed. It was upon Copernicus' work that his student Rheticus based his ideas, which soon brought trigonometry into full use. In a two-volume work, *Opus palatinum de triangulis*, Rheticus used and calculated elaborate tables for all six trigonometric functions.

Many "forgotten" women astronomers did much of the work in filling in the details for the elaborate calculations of planetary orbits. Marie Cunitz was the first woman to attempt to correct Kepler's tables of planetary motion. Her work *Urania Propitia* displayed a nice example of role reversal when she acknowledged in the preface the able assistance of her husband.

1

TRIGONOMETRIC FUNCTIONS

Contents

1.1 Angles and Degree Measure

1.2 Radian Measure of Angles

1.3 Trigonometric Functions on a Unit Circle

1.4 Trigonometric Functions of Any Angle

1.5 Trigonometric Functions of Real Numbers

1.6 Chapter 1 Summary and Review

Application for Further Study
   Angular and Linear Velocity—From Windmills to Pistons

Preview

This chapter introduces you to the basic ideas of trigonometry. We begin by talking about how to measure angles using both degrees and radians, and after defining the trigonometric functions, we spend the rest of the chapter talking about the evaluation of those functions.

Perspective

Trigonometric functions are tools in mathematics, just as are linear or quadratic functions. It is not possible to seriously study mathematics, or many of the physical sciences, without a knowledge of trigonometry. For example, in a study of the flow of traffic by Michael J. Lightthill and Gerald B. Whitham of the University of Manchester, trigo-

Progress of car in denser traffic

Progress of shockwave

Progress of car in lighter traffic

Early ⟶ Late   Early ⟶ Late

From "How to analyze the shock waves that sweep through expressway traffic," by Jearl Walker, *Scientific American*, August 1989, p. 98.

nometry played a role in describing what they called "shock waves." Other applications of trigonometry are as varied as wave motion in physics and measuring the grade of a highway in road construction. Maps, oscillations, music, surveying, and astronomy all make extensive use of trigonometry.

In this course we will study trigonometry for two important reasons: the first, as we just mentioned, is as a tool for more advanced work, and in this sense we study the trigonometric functions, their relationships, and their properties. The second reason for studying trigonometry is because of its usefulness in measurement, in particular in measuring inaccessible distances.

1.1   Angles and Degree Measure

The ancients used trigonometry in a very practical way to measure triangles as a means of surveying land. We look at this in Chapter 5. However, today trigonometry is used in a much more theoretical way, not only in mathematics, but also in electronics, engineering, and computer science. Today we can characterize trigonometry as the branch of mathematics that deals with the properties and applications of six functions: the cosine, sine, tangent, secant, cosecant, and cotangent. These functions are defined in this chapter. However, since these are functions of angles, we begin our study of trigonometry with angles.

Angles

Essential to the definition of the trigonometric functions is the idea of an angle. Recall from geometry that an **angle** is defined as two rays or line segments with a common endpoint. The rays or line segments are called the **sides** of the angle, and the common endpoint is called the **vertex.** In geometry you named angles by naming the vertex or the three points as shown in Figure 1.1. In trigonometry it is also common to use lowercase Greek letters to denote angles. A list of the common Greek letters is included as part of Figure 1.1.*

The symbol ∠ denotes angle. The given angle can be denoted in the following ways:

$$\angle ABC \qquad \angle B$$
$$\angle CBA \qquad \theta$$
↑
The vertex must be listed between the other two points.

Symbol	Name
α	alpha
β	beta
γ	gamma
δ	delta
θ	theta
λ	lambda
μ	mu
ϕ or φ	phi
ω	omega

FIGURE 1.1 Ways to denote an angle

In trigonometry and more advanced mathematics, this geometric definition of an angle is generalized.

*The Greek letter π (pi) is a lowercase letter that will not be used to represent an angle, because it denotes an irrational number approximately equal to 3.141592654.

An **angle** is formed by rotating a ray about its endpoint (called the **vertex**) from some initial position (called the **initial side**) to some terminal position (called the **terminal side**). The measure of an angle is the amount of rotation. An angle is also formed if a line segment is rotated about one of its endpoints.

If the rotation of the ray is in a counterclockwise direction, the measure of the angle is called **positive.** If the rotation is in a clockwise direction, the measure of the angle is called **negative.** A small curved arrow is used to denote the direction and amount of rotation. In geometry, you did not need to show a rotation for an angle, but now with this generalized definition, the curved arrow is essential.

Degree Measurement of Angles

FIGURE 1.2
Central angle and an arc

Several units of measurement are used for measuring angles. Historically, the most common scheme uses *degrees*. Draw a circle with any nonzero radius r and center at O. An **arc** is part of a circle, so, if P and Q are any two distinct points on the circle, then $\overset{\frown}{PQ}$ (read as arc PQ) is an arc of the circle. Angle POQ is called a **central angle** that is **subtended** by $\overset{\frown}{PQ}$ as shown in Figure 1.2.

The distance around a circle is called the **circumference** of the circle. A **degree** is an angle measurement that is defined as the size of a central angle subtended by an arc equal to $\frac{1}{360}$ of the length of the circumference. This definition of degree measure is independent of the size of the radius of the circle.

To measure angles you can use a protractor showing degree measure (see Figure 1.3).

FIGURE 1.3 Protractor for degree measure

You will not need to use a protractor in this class, but you will need to know the approximate size of various angles measured in degrees. The degree symbol is ° and is used as shown:

We use this
mark to indicate
a right angle.

$\theta = 90°$

Right angle

$0° < \theta < 90°$

Acute angle

$90° < \theta < 180°$

Obtuse angle

$\theta = 360°$

One revolution

$\theta = 180°$

Straight angle
$\frac{1}{2}$ revolution

Other common angles are shown in Example 1. When we write $\angle B = 45°$, we mean the measure of angle B is 45°, and when we write $\theta = 30°$, we mean the measure of angle θ is 30°.

EXAMPLE 1   Drawing an angle with a given degree measure

Sketch each angle from memory (without using a protractor).

a. $\alpha = 30°$ **b.** $\angle G = 45°$ **c.** $\beta = -60°$ **d.** $\gamma = 270°$

SOLUTION   **a.**   **b.**   **c.**   **d.** $\gamma = 270°$

⊗ You should remember the approximate sizes of the angles 0°, 30°, 45°, 60°, 90°, 180°, 270°, and 360° because they are frequently used in mathematics. ⊗

A degree can be divided into 60 equal parts, each called a **minute**, denoted by ′ ($1° = 60′$). Furthermore, a minute can be divided into 60 equal parts, each called a **second**, denoted by ″ ($1′ = 60″$; $1° = 60′ = 3{,}600″$). With the widespread use of calculators, it is becoming more and more common to represent minutes and seconds as decimal degrees. In this book you should use decimal degrees rather than minutes and seconds unless you are directed to do otherwise.

EXAMPLE 2   Converting degree measure to decimal degrees

a. $25°30′$ **b.** $128°14′$ **c.** $42°13′40″$

SOLUTION   **a.** Since $1° = 60′$, the number of minutes should be divided by 60 to convert it to decimal degrees.

$$30′ = \frac{30°}{60} = \frac{1°}{2} = .5° \qquad \text{Thus, } 25°30′ = 25.5°.$$

b. $14′ = \dfrac{14°}{60} = .2\overline{3}° \approx .23°$

If an angle is measured in minutes, then round the converted decimal to the nearest hundredth of a degree.* Thus, $128°14′ \approx 128.23°$.

c. Since $1° = 60′ = 3{,}600″$,

$$13′40″ = \left(\frac{13}{60}\right)° + \left(\frac{40}{3{,}600}\right)° = \left(\frac{820}{3{,}600}\right)°$$

---

*Accuracy of measurements will be important in trigonometry. With calculators it is easy to assume much more accuracy than we have. Pay attention to agreements about proper rounding as they are mentioned. Also, note the difference between the exact representation of an irrational number such as $\sqrt{2}$, π, $\frac{\pi}{2}$, and their approximations to the nearest hundredth: 1.41, 3.14, and 1.57.

By division, $\frac{820}{3,600} = .22\bar{7}$. If an angle is measured in seconds, then round the converted decimal to the nearest thousandth of a degree. Thus, $42°13'40'' \approx 42.228°$.

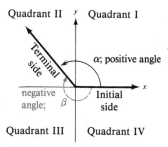

FIGURE 1.4 Standard-position angles α and β. α is a positive angle, and β is a negative angle. α and β are coterminal angles.

Standard-Position Angles

It is often useful to superimpose a Cartesian system on an angle so that the vertex is at the origin and the initial side is along the positive x-axis. The angle is then said to be in **standard position.** In this book we will assume that angles are in standard position unless otherwise noted. Angles in standard position that have the same terminal sides are **coterminal angles.** Given any angle α, there is an unlimited number of coterminal angles (some positive and some negative). In Figure 1.4, β is coterminal with α. Can you find other angles coterminal with α?

If the terminal side of θ is on one of the coordinate axes, then θ is called a **quadrantal angle.** If θ is not a quadrantal angle, then it is said to be in a certain **quadrant** if its terminal side lies in that quadrant. Thus, in Figure 1.4, both α and β would be called Quadrant II angles.

EXAMPLE 3   **Finding a positive angle coterminal with a given angle**

Finding a positive angle less than one revolution that is coterminal with the given angle. Also, classify the angle by quadrant.

a. $-30°$ **b.** $500°$ **c.** $-2,000°$

SOLUTION   **a.** $-30°$; Quadrant IV   **b.** $500°$; Quadrant II

$-30°$ and $330°$ are coterminal. $500°$ and $140°$ are coterminal.

c. $-2,000°$; Quadrant II. To find how many revolutions, divide by $360°$:

$$2,000 \div 360 = 5.\bar{5}$$

Thus, it is between 5 and 6 revolutions. To find how much more than 5 revolutions, subtract 5 revolutions $(5 \cdot 360°)$ from $2,000°$:

$$2,000° - 5(360°) = 2,000° - 1,800° = 200°$$

Now draw the negative angle (5 revolutions clockwise, plus $200°$ clockwise), as shown at the left.

$-2,000°$ and $160°$ are coterminal.

1.1* Problem Set

A

1. Name the given Greek letter.
   a. θ **b.** α **c.** ϕ **d.** ω **e.** μ
2. Name the given Greek letter.
   a. γ **b.** β **c.** δ **d.** λ **e.** φ
3. Write the symbol for the given Greek letter.
   a. delta   **b.** phi   **c.** theta   **d.** omega   **e.** mu
4. Write the symbol for the given Greek letter.
   a. alpha   **b.** beta   **c.** gamma   **d.** lambda   **e.** pi

Without using a protractor, draw an angle that approximates each angle with the measure indicated in Problems 5–10.

5. **a.** $60°$ **b.** $180°$ **c.** $45°$ **d.** $360°$ **e.** $300°$
6. **a.** $90°$ **b.** $30°$ **c.** $0°$ **d.** $270°$ **e.** $150°$
7. **a.** $120°$ **b.** $210°$ **c.** $135°$ **d.** $330°$ **e.** $240°$
8. **a.** $-30°$ **b.** $135°$ **c.** $-120°$ **d.** $-270°$ **e.** $-200°$
9. **a.** $-60°$ **b.** $315°$ **c.** $-330°$ **d.** $-45°$ **e.** $-90°$
10. **a.** $-300°$ **b.** $-210°$ **c.** $-240°$ **d.** $-360°$ **e.** $-150°$

Find a positive angle less than one revolution that is coterminal with each of the angles with the measures indicated in Problems 11–16.

11. **a.** $400°$ **b.** $540°$ **c.** $800°$ **d.** $750°$ **e.** $3,000°$
12. **a.** $370°$ **b.** $900°$ **c.** $550°$ **d.** $1,050°$ **e.** $2,500°$
13. **a.** $600°$ **b.** $730°$ **c.** $580°$ **d.** $1,000°$ **e.** $4,000°$
14. **a.** $-30°$ **b.** $-200°$ **c.** $-100°$ **d.** $-300°$ **e.** $-140°$
15. **a.** $-55°$ **b.** $-320°$ **c.** $-270°$ **d.** $-400°$ **e.** $-530°$
16. **a.** $-75°$ **b.** $-450°$ **c.** $-360°$ **d.** $-500°$ **e.** $-750°$

B *Change the given measures of the angles in Problems 17–26 to decimal degrees to the nearest hundredth.*

17. $65°40'$ 18. $146°50'$ 19. $85°20'$ 20. $127°10'$
21. $240°30'$ 22. $315°25'$ 23. $16°42'$ 24. $29°17'$
25. $143°23'$ 26. $165°58'$

Change the given measures of the angles in Problems 27–36 to decimal degrees to the nearest thousandth.

27. $128°10'40''$ 28. $13°30'50°$
29. $48°28'10''$ 30. $281°31'36''$
31. $210°40'15''$ 32. $94°21'31''$
33. $38'42''$ 34. $12'24''$
35. $6'7''$ 36. $2'6''$

37. *Historical Question* The division of one revolution (a *circle*) into 360 equal parts (called *degrees*) is no doubt due to the sexagesimal (base 60) numeration system used by the Babylonians. Several explanations have been put forward to account for the choice of this number. (For example, see Howard Eves' *In Mathematical Circles*.) One possible explanation is based on the fact that the radius of a circle can be applied exactly six times to its circumference as a chord. Illustrate this geometric fact. Notice that sexagesimal division (that is, division into 60 parts) makes this geometric construction easy. It then seems natural that each of these chords should be further divided into 60 equal parts, resulting in the division of the circle into 360 equal parts. Write a short research paper on the history of degree measure for an angle.

1.2 Radian Measure of Angles

Radian Measurement

In the last section the degree measure of an angle was discussed. There is a second commonly used measure for an angle that, instead of being based on 60 ($60'' = 1'$, $60' = 1°$, $360° = 1$ revolution), is based on a real-number measurement.

---

*The problem sets in this book are designed with equal length so that uniform problem set assignments are possible. Answers to the odd-numbered problems are in Appendix D.

Draw a circle with any positive radius r. Next, measure out an arc whose length is r. Figure 1.5a shows the case in which $r = 1$, and Figure 1.5b shows $r = 2$. Regardless of your choice of r, the size of the angle determined by this arc of length equal to the radius is the same (it is labeled θ in Figure 1.5). This fact can be proved from plane geometry, but you can easily convince yourself by drawing a few circles of *different* radii and verifying that every time the angle is the same size. Because it is the same size for every radius, we use this angle as a standard of comparison. Its measure is called one **radian.**

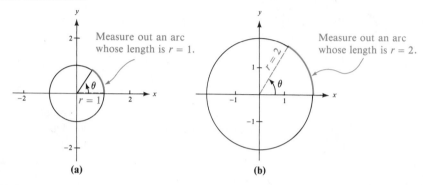

FIGURE 1.5 Angles of measure 1. The size of the angle is independent of the size of the circle.

RADIAN

One **radian** is the measure of a central angle intercepting an arc equal to the radius of the circle.

ANGLE	RADIAN MEASURE
One revolution	2π
Straight angle $\left(\dfrac{1}{2}\text{ rev}\right)$	π
Right angle $\left(\dfrac{1}{4}\text{ rev}\right)$	$\dfrac{\pi}{2}$

⧖ Note the use of degree and radian measures in this book. ⊗

Notice that the circumference generates an angle of one revolution. Since $C = 2\pi r$, and since the radian is measured in terms of r, **an angle of one revolution has a radian measure of 2π.**

Notice that, when measuring angles in radians, we are using *real numbers*. Because radian measure is used so frequently in more advanced work, we agree that radian measure is understood when *no units* of measure for an angle are indicated.

In this book, when measuring in degrees we will always say *degrees* or use the degree symbol, as with $\theta = 5°$. When measuring in radians we will **omit** the word *radians*. That is, *when no unit is given, radian measure is understood.* For example,

$$\theta = 5$$

means the measure of θ is radians.

Figure 1.6 shows a protractor for measuring angles using radian measure. Notice that the maximum *direct* measurement you can make with this protractor is a straight angle (π radians). Therefore, if the angle you wish to measure is more than π radians, you must subtract the measure of a straight angle (see Example 1c).

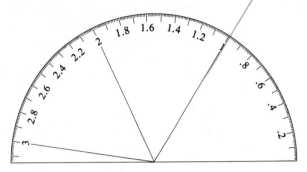

FIGURE 1.6 Protractor for radian measure

Just as with degree measure, you should have the size of 1 radian memorized so that you can draw angles of various sizes from memory with a reasonable amount of accuracy. You will have to practice thinking in terms of radian measure. You should memorize the approximate size of an angle of measure 1 in much the same way you have memorized the approximate size of an angle of 45°.

EXAMPLE 1   Drawing an angle with a given radian measure

Use Figure 1.6 to draw the angles whose measures are given. Do not attempt to change the measures in these problems to degrees. Work them directly in radians; do not make comparisons back to degree measure.

a. $\theta = 2$ **b.** $\theta = -3$ **c.** $\theta = 4$ **d.** $\theta = -\frac{\pi}{4}$

SOLUTION   **a.** $\theta = 2$

b. $\theta = -3$

c. $\theta = 4$
You cannot measure values greater than π (about 3.14) *directly* because a protractor measures angles up to a straight angle. Thus, **subtract the measure of a straight angle** (or a multiple) to give a residual less than π; use the protractor to measure the residual:

$$4 - \pi \approx .86$$

Thus, $$4 \approx \pi + .86$$

d. $\theta = -\frac{\pi}{4}$

$-\frac{\pi}{4}$ is called an *exact measurement* of the angle.

To use a protractor, write an approximation for the measure of the angle:

$$-\frac{\pi}{4} \approx -.79$$

Degree–Radian Conversions

We are led to the following relationships:

1 revolution is measured by 360°.

1 revolution is measured by 2π.

Thus,

$$\text{number of revolutions} = \frac{\text{angle in degrees}}{360}$$

$$\text{number of revolutions} = \frac{\text{angle in radians}}{2\pi}$$

Therefore, we can state the following equivalency:

RELATIONSHIP BETWEEN DEGREE AND RADIAN MEASURE

$$\frac{\text{angle in degrees}}{360} = \frac{\text{angle in radians}}{2\pi}$$

To change from degree measure to radian measure, let θ be the angle measure in radians. Multiply both sides by 2π to obtain

$$\theta = \frac{2\pi}{360} (\text{angle in degrees})$$

$$= \frac{\pi}{180} (\text{angle in degrees})$$

DEGREE TO RADIAN

Multiply the angle in degrees by the constant $\frac{\pi}{180}$ to obtain the radian measure.

EXAMPLE 2   **Change degree measure to radian measure**

    a. 45°   **b.** 125°   **c.** 1°

SOLUTION   **a.** $\theta = \dfrac{\pi}{180}(45)$

An alternate method for converting this angle measure is to remember that π radians is 180°. That is, because 45° is $\frac{1}{4}$ of 180°, you know that the radian measure is $\frac{\pi}{4}$.

$$= \dfrac{\pi}{4} \quad \text{Exact value}$$

b. $\theta = \dfrac{\pi}{180}(125)$

$$= \dfrac{25\pi}{36} \quad \text{Exact}$$

If you want a calculator approximation of θ:

$$125° \approx 2.1817 \qquad \text{Approximate}$$

c. $\theta = \dfrac{\pi}{180}(1)$

$$= \dfrac{\pi}{180} \quad \text{Exact}$$

If you want a calculator approximation of θ:

$$1° \approx .0175 \qquad \text{Approximate}$$

To change from radian measure to degree measure, let θ be the angle measured in degrees. Multiply both sides of the equivalency formula by 360 to obtain

$$\theta = \dfrac{360}{2\pi} \text{(angle in radians)}$$

$$= \dfrac{180}{\pi} \text{(angle in radians)}$$

RADIAN TO DEGREE | **Multiply the angle in radians by the constant $\frac{180}{\pi}$ to obtain the degree measure.**

EXAMPLE 3   **Change radian measure to degree measure**

    a. $\frac{\pi}{9}$ **b.** 1   **c.** 2.30

SOLUTION   **a.** $\theta = \dfrac{180}{\pi}\left(\dfrac{\pi}{9}\right)$

$$= 20$$

Thus, $\frac{\pi}{9}$ is an angle with measure of 20°.

b. $\theta = \dfrac{180}{\pi}(1)$

$$= \dfrac{180}{\pi}$$

On a calculator, $\theta \approx 57.29577951°$.

c. $\theta = \dfrac{180}{\pi}(2.30)$

$$\approx (57.296)(2.3)$$

$$= 131.7803 \qquad \text{To the nearest hundredth, the angle is } 131.78°.$$

If you have a calculator, you can obtain a much more accurate answer by using a better approximation for $\frac{180}{\pi}$. For example, if 2.3 is exact, then by calculator:

DISPLAY: 131.7802929

CALCULATOR COMMENT

Many calculators are programmable. For example, on the TI-81 you can press ⎡PRGM⎤ and then write a two line program to convert from

<table>
<tr><td>radian to degree measure:</td><td>degree to radian measure:</td></tr>
<tr><td>:180/π*R</td><td>:π/180*D</td></tr>
<tr><td>:Disp Ans</td><td>:Disp Ans</td></tr>
</table>

If you are changing from radian, store the radian measure in ⎡R⎤; if you are changing from degree, store the degree measure in ⎡D⎤. Then execute the appropriate program.

Even if you do not have a programmable calculator, you can store the constant $180/\pi$ to change a set of numbers from radian to degree measure. On the other hand, if you want to change a set of numbers from degree to radian measure, store the constant $\pi/180$.

For the more common measures of angles, it is a good idea to memorize the equivalent degree and radian measures shown in Table 1.1. If you keep in mind that 180° equals π in radian measure, the rest of the table will be very easy to remember.

TABLE 1.1 Relationship between degree and radian measure of angles

DEGREES	RADIANS
0°	0
30°	$\frac{\pi}{6}$
45°	$\frac{\pi}{4}$
60°	$\frac{\pi}{3}$
90°	$\frac{\pi}{2}$
180°	π
270°	$\frac{3\pi}{2}$
360°	2π

Quadrantal angles

Common first quadrant angles

$\leftarrow$

If you remember

$$180° = \pi$$

you will be able to easily remember the other entries in this table.

Arc Length

In the preceding discussion we used an arc to find the measure of an angle, but we can also use the measure of an angle to find the length of an arc of a circle, called **arc length**. To do this, use the formula for the circumference of a circle with radius *r*.

ARC LENGTH

The **arc length** of a circle of radius r and central angle θ (measured in radians) is denoted by s and is found by

$$s = r\theta$$

To see where this formula comes from, recall the definition of radian measure to notice that the angle is 2π times that portion of the circle by the angle θ. For example,

$$1 \text{ revolution}, \frac{s}{C} = 1, \text{ so } \theta = 1(2\pi) = 2\pi$$

$$\frac{1}{2} \text{ revolution}, \frac{s}{C} = \frac{1}{2}, \text{ so } \theta = \frac{1}{2}(2\pi) = \pi$$

$$\frac{1}{4} \text{ revolution}, \frac{s}{C} = \frac{1}{4}, \text{ so } \theta = \frac{1}{4}(2\pi) = \frac{\pi}{2}$$

In general $\theta = \dfrac{s}{C}(2\pi)$. Now, algebraically solve for s: $s = \dfrac{\theta C}{2\pi}$. But $C = 2\pi r$, so

$$s = \frac{\theta(2\pi r)}{2\pi}$$

$$s = r\theta$$

EXAMPLE 4   **Finding an arc length when central angle is measured in radians**

What is the length of the arc of a circle of radius 10 cm if the central angle is 2?

SOLUTION
$$\begin{aligned} s &= r\theta \\ &= (10 \text{ cm})(2) \\ &= 20 \text{ cm} \end{aligned}$$

EXAMPLE 5   **Finding an arc length when central angle is measured in degrees**

Find the arc length if the central angle is 36° and the radius of the circle is 20 centimeters (cm).

SOLUTION   First, change the degree measure to radian measure:

$$\theta = \frac{\pi}{180}(36) = \frac{\pi}{5} \quad \text{where } \theta \text{ is the radian measure of the angle}$$

Thus,
$$\begin{aligned} s &= r\theta \\ &= 20\left(\frac{\pi}{5}\right) \\ &= 4\pi \end{aligned}$$

The length of the arc is 4π cm or about 12.6 cm.

Locations on the earth are given by *latitude* and *longitude*. Latitude is measured north or south of the equator by angles between 0° and 90°. Longitude is measured by angles between 0° and 180° either west or east of the Greenwich meridian.

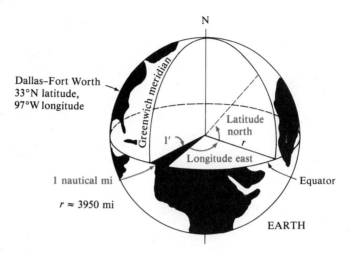

EXAMPLE 6   Using arc length to find distances between cities

The Dallas–Fort Worth airport is located at approximately 33°N latitude and 97°W longitude. How far north (to the nearest hundred miles) is it from the equator?

SOLUTION   The location is 33°N latitude, so first convert 33° to radians:

$$\theta \approx 33° = 33\left(\frac{\pi}{180}\right) \approx .576$$

Next, use the arc length formula:

$$s = r\theta$$
$$\approx 3{,}950\theta$$
$$\approx 2{,}300$$

The Dallas–Fort Worth airport is about 2,300 miles north of the equator.

1.2   Problem Set

A

1. From memory, give the radian measure for each of the angles whose degree measure is stated.

  a. 30°   **b.** 90°   **c.** 270°   **d.** 45°

  e. 360°   **f.** 60°   **g.** 180°

2. From memory, give the degree measure for each of the angles whose radian measure is stated.

  a. π **b.** $\frac{\pi}{4}$ **c.** $\frac{\pi}{3}$ **d.** 2π

  e. $\frac{\pi}{2}$ **f.** $\frac{3\pi}{2}$ **g.** $\frac{\pi}{6}$

Use the radian protractor in Figure 1.6 to help you sketch each of the angles with the measures indicated in Problems 3–6. Do not change them to degrees, but try to think in radians.

3. a. $\frac{\pi}{2}$ **b.** $\frac{\pi}{6}$ **c.** 2.5   **d.** 6

4. a. $\frac{2\pi}{3}$ **b.** $\frac{3\pi}{4}$ **c.** -1 **d.** 5

5. a. $-\frac{\pi}{4}$ **b.** $\frac{7\pi}{6}$ **c.** -2.76 **d.** $\sqrt{17}$

6. a. $-\frac{3\pi}{2}$ **b.** $\frac{13\pi}{3}$ **c.** -1.2365 **d.** $\sqrt{23}$

Find a positive angle less than one revolution that is coterminal with each of the angles with the measures indicated in Problems 7–10.

7. a. 3π **b.** $\frac{13\pi}{6}$ **c.** $-\pi$ **d.** 7

8. a. $-\frac{\pi}{4}$ **b.** $\frac{17\pi}{4}$ **c.** $\frac{11\pi}{3}$ **d.** -2

9. a. 9   **b.** -5 **c.** $\sqrt{50}$ **d.** -6

10. a. 6.2832   **b.** -3.1416 **c.** 30   **d.** $3\sqrt{5}$

Change each of the angles with the indicated measures in Problems 11–13 to radians using exact values.

11. a. 40°   **b.** 20°

12. a. $-64°$ **b.** $-220°$

13. a. 254°   **b.** 85°

Change each of the angles with the indicated measures in Problems 14–16 to radians correct to the nearest hundredth.

14. a. 112°   **b.** 314°

15. a. $-62.8°$ **b.** 350°

16. a. $-205.2°$ **b.** 535°

Change each of the angles with the indicated measures in Problems 17–19 to degrees correct to the nearest hundredth.

17. a. $\frac{2\pi}{9}$ **b.** $\frac{\pi}{10}$

18. a. $-\frac{11\pi}{12}$ **b.** 2

19. a. -3 **b.** $-.25$

In problems 20–25 find the length of each intercepted arc to the nearest hundredth, if the measure of the central angle and radius are as given.

20. Angle 1, radius 1 m

21. Angle 2.34, radius 6 cm

22. Angle $\frac{\pi}{3}$, radius 4 m

23. Angle $\frac{3\pi}{2}$, radius 15 cm

24. Angle 40°, radius 7 km

25. Angle 72°, radius 10 dm

26. How far (to the nearest hundredth) does the tip of an hour hand measuring 2.00 cm move in three hours?

27. A 50-cm pendulum on a clock swings through an angle of 100°. How far (to the nearest centimeter) does the tip travel in one arc?

28. GEOGRAPHY   If New York City is at 40°N latitude and 74°W longitude, what is its approximate distance (to the nearest hundred miles) from the equator?

29. GEOGRAPHY   If Chicago is located at 42°N latitude and 88°W longitude, what is its approximate distance (to the nearest hundred miles) from the equator?

30. GEOGRAPHY   If Columbia, South Carolina, is located at 34°N latitude and 81°W longitude, what is its approximate distance (to the nearest hundred miles) from Disneyland, which is located at 34°N latitude and 118°W longitude? [*Hint:*   The radius at 34° is approximately 3,250 mi.]

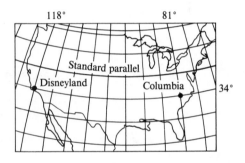

*31. PHYSICS   Through how many radians does a pulley with a 20.0-cm diameter turn when 1.00 m of cable is pulled through it without slippage? (1 m = 100 cm)

*32. PHYSICS   Answer the question in Problem 31 for a pulley with a 5.0-cm diameter.

---

*See the Application for Further Study at the end of this chapter for further applications involving linear velocity and angular velocity.

33. *Historical Question* In about 230 B.C. a mathematician named Eratosthenes estimated the radius of the earth using the following information. Syene and Alexandria in Egypt are on the same line of longitude. They are also 800 km apart. At noon on the longest day of the year, when the sun was directly overhead in Syene, Eratosthenes measured the sun to be 7.2° from the vertical in Alexandria. Because of the distance of the sun from the earth, he assumed that the sun's rays were parallel. Thus he concluded that the central angle subtending rays from the center of the earth to Syene and Alexandria was also 7.2°. Using this information, find the approximate radius of the earth.

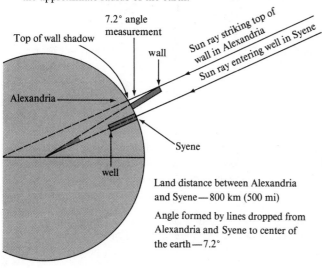

7.2° angle measurement

Top of wall shadow

wall

Sun ray striking top of wall in Alexandria

Sun ray entering well in Syene

Alexandria

Syene

well

Land distance between Alexandria and Syene—800 km (500 mi)

Angle formed by lines dropped from Alexandria and Syene to center of the earth—7.2°

34. SURVEYING Omaha, Nebraska, is located at approximately 97°W longitude, 41°N latitude; Wichita, Kansas, is located at approximately 97°W longitude, 37°N latitude. Notice that these two cities have about the same longitude. If we know that the radius of the earth is about 6,370 kilometers (km), what is the distance between these cities to the nearest 10 km?

35. SURVEYING Entebbe, Uganda, is located at approximately 33°E longitude, and Stanley Falls in Zaire is located at 25°E longitude. Also, both these cities lie approximately on the equator. If we know that the radius of the earth is about 6,370 km, what is the distance between the cities to the nearest 10 km?

36. ASTRONOMY Suppose it is known that the moon subtends an angle of 45.75′ at the center of the earth. It is also known that the center of the moon is 384,417 km from the surface of the earth. If you assume that the radius of the earth is 6,370 km, what is the diameter of the moon to the nearest 10 km?

Earth

Moon

Intercepted arc

45.75′

Diameter of moon
Chord ≈ intercepted arc

For small central angles with large radii, the intercepted arc is approximately equal to its chord.

1.3 Trigonometric Functions on a Unit Circle

Unit Circle

To introduce you to the trigonometric functions, we will consider a relationship between angles and circles. The **unit circle** is the circle centered at the origin with radius one. The equation of the unit circle is

$$x^2 + y^2 = 1$$

Draw a unit circle with an angle θ in standard position, as shown in Figure 1.7.

Terminal side

$A(a, b)$

Initial side

$B(1, 0)$

FIGURE 1.7 The unit circle

Cosine and Sine Functions

The initial side of θ intersects the unit circle at $(1, 0)$, and the terminal side intersects the unit circle at (a, b). Define functions of θ as follows:

$$c(\theta) = a \quad \text{and} \quad s(\theta) = b$$

EXAMPLE 1   **Evaluating c and s functions**

  a. $s(90°)$ **b.** $c(90°)$ **c.** $s(-270°)$ **d.** $s\left(-\frac{\pi}{2}\right)$ **e.** $c(-3.1416)$

SOLUTION   **a.** $s(90°)$; this is the second component of the ordered pair (a, b), where (a, b) is the point of intersection of the terminal side of a 90° standard-position angle and the unit circle. By inspection, it is 1. Thus, $s(90°) = 1$.

  b. $c(90°) = 0$

  c. $s(-270°) = 1$; this is the same as part **a** because the angles 90° and $-270°$ are coterminal.

  d. $s\left(-\frac{\pi}{2}\right) = -1$ **e.** $c(-3.1416) \approx c(-\pi) = -1$

Trigonometric Functions

The function $c(\theta)$ is called the **cosine function**, and the function $s(\theta)$ is called the **sine function**. These functions, along with four additional functions, make up what are called the **trigonometric functions** or **trigonometric ratios.**

UNIT CIRCLE DEFINITION OF THE TRIGONOMETRIC FUNCTIONS

Let θ be an angle in standard position with the point (a, b) the interesection of the terminal side of θ and the unit circle. Then the six trigonometric functions are defined as follows:

cosine: $\cos \theta = a$	**secant:** $\sec \theta = \dfrac{1}{a} \quad a \neq 0$	
sine: $\sin \theta = b$	**cosecant:** $\csc \theta = \dfrac{1}{b} \quad b \neq 0$	
tangent: $\tan \theta = \dfrac{b}{a} \quad a \neq 0$	**cotangent:** $\cot \theta = \dfrac{a}{b} \quad b \neq 0$	

⊘ cos, sin, tan, . . . are not variables. They cannot stand alone, and must be associated with an angle. ⊘

The angle θ is an important part of these definitions. The words *cos, sin, tan,* and so on as defined in the preceding box are meaningless without the θ. You can speak of the *cosine function* or of *cos θ*, but you cannot simply say *cos.*

Notice the condition on the tangent, secant, cosecant, and cotangent functions. These exclude division by 0; for example, $a \neq 0$ means that θ cannot be 90°, 270°, or any angle coterminal to these angles. We summarize this condition by saying that the *tangent and secant are not defined for* 90° *or* 270°. If $b \neq 0$, then $\theta \neq$ 0°, 180°, or any angle coterminal to these angles; thus, *cosecant and cotangent are not defined for* 0° *or* 180°. In this book we will assume all values that cause division by 0 are excluded from the domain.

Fundamental Identities

There are certain relationships among the trigonometric functions that you will need to evaluate these functions. These relationships are commonly referred to as the **fundamental identities**. Recall from algebra that an **identity** is an open equation (has at least one variable) that is true for all values in the domain.

The first three identities are called the **reciprocal identities**. Remember from arithmetic that 5 and $\frac{1}{5}$ are reciprocals, as are $\frac{2}{3}$ and $\frac{3}{2}$; in general, two numbers are reciprocals if their product is 1.

RECIPROCAL IDENTITIES

1. $\sec \theta = \dfrac{1}{\cos \theta}$ The cosine and secant are reciprocal functions.

2. $\csc \theta = \dfrac{1}{\sin \theta}$ The sine and cosecant are reciprocal functions.

3. $\cot \theta = \dfrac{1}{\tan \theta}$ The tangent and cotangent are reciprocal functions.

It is easy to prove that these are identities by looking at the definition of secant, cosecant, and cotangent. Also, notice that these can be rewritten in other forms:

$$\cos \theta \sec \theta = 1 \quad \text{or} \quad \cos \theta = \frac{1}{\sec \theta}$$

EXAMPLE 2   **Recognizing different algebraic representations of an identity**

Write the second identity two different ways.

SOLUTION $\sin \theta \csc \theta = 1 \quad \text{and} \quad \sin \theta = \dfrac{1}{\csc \theta}$

EXAMPLE 3   **Using reciprocal identities**

Write each given function in terms of its reciprocal function.

a. $\dfrac{1}{\sin 81°}$ **b.** $\sec 2$ **c.** $\tan \frac{\pi}{4}$ **d.** $\cot 35°$ **e.** $\csc 3$

SOLUTION   **a.** $\dfrac{1}{\sin 81°} = \csc 81°$ **b.** $\sec 2 = \dfrac{1}{\cos 2}$ **c.** $\tan \frac{\pi}{4} = \dfrac{1}{\cot \frac{\pi}{4}}$

d. $\cot 35° = \dfrac{1}{\tan 35°}$ **e.** $\csc 3 = \dfrac{1}{\sin 3}$

The next two identities are called the **ratio identities**; they express the tangent and cotangent functions using sine and cosine functions.

RATIO IDENTITIES

$$4. \quad \tan \theta = \frac{\sin \theta}{\cos \theta} \qquad 5. \quad \cot \theta = \frac{\cos \theta}{\sin \theta}$$

To prove these identities, we also use the definition:

$$\tan \theta = \frac{b}{a} \qquad \text{Definition of tangent}$$

$$= \frac{\sin \theta}{\cos \theta} \qquad \begin{array}{l} \text{Substitution of values from the} \\ \text{definition: } \sin \theta = b \text{ and } \cos \theta = a \end{array}$$

The ratio identity for the cotangent is proved in exactly the same way.

EXAMPLE 4   Using ratio identities

If $\sin \theta = -\frac{3}{5}$ and $\cos \theta = \frac{4}{5}$, find the values of $\tan \theta$ and $\cot \theta$.

SOLUTION

$$\tan \theta = \frac{\sin \theta}{\cos \theta} = \frac{-\frac{3}{5}}{\frac{4}{5}} = -\frac{3}{5} \cdot \frac{5}{4} = -\frac{3}{4} \qquad \cot \theta = \frac{1}{\tan \theta} = -\frac{4}{3}$$

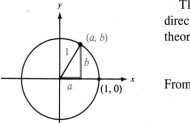

The last three identities are called the **Pythagorean identities** because they follow directly from the Pythagorean theorem. Since (a, b) is on a unit circle, the Pythagorean theorem tells us

$$a^2 + b^2 = 1$$

From the definition of sine and cosine (and by substitution), we obtain

$$(\cos \theta)^2 + (\sin \theta)^2 = 1$$

In trigonometry, we simplify the notation involving the square of a trigonometric function by writing the previous equation as follows:

$$\cos^2 \theta + \sin^2 \theta = 1$$

PYTHAGOREAN IDENTITIES

$$6. \quad \cos^2 \theta + \sin^2 \theta = 1 \qquad 7. \quad 1 + \tan^2 \theta = \sec^2 \theta \qquad 8. \quad \cot^2 \theta + 1 = \csc^2 \theta$$

To prove identities 7 and 8, we will need only to divide both sides of $a^2 + b^2 = 1$ by a^2 (for identity 7) or b^2 (for identity 8):

$$\frac{a^2}{a^2} + \frac{b^2}{a^2} = \frac{1}{a^2} \qquad \text{Dividing both sides by } a^2 \ (a \neq 0)$$

$$1 + \left(\frac{b}{a}\right)^2 = \left(\frac{1}{a}\right)^2 \qquad \text{Properties of exponents}$$

$$1 + (\tan \theta)^2 = (\sec \theta)^2 \qquad \text{Definition of } \tan \theta \text{ and } \sec \theta, \text{ and substitution}$$

$$1 + \tan^2 \theta = \sec^2 \theta$$

The proof of identity 8 is left as an exercise.

These Pythagorean identities are frequently written in a different form. For example, we can solve identity 7 for $\tan \theta$:

$$1 + \tan^2\theta = \sec^2\theta$$ Identity 7

$$\tan^2\theta = \sec^2\theta - 1$$ Subtract 1 from both sides.

$$\tan \theta = \pm\sqrt{\sec^2\theta - 1}$$ Do not forget the $\pm$ using the square root property from algebra.

EXAMPLE 5   Using fundamental identities

Write all the trigonometric functions in terms of $\sin \theta$ by using the eight fundamental identities.

SOLUTION   **a.** $\sin \theta = \sin \theta$

b. $\cos \theta$:   Use identity 6:

$$\cos^2\theta + \sin^2\theta = 1$$
$$\cos^2\theta = 1 - \sin^2\theta$$
$$\cos \theta = \pm\sqrt{1 - \sin^2\theta}$$

c. $\tan \theta = \dfrac{\sin \theta}{\cos \theta}$ From identity 4

$$= \frac{\sin \theta}{\pm\sqrt{1 - \sin^2\theta}}$$ From part b*

d. $\cot \theta = \dfrac{1}{\tan \theta}$ From identity 3

$$= \frac{1}{\dfrac{\sin \theta}{\pm\sqrt{1 - \sin^2\theta}}}$$ From part c

$$= \frac{\pm\sqrt{1 - \sin^2\theta}}{\sin \theta}$$ Invert and multiply.

e. $\csc \theta = \dfrac{1}{\sin \theta}$ Identity 2

f. $\sec \theta = \dfrac{1}{\cos \theta}$ Identity 1

$$= \frac{1}{\pm\sqrt{1 - \sin^2\theta}}$$ From part b

---

*In algebra you needed to rationalize the denominator; however, since we work with many reciprocals in trigonometry, we generally relax the requirement that expressions with a radical in a denominator need to be rationalized.

Evaluation of Trigonometric Functions

In many applications, you will know the angle measure and will want to find one or more of its trigonometric functions. To do this you will carry out a process called **evaluation of the trigonometric functions**. In order to help you see the relationship between the angle and the function, consider Figure 1.8 and Example 6.

Evaluation Using the Definition

EXAMPLE 6   **Evaluating trigonometric functions using the definition**

Evaluate the trigonometric functions of $\theta = 100°$ by drawing the unit circle and approximating the point (a, b).

SOLUTION   Draw a circle with $r = 1$ as shown in Figure 1.8.

FIGURE 1.8 Approximate values of circular functions

Next, draw the terminal side of the angle 110°. Estimate (by counting squares) $a \approx -.35$ and $b \approx .95$. Thus,

$$\cos 110° \approx -.35 \qquad \sec 110° \approx \frac{1}{-.35} \approx -2.9$$

$$\sin 110° \approx .95 \qquad \csc 110° \approx \frac{1}{.95} \approx 1.1$$

$$\tan 110° \approx \frac{.95}{-.35} \approx -2.7 \qquad \cot 110° \approx \frac{-.35}{.95} \approx -.37$$

In practice you will not use the definition and Figure 1.8 to evaluate trigonometric functions, but this exercise is important because it will help you to understand what the values you obtain from a calculator mean.

Using Figure 1.8, it is easy to derive a very important property of the trigonometric functions. Notice if θ is in Quadrant I, then all of the trigonometric functions are positive since a and b are both positive.

In Quadrant II, $a < 0$ and $b > 0$, so

$$\cos \theta = a < 0$$

$$\sin \theta = b > 0$$

$$\tan \theta < 0$$

Do you see why tangent is negative in Quadrant II? If not, use the definition: $\tan \theta = \dfrac{b}{a}$, which is a positive divided by a negative.

The signs of sec θ, csc θ, and cot θ are the same as their respective reciprocals. In Quadrant III, $a < 0$ and $b < 0$, so

$$\begin{cases} \cos \theta < 0 \\ \sin \theta < 0 \\ \tan \theta > 0 \end{cases}$$

Remember, a negative divided by a negative is positive.

$$\begin{cases} \sec \theta < 0 \\ \csc \theta < 0 \\ \cot \theta > 0 \end{cases}$$

Finally, in Quadrant IV, $a > 0$ and $b < 0$, so

$$\begin{cases} \cos \theta > 0 \\ \sin \theta < 0 \\ \tan \theta < 0 \end{cases}$$

$$\begin{cases} \sec \theta > 0 \\ \csc \theta < 0 \\ \cot \theta < 0 \end{cases}$$

It will be important in your work in trigonometry to remember which of the trigonometric functions are positive in each of the quadrants. However, this information is difficult to remember using the above notation, so it is summarized below. Remember that the values of sec θ, csc θ, and cot θ have the same signs as their reciprocals.

SIGNS OF THE TRIGONOMETRIC FUNCTIONS   EASY-TO-REMEMBER FORM:

SINE POSITIVE	**A**LL POSITIVE	**S**MART	**A**
TANGENT POSITIVE	**C**OSINE POSITIVE	**T**RIG	**C**LASS

EXAMPLE 7   **Finding the quadrants of a trigonometric function by looking at its sign**

State the quadrant or quadrants in which θ may lie to make each of the expressions true.

   a. $\tan \theta = -.9$ **b.** $\sec \theta = -\frac{5}{3}$ **c.** $\cot \theta = .5$

   d. $\sin \theta = .3$ and $\tan \theta < 0$ **e.** $\tan \theta = -2$ and $\sin \theta > 0$

   f. $\sin \theta = \sqrt{1 - \cos^2\theta}$ **g.** $\csc \theta = -\sqrt{\cot^2\theta + 1}$

SOLUTION   **a.** Quadrants II and IV (tangent is negative)

   b. The reciprocal of secant is cosine and cosine is negative in Quadrants II and III, so the secant is negative in Quadrants II and III.

   c. Quadrant I or III   (Think about where the tangent is positive.)

   d. The function $\sin \theta$ is positive in Quadrants I and II and the tangent is negative in Quadrants II and IV, so there is one quadrant that makes both statements true; θ is in Quadrant II.

   e. $\tan \theta$ is negative in Quadrants II and IV, and $\sin \theta$ is positive in Quadrants I and II; answer is Quadrant II.

   f. Radical symbols are positive, so sine is positive in Quadrants I and II.

   g. Since radicals are positive, opposites of radicals must be negative, so θ is in Quadrant III or IV (think about the reciprocal of cosecant, namely, sine).

EXAMPLE 8   **Finding the values of the trigonometric functions given one value**

If $\sin \theta = -.75$ and θ is in Quadrant III, use the results of Example 5 to find the values (correct to the nearest hundredth) of the other trigonometric functions.

SOLUTION

$$\cos \theta = -\sqrt{1 - \sin^2\theta} \qquad \text{Cosine is negative in Quadrant III.}$$
$$= -\sqrt{1 - (-.75)^2}$$
$$\approx -.66$$

 You might use the identity $\tan \theta = \dfrac{\sin \theta}{\cos \theta} \approx \dfrac{-.75}{-.66} \approx 1.14$; this answers includes a **round-off error.** You should not work with rounded results (excepting the limits of your calculator); round your answers only once, when you are ready to state the final answer. ⊗

$$\tan \theta = \frac{\sin \theta}{-\sqrt{1 - \sin^2\theta}} = \frac{-.75}{-\sqrt{1 - (-.75)^2}} \approx 1.13$$

$$\cos \theta = \frac{1}{\sin \theta} = \frac{1}{-.75} \approx 1.33$$

$$\sec \theta = \frac{1}{-\sqrt{1 - \sin^2\theta}} = \frac{1}{-\sqrt{1 - (-.75)^2}} \approx 1.51$$

$$\cot \theta = \frac{-\sqrt{1 - \sin^2\theta}}{\sin \theta} = \frac{-\sqrt{1 - (-.75)^2}}{-.75} \approx .88$$

Evaluation Using a Calculator

Evaluating the trignometric ratios by calculator is very straightforward, but you must note several details.

CALCULATOR EVALUATION OF THE TRIGONOMETRIC FUNCTION

1. *Unit of measure used.* Most calculators measure angles in both degrees and radians. You must make certain that your calculator is in the *proper mode*. Some calculators simply have a switch (often labeled R–D) that changes the angle-measure mode. However, there are other ways, so you should consult the owner's manual to find out how to make sure your calculator is in either degree or radian mode as needed. From now on, we will assume that you are working in the appropriate angle-measure mode. Remember, if later in the course you suddenly start obtaining "strange" answers and have no idea what you are doing wrong, double-check to make sure your calculator is in the proper mode.

2. With some calculators you *enter the angle first* and then press the button corresponding to the trigonometric function.

3. Your calculator has keys for $\boxed{\cos}$, $\boxed{\sin}$, and $\boxed{\tan}$, but not for secant, cosecant, or cotangent. Keys for the latter three functions are unnecessary as they are readily obtained using the reciprocal identities and the $\boxed{1/x}$ or $\boxed{x^{-1}}$ key on your calculator. If $x \neq 0$, you can enter x, press the reciprocal button, and see the reciprocal of x displayed in decimal form.

EXAMPLE 9   Evaluating trigonometric functions using a calculator

Find the values of the trigonometric functions for $\theta = 37.8°$. Round to four decimal places.

SOLUTION

ALGEBRAIC	OR	GRAPHING	DISPLAY	VALUE
$\boxed{37.8}$ $\boxed{\cos}$	or	$\boxed{\cos}$ $\boxed{37.8}$ $\boxed{\text{ENTER}}$	.7901550124*	$\cos 37.8° \approx .7902$
$\boxed{37.8}$ $\boxed{\sin}$	or	$\boxed{\sin}$ $\boxed{37.8}$ $\boxed{\text{ENTER}}$	.6129070537	$\sin 37.8° \approx .6129$
$\boxed{37.8}$ $\boxed{\tan}$	or	$\boxed{\tan}$ $\boxed{37.8}$ $\boxed{\text{ENTER}}$	.7756795108	$\tan 37.8° \approx .7757$
$\boxed{37.8}$ $\boxed{\cos}$ $\boxed{1/x}$	or	$\boxed{\cos}$ $\boxed{37.8}$ $\boxed{x^{-1}}$ $\boxed{\text{ENTER}}$	1.265574456	$\sec 37.8° \approx 1.2656$
$\boxed{37.8}$ $\boxed{\sin}$ $\boxed{1/x}$	or	$\boxed{\sin}$ $\boxed{37.8}$ $\boxed{x^{-1}}$ $\boxed{\text{ENTER}}$	1.631568758	$\csc 37.8° \approx 1.6316$
$\boxed{37.8}$ $\boxed{\tan}$ $\boxed{1/x}$	or	$\boxed{\tan}$ $\boxed{37.8}$ $\boxed{x^{-1}}$ $\boxed{\text{ENTER}}$	1.289192232	$\cot 37.8° \approx 1.2892$

1.3 Problem Set

A

1. State the unit circle definition of the trigonometric functions, from memory.

2. Name the reciprocal of each given trigonometric function.

    a. cosine **b.** cotangent **c.** sine **d.** secant **e.** cosecant

3. Solve $\cos^2\theta + \sin^2\theta = 1$ for $\cos\theta$.

4. Solve $\cos^2\theta + \sin^2\theta = 1$ for $\sin\theta$.

5. Solve $\cot^2\theta + 1 = \csc^2\theta$ for $\cot\theta$.

*Your calculator may have fewer digits in the display than are shown here. Also, if you obtained .9949151, your calculator is set to radian mode.

Use Figure 1.9 *and the definition of the trigonometric functions to evaluate the functions given in Problems 6–8 to one decimal place.*

FIGURE 1.9 Using a unit circle definition of the trigonometric functions

6. a. $\cos 50°$ **b.** $\cos 2$ **c.** $\sec 70°$ **d.** $\csc(-150°)$

7. a. $\sin 200°$ **b.** $\cos 250°$ **c.** $\tan(-20°)$ **d.** $\cot 3$

8. a. $\cos 4$ **b.** $\sec 3$ **c.** $\cot 6$ **d.** $\csc 3$

Use a calculator to find the values of the trigonometric functions given in Problems 9–11. Round to four decimal places.

9. a. $\sin 34°$ **b.** $\sec 56°$ **c.** $\tan 15°$ **d.** $\sin 61.5°$

10. a. $\sec 18.6°$ **b.** $\cot 75.4°$ **c.** $\sin 1$ **d.** $\sec 1.5$

11. a. $\cos .5$ **b.** $\csc .25$ **c.** $\cot .43$ **d.** $\cot .65$

In Problems 12–14 use the definition of the trigonometric functions to write each given function in terms of its reciprocal function.

12. a. $\dfrac{1}{\csc 25°}$ **b.** $\dfrac{1}{\cos 125°}$ **c.** $\dfrac{1}{\cot 320°}$

   d. $\dfrac{1}{\sin 210°}$ **e.** $\dfrac{1}{\tan 18°}$ **f.** $\dfrac{1}{\sec 35°}$

13. a. $\dfrac{1}{\cos 3}$ **b.** $\dfrac{1}{\sec 1}$ **c.** $\dfrac{1}{\cot 2}$

   d. $\sin 19°$ **e.** $\sec 181°$ **f.** $\cot 210°$

14. a. $\sin 3$ **b.** $\cos 1$ **c.** $\cot 6$

   d. $\tan 2$ **e.** $\sec 3$ **f.** $\csc 4$

In each of Problems 15–17 state the quadrant or quadrants in which θ may lie to make the expression true.

15. a. $\cos \theta = .2345$ **b.** $\sin \theta = -.43$

   c. $\sin \theta = \sqrt{1 - \cos^2\theta}$ **d.** $\cos \theta = -\sqrt{1 - \sin^2\theta}$

16. a. $\tan \theta = 2$

   b. $\sec \theta = 4.1258$

   c. $\cot \theta = \sqrt{\csc^2\theta - 1}$

   d. $\sin \theta = -\sqrt{1 - \cos^2\theta}$ and $\tan \theta > 0$

17. a. $\csc \theta = \frac{3}{2}$ and $\tan \theta < 0$

   b. $\sec \theta = -\sqrt{1 + \tan^2\theta}$ and $\sin \theta > 0$

   c. $\cot \theta = \sqrt{5}$ and $\sin \theta < 0$

   d. $\csc \theta = \sqrt{1 + \cot^2\theta}$ and $\tan \theta < 0$

Write each of the expressions in Problems 18–20 as a single trigonometric function of some angle by using one of the eight fundamental identities.

18. a. $\dfrac{\sin 50°}{\cos 50°}$ **b.** $1 - \cos^2 18°$ **c.** $-\sqrt{1 - \cos^2 210°}$

19. a. $\dfrac{\cos(A + B)}{\sin(A + B)}$ **b.** $\dfrac{1}{\sec 75°}$ **c.** $\sqrt{1 + \tan^2 210°}$

20. a. $\dfrac{1}{\cot \frac{\pi}{15}}$ **b.** $1 - \tan^2 \frac{\pi}{12}$ **c.** $-\sqrt{1 - \sin^2 127°}$

21. Evaluate each expression using a fundamental identity.
   a. $\cos 128° \sec 128°$ **b.** $\sin^2 53° + \cos^2 53°$
   c. $\tan^2 135° - \sec^2 135°$

22. Evaluate each expression using a fundamental identity.
   a. $\sec^2 \frac{\pi}{12} - \tan^2 \frac{\pi}{12}$ **b.** $\csc 85° \sin 85°$
   c. $\cot^2 37° - \csc^2 37°$

B *In Problems 23–27 write all the trigonometric functions in terms of the given functions. You do not need to rationalize denominators.*

23. $\cos \theta$ **24.** $\tan \theta$ **25.** $\cot \theta$ **26.** $\sec \theta$ **27.** $\csc \theta$

In Problems 28–31, use the fundamental identities to find the other trigonometric functions of θ using the given information. State the answers to Problems 30 and 31 correct to the nearest hundredth.

28. $\cos \theta = \frac{5}{13}$; $\tan \theta < 0$ **29.** $\tan \theta = \frac{5}{12}$; $\sin \theta > 0$

30. $\sin \theta = .65$; $\sec \theta > 0$ **31.** $\cot \theta = 1.25$; $\cos \theta < 0$

In Problems 32–35, prove the identity.

32. $\csc \theta = \dfrac{1}{\sin \theta}$ **33.** $\cot \theta = \dfrac{\cos \theta}{\sin \theta}$

34. $1 + \tan^2\theta = \sec^2\theta$ **35.** $\cot^2\theta + 1 = \csc^2\theta$

36. *Historical Question*   There is evidence (see Eves' *In Mathematical Circles*) that the Babylonians had constructed tables of the trigonometric functions as early as 1900 to 1600 B.C. A tablet known as the Plimpton 322 can be used as a table to secant values from 45° to 31°. There were probably companion tablets for angles ranging from 30° to 16° and from 15° to 1°. The Babylonians used primitive Pythag-

orean triplets and a right triangle to build their table. Today, we would use a computer. Write a computer program that will output a table of trigonometric functions for the sine, cosine, and tangent for every degree from 0° to 90°.

37. *Historical Question*   Do some research on primitive Pythagorean triplets (see Problem 36), and write a paper on this topic.

1.4 Trigonometric Functions of Any Angle

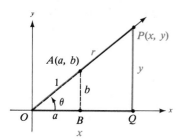

FIGURE 1.10 Similar triangles

It is not always convenient to work with the unit circle definition of the trigonometric functions developed in the last section. In this section we will make use of a result from geometry concerning similar triangles that will allow us to generalize the unit circle definition.

Similar Triangles

Suppose we choose *any* point $P(x, y)$ on the terminal side of a standard-position angle θ so that its distance r from the origin is not zero, as shown in Figure 1.10.

As before, let $A(a, b)$ be the point of intersection of the terminal side and a unit circle. If we drop perpendicular lines from A and P to the x-axis and label the points B and Q as shown in Figure 1.10, two right triangles AOB and POQ are formed. Because $\triangle AOB$ and $\triangle POQ$ are **similar triangles**, it follows that the corresponding sides of the triangles are proportional:

$$\frac{a}{1} = \frac{x}{r} \qquad \frac{b}{1} = \frac{y}{r} \qquad \frac{b}{a} = \frac{y}{x}$$

$$\frac{1}{a} = \frac{r}{x} \qquad \frac{1}{b} = \frac{r}{y} \qquad \frac{a}{b} = \frac{x}{y}$$

Generalized Definition of the Trigonometric Functions

Using these ratios, we can now state the generalized definition of the trigonometric functions that does not depend on a unit circle.

GENERALIZED DEFINITION OF THE TRIGONOMETRIC FUNCTIONS

Let θ be any angle in standard position with a point $P(x, y)$ on the terminal side a distance of r from the origin ($r \neq 0$). Then the six trigonometric functions are defined as follows:

$$\cos \theta = \frac{x}{r} \qquad\qquad \sec \theta = \frac{r}{x} \quad (x \neq 0)$$

$$\sin \theta = \frac{y}{r} \qquad\qquad \csc \theta = \frac{r}{y} \quad (y \neq 0)$$

$$\tan \theta = \frac{y}{x} \quad (x \neq 0) \qquad \cot \theta = \frac{x}{y} \quad (y \neq 0)$$

This definition of the trigonometric functions should be memorized because a great deal of your success in this course will depend on a quick recall of this definition. Note from the Pythagorean theorem that r is determined by x and y:*

$$r^2 = x^2 + y^2$$

or

$$r = \sqrt{x^2 + y^2}$$

EXAMPLE 1   Evaluating trigonometric functions given a point on the terminal side

Find the values of the trigonometric functions for an angle θ in standard position whose terminal side passes through $(-5, 5)$.

SOLUTION $x = -5$ and $y = 5$ (given); $r = \sqrt{x^2 + y^2} = \sqrt{(-5)^2 + (5)^2} = 5\sqrt{2}$.

$$\cos\theta = \frac{-5}{5\sqrt{2}} \qquad \sin\theta = \frac{5}{5\sqrt{2}} \qquad \tan\theta = \frac{5}{-5}$$

$$= \frac{-\sqrt{2}}{2} \qquad\qquad = \frac{\sqrt{2}}{2} \qquad\qquad = -1$$

$$\sec\theta = \frac{5\sqrt{2}}{-5} \qquad \csc\theta = \frac{5\sqrt{2}}{5} \qquad \cot\theta = \frac{-5}{5}$$

$$= -\sqrt{2} \qquad\qquad = \sqrt{2} \qquad\qquad = -1$$

It is also important to understand that, even though the trigonometric functions are defined in terms of x, y, or r, they are indeed functions of θ. The relationship between x, y, r, and θ is shown in Figure 1.11.

x is pos.
y is pos.
r is pos.

θ in Quadrant I

x is neg.
y is pos.
r is pos.

θ in Quadrant II

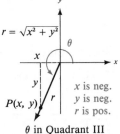

x is neg.
y is neg.
r is pos.

θ in Quadrant III

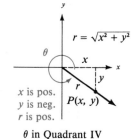

x is pos.
y is neg.
r is pos.

θ in Quadrant IV

FIGURE 1.11 Relationship between θ and x, y, and r

Reference Angles and Reduction Principle

Related to every angle is another angle that is of utmost importance in trigonometry. It is called the reference angle, and, given some angle θ, you will need to be able to find

---

*From algebra the square root property says $r = \pm\sqrt{x^2 + y^2}$ but we include only the positive value because r represents a distance.

its **reference angle**, called θ', routinely and effortlessly.* For this reason, you should consider the following definition with care.

REFERENCE ANGLE

Given any angle θ, the **reference angle** θ' is defined as the acute angle the terminal side of θ makes with the x-axis. This means that θ' is positive and is between 0° and 90° or between 0 and $\frac{\pi}{2}$. (Note that $\frac{\pi}{2} \approx 1.57$.)

The procedure for finding the reference angle varies, depending on the quadrant of θ, as shown in Figure 1.12. Notice that the reference angle θ', shown in color, is always drawn to the x-axis and *never* to the y-axis.

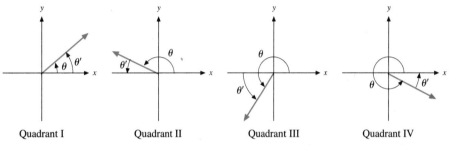

FIGURE 1.12 Reference angles

EXAMPLE 2   Finding reference angles

Find the reference angle, and draw both the given angle and the reference angle.

a. 210°   **b.** 150°

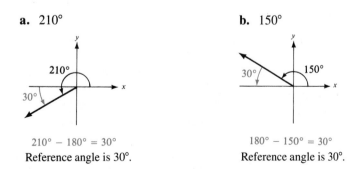

210° − 180° = 30°   180° − 150° = 30°
Reference angle is 30°.   Reference angle is 30°.

---

*We will use θ' to denote the reference angle of θ, α' to denote the reference angle of α, and so on throughout this book. That is, if you see the notation β', you should know that it is the reference angle for β.

c. $-\frac{5\pi}{3}$

$$2\pi - \frac{5\pi}{3} = \frac{6\pi}{3} - \frac{5\pi}{3} = \frac{\pi}{3}$$

Reference angle is $\frac{\pi}{3}$

d. 812°
If the angle is more than one revolution, first find a nonnegative coterminal angle less than one revolution. 812° is coterminal with 92°.

$$180° - 92° = 88°$$

Reference angle is 88°.

e. 2.5

$$\pi - 2.5 \approx 3.14 - 2.5 = .64$$

Reference angle is .64.

f. 30
30 is coterminal with 4.867.
$$(30 - 8\pi \approx 4.867)$$

$$2\pi - 4.867 \approx 6.282 - 4.867 = 1.416$$

Reference angle is 1.416.

Reference angles are used in a procedure that reduces the value of any trigonometric function to a function of its reference angle. To carry out this procedure, called the **reduction principle**, you will also need to remember in which quadrants the trigonometric functions are positive and in which quadrants they are negative.

REDUCTION PRINCIPLE   A trigonometric function of an angle θ equals that function of the reference angle θ' with the sign (positive or negative) appropriate to the quadrant of θ.

E X A M P L E 3   Writing trigonometric functions as functions of acute angles

Rewrite each function so that it is a function of an acute angle.

a. tan 210°   **b.** sec 210°   **c.** cos 405°

SOLUTION   **a.** tan 210°   =   +   tan 30°

Quadrant III: Reference
tangent positive,   angle
secant negative

b. sec 210°   =   −   sec 30°

c. cos 405° = cos 45°

Quadrant I:
cosine positive

1.4

A *Find the reference angle in each of Problems 1–6.*

1. a. 30°   **b.** 60°   **c.** 120°   **d.** 240°   **e.** 50°
2. a. 300°   **b.** 330°   **c.** 390°   **d.** 400°   **e.** 450°
3. a. 140°   **b.** 70°   **c.** 200°   **d.** 310°   **e.** 510°
4. a. −60°   **b.** −120°   **c.** −150°   **d.** −180°   **e.** −75°
5. a. −210°   **b.** −240°   **c.** −270°   **d.** −300°   **e.** −400°
6. a. −315°   **b.** −268°   **c.** −418°   **d.** −910°   **e.** −520°

Use a calculator to find the values of the trigonometric functions in Problems 7–12 correct to six decimal places.

7. a. cos 50° **b.** sin 20° **c.** tan 440° **d.** sin 148°   **e.** cos 85°
8. a. sin 70° **b.** tan 28° **c.** cos 409° **d.** sin 185°   **e.** cos 215°
9. a. sec 35° **b.** csc 48° **c.** csc 13.3° **d.** sec 794.6° **e.** cot 18°
10. a. cot 25°   **b.** sec 75°   **c.** csc 52.8°
    d. cot 399.8°   **e.** csc 183.2°
11. a. tan(−20°)   **b.** sin(−40°)   **c.** cos(−50°)
    d. sec(−18°)   **e.** cot(−38°)
12. a. sin(−128°)   **b.** csc(−213°)   **c.** cot(−170°)
    d. sec(−335°)   **e.** cot(−415°)

Rewrite each function in Problems 13–18 so that it is a function of an acute angle.

13. a. cos 250°   **b.** sin 220°   **c.** tan 440°
    d. sin 148°   **e.** cos 175°
14. a. sin 370°   **b.** tan 328°   **c.** cos 409°
    d. sin 185°   **e.** cos 365°
15. a. cos 460°   **b.** tan 452°   **c.** sin 276°
    d. cos 498°   **e.** tan 389°
16. a. sec 335°   **b.** csc 348°   **c.** csc 413.3°
    d. sec 794.6°   **e.** cot 125.3°
17. a. cot 625°   **b.** sec 715°   **c.** csc 388°
    d. cot 165°   **e.** sec 193°
18. a. tan(−20°)   **b.** sin(−40°)   **c.** cos(−50°)
    d. sec(−18°)   **e.** cot(−80°)

B *Find the values of the six trigonometric functions for an angle θ in standard position whose terminal side passes through the points in Problems 19–24. Draw a picture showing θ and the reference angle θ'.*

19. (3, 4)   **20.** (3, −4)   **21.** (−5, −12)
22. (−5, 12)   **23.** (−6, 1)   **24.** (−2, −3)

Use your calculator to evaluate the expressions given in Problems 25–30, and round to four decimal places.

25. a. $\cos(2 \cdot 30°)$ **b.** $2 \cos 30°$

  c. $\dfrac{\cos(2 \cdot 60°)}{2}$ **d.** $\dfrac{2 \cos 60°}{2}$

26. a. $(\cos 30°)^2$ **b.** $(\sin 30°)^2$

  c. $\csc\left(\tfrac{1}{2} \cdot 60°\right)$ **d.** $\dfrac{\csc 60°}{2}$

27. a. $\sin 17° \csc 17°$ **b.** $\cos 38° \sec 38°$

  c. $\tan 25° \cot 25°$

28. a. $\cos(80° - 50°)$ **b.** $\cos 80° - \cos 50°$

  c. $\cos 80° \cos 50° + \sin 80° \sin 50°$

29. a. $\sin(70° - 40°)$ **b.** $\sin 70° - \sin 40°$

  c. $\sin 70° \cos 40° - \cos 70° \sin 40°$

30. a. $\sin 18.6° \csc 18.6°$ **b.** $\tan 134.2° \cot 134.2°$

  c. $\cos 214.3° \sec 214.3°$

31. *Historical Question* In this book we have measured angles in degrees and in radians. Another measurement, called a *grad,* was proposed in France and is used in Europe. A grad, also called a *new degree,* is found on many calculators and is defined as a unit of measurement that divides one revolution into 400 equal parts. These new degrees are denoted by 1^g. Notice that 100^g is a right angle and 200^g is a straight angle. Write a formula relating grads and degrees.

32. Use Problem 31 to evaluate (correct to four decimal places) the following functions.

  a. $\cos 50^g$ **b.** $\sin 150^g$ **c.** $\tan 25^g$

33. Use Problem 31 to evaluate (correct to four decimal places) the following functions.

  a. $\sin 14^g$ **b.** $\cos 83^g$ **c.** $\tan 261^g$

1.5 Trigonometric Functions of Real Numbers

Trigonometric Functions

In Section 1.3, the trigonometric functions were defined using a unit circle, and this definition was extended in Section 1.4 to include any angle. In this section we will extend the definition of the trigonometric functions to include real-number domains.

The extension of the definition is easy because radian measure is already in terms of real numbers. That is, for

$$\cos \frac{\pi}{2} \quad \text{or} \quad \cos 2$$

it does not matter whether $\frac{\pi}{2}$ and 2 are considered as radian measures of angles or simply as real numbers—the functional values are the same.

TRIGONOMETRIC FUNCTIONS OF REAL NUMBERS

For any real number t, let $t = \theta$, where θ is a standard-position angle measured in radians. Then

$$\cos t = \cos \theta \qquad \sin t = \sin \theta \qquad \tan t = \tan \theta$$
$$\sec t = \sec \theta \qquad \csc t = \csc \theta \qquad \cot t = \cot \theta$$

EXAMPLE 1   Functional values of trigonometric functions of real numbers

  a. $\cos(-2)$ **b.** $\sin \frac{\pi}{12}$ **c.** $\sec 2.68$ **d.** $\cot \sqrt{3}$

SOLUTION   Switch to radian mode (see your owner's manual if you are not sure how to do this).

    a. DISPLAY: $-.4161468365$

       Note the negative value because the angle -2 is in Quadrant III (cosine negative).

    b. DISPLAY: $.2588190451$

       Note the positive value because $\frac{\pi}{12}$ is in Quadrant I (sine positive).

    c. DISPLAY: -1.116888769

       Note the negative value because 2.68 is in Quadrant II (secant negative).

    d. DISPLAY: $-.1626668737$

       Note the negative value because $\sqrt{3} \approx 1.71$ is in Quadrant II
       (cotangent negative).

Exact Values

Calculators and tables give most values for the trigonometric functions as approximations. However, many applications in this course, as well as in calculus and more advanced courses, require **exact values**.

EXAMPLE 2   **Exact values of trigonometric functions of $\frac{\pi}{4} = 45°$**

Find the exact value of

    a. $\cos \dfrac{\pi}{4}$ **b.** $\sin \dfrac{\pi}{4}$ **c.** $\tan \dfrac{\pi}{4}$ **d.** $\sec \dfrac{\pi}{4}$ **e.** $\csc \dfrac{\pi}{4}$ **f.** $\cot \dfrac{\pi}{4}$

SOLUTION   **a.** $\cos \dfrac{\pi}{4} = \dfrac{x}{r}$ From the definition (this is true for any angle)

           $= \dfrac{x}{\sqrt{x^2 + y^2}}$ By substitution $(r = \sqrt{x^2 + y^2})$

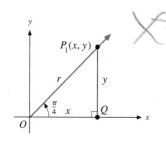

If $\theta = \frac{\pi}{4}$, then $x = y$ since $\frac{\pi}{4}$ bisects Quadrant I. By substitution,

$$\cos \frac{\pi}{4} = \frac{x}{\sqrt{x^2 + x^2}}$$

$$= \frac{x}{\sqrt{2x^2}}$$

$$= \frac{x}{x\sqrt{2}}\qquad\qquad \sqrt{x^2} = |x| = x \quad \text{since } x \text{ is positive in Quadrant I}$$

$$= \frac{1}{\sqrt{2}}$$

$$= \frac{1}{2}\sqrt{2}\qquad\qquad \frac{1}{\sqrt{2}} = \frac{1}{\sqrt{2}} \cdot \frac{\sqrt{2}}{\sqrt{2}} = \frac{\sqrt{2}}{2} = \frac{1}{2}\sqrt{2}$$

b. $\sin \dfrac{\pi}{4} = \dfrac{y}{\sqrt{x^2 + y^2}} = \dfrac{x}{\sqrt{x^2 + x^2}} = \dfrac{\sqrt{2}}{2}$ From part **a** and the fact that $x = y$

c. $\tan \dfrac{\pi}{4} = \dfrac{\sin \frac{\pi}{4}}{\cos \frac{\pi}{4}}$ Ratio identity

$= \dfrac{\frac{\sqrt{2}}{2}}{\frac{\sqrt{2}}{2}}$ Parts **a** and **b**

$= 1$

d. $\sec \dfrac{\pi}{4} = \dfrac{1}{\cos \frac{\pi}{4}}$ Reciprocal identity

$= \dfrac{1}{\frac{1}{2}\sqrt{2}}$ Part **a**

$= \sqrt{2}$

e. $\csc \dfrac{\pi}{4} = \dfrac{1}{\sin \frac{\pi}{4}} = \sqrt{2}$ **f.** $\cot \dfrac{\pi}{4} = \dfrac{1}{\tan \frac{\pi}{4}} = 1$

E X A M P L E   3   Exact values of trigonometric functions of $30° = \frac{\pi}{6}$

SOLUTION   Consider not only the standard-position angle 30°, but also the standard-position angle $-30°$ as shown in Figure 1.13. Choose $P_1(x, y)$ and $P_2(x, -y)$, respectively, on the terminal sides.

Angles OQP_1 and OQP_2 are right angles, so $\angle OP_1Q = 60°$ and $\angle OP_2Q = 60°$. Thus, $\triangle OP_1P_2$ is an equiangular triangle (all angles measure 60°). From geometry, an equiangular triangle has sides that are the same length. Thus, $2y = r$. Notice the following relationship between x and y:

$$r^2 = x^2 + y^2 \quad \text{Pythagorean theorem}$$
$$(2y)^2 = x^2 + y^2 \quad \text{Since } r = 2y$$
$$3y^2 = x^2 \quad \text{Subtract } y^2 \text{ from both sides.}$$
$$\sqrt{3}|y| = |x|$$

For 30°, x and y are both positive, so $x = \sqrt{3}y$.

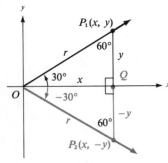

FIGURE 1.13 30°–60°–90° triangle

THE RECIPROCAL FUNCTIONS ARE EASIER:

$\cos 30° = \dfrac{x}{r} = \dfrac{\sqrt{3}y}{2y} = \dfrac{\sqrt{3}}{2}$ $\sec 30° = \dfrac{2}{\sqrt{3}} = \dfrac{2}{3}\sqrt{3}$

$\sin 30° = \dfrac{y}{r} = \dfrac{y}{2y} = \dfrac{1}{2}$ $\csc 30° = 2$

$\tan 30° = \dfrac{y}{x} = \dfrac{y}{\sqrt{3}y} = \dfrac{1}{\sqrt{3}} = \dfrac{\sqrt{3}}{3}$ $\cot 30° = \sqrt{3}$

EXAMPLE 4   **Exact values of trigonometric functions of $\frac{\pi}{3} = 60°$**

Find the exact values for the trigonometric functions of $\dfrac{\pi}{3}$.

SOLUTION   For an angle of 60° instead of 30° (see Figure 1.13), it can be shown that $r = 2x$, so that

$$r^2 = x^2 + y^2 \qquad \text{Pythagorean theorem}$$
$$(2x)^2 = x^2 + y^2 \qquad \text{Since } r = 2x$$
$$3x^2 = y^2 \qquad \text{Subtract } x^2 \text{ from both sides.}$$
$$\sqrt{3}|x| = y$$

For 60°, x and y are both positive so $y = \sqrt{3}\,x$:

$$\cos 60° = \frac{x}{r} = \frac{x}{2x} = \frac{1}{2} \qquad\qquad \sec 60° = 2$$

$$\sin 60° = \frac{y}{r} = \frac{\sqrt{3}\,x}{2x} = \frac{\sqrt{3}}{2} \qquad \csc 60° = \frac{2}{\sqrt{3}} = \frac{2}{3}\sqrt{3}$$

$$\tan 60° = \frac{y}{x} = \frac{\sqrt{3}\,x}{x} = \sqrt{3} \qquad \cot 60° = \frac{1}{\sqrt{3}} = \frac{1}{3}\sqrt{3}$$

The results of Examples 2–4 are summarized in Table 1.2. You should now be able to verify all the entries in this table.

TABLE 1.2   Exact Trigonometric Values

ANGLE θ FUNCTION	$0 = 0°$	$\dfrac{\pi}{6} = 30°$	$\dfrac{\pi}{4} = 45°$	$\dfrac{\pi}{3} = 60°$	$\dfrac{\pi}{2} = 90°$	$\pi = 180°$	$\dfrac{3\pi}{2} = 270°$
$\cos\theta$	1	$\dfrac{\sqrt{3}}{2}$	$\dfrac{\sqrt{2}}{2}$	$\dfrac{1}{2}$	0	-1	0
$\sin\theta$	0	$\dfrac{1}{2}$	$\dfrac{\sqrt{2}}{2}$	$\dfrac{\sqrt{3}}{2}$	1	0	-1
$\tan\theta$	0	$\dfrac{\sqrt{3}}{3}$	1	$\sqrt{3}$	undef.	0	undef.
$\sec\theta$	1	$\dfrac{2}{\sqrt{3}} = \dfrac{2}{3}\sqrt{3}$	$\dfrac{2}{\sqrt{2}} = \sqrt{2}$	$\dfrac{2}{1} = 2$	undef.	$\dfrac{1}{-1} = -1$	undef.
$\csc\theta$	undef.	$\dfrac{2}{1} = 2$	$\dfrac{2}{\sqrt{2}} = \sqrt{2}$	$\dfrac{2}{\sqrt{3}} = \dfrac{2}{3}\sqrt{3}$	1	undef.	-1
$\cot\theta$	undef.	$\dfrac{3}{\sqrt{3}} = \sqrt{3}$	1	$\dfrac{1}{\sqrt{3}} = \dfrac{\sqrt{3}}{3}$	0	undef.	0

$\sec\theta$, $\csc\theta$, and $\cot\theta$ are the reciprocals (which is why the exact values are given in reciprocal form as well as in simplified form).

This table of values is used extensively, and you should memorize it as you did multiplication tables in elementary school. It is used along with the reduction principle as shown in Example 5.

EXAMPLE 5   Exact values by using the reduction principle

  a. $\cos 135°$ **b.** $\tan 210°$ **c.** $\sin\left(-\dfrac{7\pi}{6}\right)$

  d. $\tan\dfrac{5\pi}{3}$ **e.** $\csc\left(-\dfrac{5\pi}{3}\right)$ **f.** $\cot\left(\dfrac{-11\pi}{6}\right)$

SOLUTION   a. $\cos 135° = -\cos 45° = -\dfrac{\sqrt{2}}{2}$ �месReference angle

Quadrant II, so
cosine is negative

  b. $\tan 210° = +\tan 30° = \dfrac{\sqrt{3}}{3}$ (Quadrant III)

  c. $\sin\left(-\dfrac{7\pi}{6}\right) = +\sin\dfrac{\pi}{6} = \dfrac{1}{2}$ (Quadrant II)

  d. $\tan\dfrac{5\pi}{3} = -\tan\dfrac{\pi}{3} = -\sqrt{3}$ (Quadrant IV)

  e. $\csc\left(-\dfrac{5\pi}{3}\right) = +\csc\dfrac{\pi}{3} = \dfrac{2}{\sqrt{3}} = \dfrac{2}{3}\sqrt{3}$ (Quadrant I)

  f. $\cot\left(-\dfrac{11\pi}{6}\right) = +\cot\dfrac{\pi}{6} = \dfrac{3}{\sqrt{3}} = \sqrt{3}$ (Quadrant I)

1.5 Problem Set

A *Find the functional values correct to four decimal places in Problems 1–8 by calculator.*

1. a. $\sin 1$ **b.** $\cos .2$ **c.** $\tan .3$
   d. $\sec .5$ **e.** $\csc .8$

2. a. $\cos 1.5$ **b.** $\sin .75$ **c.** $\tan .6$
   d. $\cot 1.2$ **e.** $\sec .95$

3. a. $\cos .45$ **b.** $\tan 1.35$ **c.** $\sin 1.08$
   d. $\sec .66$ **e.** $\cot .52$

4. a. $\sin 2.5$ **b.** $\cos 4$ **c.** $\cos 5$
　　d. $\sec 4$ **e.** $\csc 6$

5. a. $\tan(-2)$ **b.** $\sin(-1.5)$ **c.** $\cos(-.85)$
　　d. $\cot(-.55)$ **e.** $\csc(-.25)$

6. a. $\tan(-2.4)$ **b.** $\cos(-7)$ **c.** $\sin(-8)$
　　d. $\cot(-15)$ **e.** $\sec \sqrt{2}$

7. a. $\csc \frac{\pi}{15}$ **b.** $\cos 5.8$ **c.** $\sin 10$
　　d. $\tan 23$ **e.** $\sec\left(-\frac{7\pi}{5}\right)$

8. a. $\sec\left(-\frac{3\pi}{10}\right)$ **b.** $\sin \frac{5\pi}{3}$ **c.** $\cos \frac{2\pi}{7}$
　　d. $\tan \frac{\pi}{15}$ **e.** $\cot\left(-\frac{3\pi}{11}\right)$

In Problems 9–15 give the exact functional value of each from memory.

9. a. $\tan \frac{\pi}{4}$ **b.** $\cos 0$ **c.** $\sin 60°$
　　d. $\csc \frac{\pi}{6}$ **e.** $\csc 0$

10. a. $\cos 270°$ **b.** $\cos 30°$ **c.** $\sec \frac{\pi}{4}$
　　d. $\tan 180°$ **e.** $\sin 45°$

11. a. $\sin \frac{\pi}{2}$ **b.** $\sec 0°$ **c.** $\tan 0$
　　d. $\sec \frac{\pi}{6}$ **e.** $\cos \frac{\pi}{4}$

12. a. $\cot \pi$ **b.** $\sec \frac{\pi}{4}$ **c.** $\tan 90°$
　　d. $\tan 60°$ **e.** $\cos \frac{\pi}{3}$

13. a. $\cot 45°$ **b.** $\cos \pi$ **c.** $\sin \frac{3\pi}{2}$
　　d. $\sin 0°$ **e.** $\sec \frac{\pi}{2}$

14. a. $\csc \frac{\pi}{2}$ **b.** $\cos 90°$ **c.** $\sec \pi$
　　d. $\tan 270°$ **e.** $\sin \frac{\pi}{6}$

15. a. $\tan \frac{\pi}{6}$ **b.** $\sin \pi$ **c.** $\cot 90°$
　　d. $\sec \frac{\pi}{3}$ **e.** $\sec 270°$

B *Use Table* 1.2 *and the reduction principle to find the exact functional value in Problems* 16–21.

16. a. $\cos 300°$ **b.** $\sin 120°$ **c.** $\tan 120°$
　　d. $\cot 240°$ **e.** $\sec 120°$

17. a. $\tan 135°$ **b.** $\sin(-135°)$ **c.** $\cos 225°$
　　d. $\cot 225°$ **e.** $\tan 315°$

18. a. $\sin 240°$ **b.** $\cos(-300°)$ **c.** $\tan(-120°)$
　　d. $\sin 300°$ **e.** $\cos 120°$

19. a. $\csc \frac{3\pi}{2}$ **b.** $\sin 390°$ **c.** $\sin \frac{17\pi}{4}$
　　d. $\cos(-6\pi)$ **e.** $\sin(-675°)$

20. a. $\cos\left(-\frac{5\pi}{6}\right)$ **b.** $\sin \frac{5\pi}{6}$ **c.** $\cot \frac{5\pi}{6}$
　　d. $\tan \frac{7\pi}{6}$ **e.** $\sec\left(-\frac{7\pi}{6}\right)$

21. a. $\sin \frac{7\pi}{3}$ **b.** $\cos\left(-\frac{11\pi}{3}\right)$ **c.** $\tan\left(-\frac{7\pi}{3}\right)$
　　d. $\sec \frac{5\pi}{3}$ **e.** $\sec\left(-\frac{5\pi}{3}\right)$

Simplify each of the expressions in Problems 22–31 by substituting the exact values for the trigonometric functions and then algebraically simplifying.

22. a. $\sin \frac{\pi}{6} \csc \frac{\pi}{6}$ **b.** $\csc \frac{\pi}{2} \sin \frac{\pi}{2}$

23. a. $\sin^2 \frac{\pi}{3} + \cos^2 \frac{\pi}{3}$ **b.** $\sin^2 \frac{\pi}{6} + \cos^2 \frac{\pi}{3}$

24. a. $\sin\left(\frac{\pi}{4} - \frac{\pi}{2}\right)$ **b.** $\sin \frac{\pi}{4} - \sin \frac{\pi}{2}$

25. a. $\tan(2 \cdot 30°)$ **b.** $2 \tan 30°$

26. a. $\csc\left(\frac{1}{2} \cdot 60°\right)$ **b.** $\dfrac{\csc 60°}{2}$

27. a. $\cos\left(\frac{\pi}{2} - \frac{\pi}{6}\right)$ **b.** $\cos \frac{\pi}{2} \cos \frac{\pi}{6} + \sin \frac{\pi}{2} \sin \frac{\pi}{6}$

28. a. $\tan(2 \cdot 60°)$ **b.** $\dfrac{2 \tan 60°}{1 - \tan^2 60°}$

29. a. $\cos\left(\frac{1}{2} \cdot 60°\right)$ **b.** $\sqrt{\dfrac{1 + \cos 60°}{2}}$

30. a. $\sin\left(\frac{1}{2} \cdot 60°\right)$ **b.** $\sqrt{\dfrac{1 - \cos 60°}{2}}$

31. a. $\sin\left(\frac{1}{2} \cdot 120°\right)$ **b.** $\sqrt{\dfrac{1 - \cos 120°}{2}}$

32. You will learn in calculus that (for x near 0)

$$\sin x = x - \frac{x^3}{3!} + \frac{x^5}{5!} - \frac{x^7}{7!} + \cdots$$

where $n! = n(n - 1)(n - 2) \cdots \cdot 3 \cdot 2 \cdot 1$. Find $\sin 1$ correct to four decimal places by using this equation.

33. You will learn in calculus that

$$\cos x = 1 - \frac{x^2}{2!} + \frac{x^4}{4!} - \frac{x^6}{6!} + \cdots$$

where $n! = n(n - 1)(n - 2) \cdots \cdot 3 \cdot 2 \cdot 1$. Find $\cos 1$ correct to four decimal places by using this equation.

1.6 Summary and Review

Objectives	*Pages / Examples / Problems*

1.1 Angles and Degree Measure

1. Know the Greek letters used to name angles in trigonometry.

p. 4; Figure 1.1; Problems 1–4.

2. Draw angles whose measures are given in degrees; know the approximate size of angles 0°, 30°, 45°, 60°, 90°, and multiples of these angles.

pp. 5–6; Example 1; Problems 5–10.

3. Convert degree–minutes–seconds to decimal degrees.

pp. 6–7; Example 2; Problems 17–36.

4. Find a positive angle less than one revolution that is coterminal with a given angle.

p. 7; Example 3; Problems 11–16. Also, Section 1.2, Problem 7–10.

1.2 Radian Measure of Angles

5. Draw angles whose measures are given in radians.

pp. 9–11; Example 1; Problems 3–6.

6. Convert degree measure to radian measure.

pp. 11–12; Example 2; Problems 1, 11–16.

7. Convert radian measure to degree measure.

pp. 12–13; Example 3; Problems 2, 17–19.

8. Find arc length.

pp. 13–15; Examples 4–6; Problems 20–32.

1.3 Trigonometric Functions on a Unit Circle

9. Know the unit circle definition of the trigonometric functions.

p. 18; Example 1; Problem 1.

10. State the eight fundamental identities.

pp. 19–21; Example 5; Problems 2–5; 23–27.

11. Prove the eight fundamental identities.

pp. 19–21; Problems 32–35.

12. Use the fundamental identities to simplify trigonometric expressions and to evaluate trigonometric functions.

pp. 19–21; Examples 2–4; Problems 3–5, 12–14, 18–31.

13. Find the approximate values of the trigonometric functions using a unit circle and the definition.

p. 22; Examples 6 and 8; Problems 6–8.

14. Know the signs of the trigonometric functions by quadrant.

pp. 23–24; Example 7; Problems 15–17.

15. Evaluate the trigonometric functions using a calculator.

p. 25; Example 9; Problems 9–11, 21–22. Also Section 1.4; Problems 7–12, 25–30.

1.4 Trigonometric Functions of Any Angle

16. Know the generalized definition of the trigonometric functions.

p. 27.

17. Evaluate the trigonometric functions given a point on the terminal side.

p. 28; Example 1; Problems 19–24.

18. Find the reference angle for any given angle.

pp. 28–29; Example 2; Problems 1–6.

19. Use the reduction principle to write trigonometric functions as functions of acute angles.

pp. 30–31; Example 3; Problems 13–18.

1.5 **Trigonometric Functions of Real Numbers**

20. Evaluate trigonometric functions of real numbers by calculator.   pp. 32–33; Example 1; Problems 1–8.

21. Know the exact values for the trigonometric functions of $0, \frac{\pi}{6}, \frac{\pi}{4}, \frac{\pi}{3}, \frac{\pi}{2}, \pi,$ and $\frac{3\pi}{2}$.   pp. 33–35; see Table 1.2 and Examples 2–5; Problems 9–31.

TERMS

Acute angle [1.1]   Coterminal angles [1.1]   Omega (ω) [1.1]   Side of an angle [1.1]
Alpha (α) [1.1]   Degree [1.1]   Phi (ϕ) [1.1]   Similar triangles [1.4]
Angle [1.1]   Delta (δ) [1.1]   Positive angle [1.1]   Sine [1.3]
Arc [1.1]   Evaluation [1.3]   Pythagorean identities [1.3]   Standard-position
Arc length [1.2]   Exact values [1.5]   Quadrant [1.1]   angle [1.1]
Beta (β) [1.1]   Fundamental identity [1.3]   Quadrantal angle [1.1]   Straight angle [1.1]
Central angle [1.1]   Gamma (γ) [1.1]   Radian [1.2]   Subtended angle [1.1]
Circular functions [1.5]   Identity [1.3]   Ratio identities [1.3]   Tangent [1.3]
Circumference [1.1]   Initial side [1.1]   Reciprocal identities [1.3]   Terminal side [1.1]
Complementary   Lambda (λ) [1.1]   Reduction principle [1.4]   Theta (θ) [1.1]
   relationships [1.3]   Minute [1.1]   Reference angle [1.4]   Trigonometric functions [1.3]
Cosecant [1.3]   Mu (μ) [1.1]   Right angle [1.1]   Unit circle [1.3]
Cosine [1.3]   Negative angle [1.1]   Secant [1.3]   Vertex [1.1]
Cotangent [1.3]   Obtuse angle [1.1]   Second [1.1]

STUDY HINTS

In the following review of objectives, there is one question for each objective in this chapter. If you have forgotten how to do a particular problem or you miss a review problem, you can refer to the pages given in this section to find discussion, examples, and similar problems. (*All* answers for the chapter review of objectives are provided in the answer section.)

CHAPTER 1 REVIEW OF OBJECTIVES

1. Name the following Greek letters.
   a. α **b.** θ **c.** γ **d.** λ **e.** β

2. Draw from memory the angle whose measure is given.
   a. 30°   **b.** 45°   **c.** 60°   **d.** 180°   **e.** 300°

3. Change 50°36′ to decimal degrees correct to the nearest hundredth degree.

4. Find the positive angle less than one revolution that is coterminal with the given angle.
   a. 400°   **b.** −150°   **c.** 7   **d.** $\frac{-5\pi}{6}$ **e.** −π

5. Draw from memory the angle whose measure is given.
   a. 1   **b.** 3   **c.** $\frac{3\pi}{4}$ **d.** $\frac{2\pi}{3}$ **e.** $\frac{5\pi}{6}$

6. Change 500° to radians correct to the nearest hundredth.

7. Change −3.8 to degrees correct to the nearest degree.

8. A curve on a highway is laid out as the arc of a circle of radius 500 m. If the curve subtends a central angle of 18.0°, what is the distance around this section of road? Give the exact answer and an answer rounded to the nearest meter.

9. If α is an angle in standard position, what are the coordinates of P_α, which is the point of intersection of the terminal side of α and the unit circle?

10. State from memory the eight fundamental identities.

11. Prove $1 + \tan^2\theta = \sec^2\theta$ without using any of the fundamental identities.

12. Find the exact value of $\tan^2 114° - \sec^2 114°$ without using tables or a calculator.

13. Use the unit circle definition to approximate the values of the trigonometric functions of 340°. Use Figure 1.9 on page 26.

14. Classify each of the six trigonometric functions as positive or negative for each of the quadrants.

	Quadrant I	Quadrant II	Quadrant III	Quadrant IV
Positive	**a.**	**c.**	**e.**	**g.**
Negative	**b.**	**d.**	**f.**	**h.**

15. Find the values of the trigonometric functions for an angle with measure 42.5°. Round to four decimal places.

16. State the generalized definition of the trigonometric functions.

17. Using the definition of the trigonometric functions, find their exact values for an angle δ whose terminal side passes through $(-1, \sqrt{8})$.

18. Find the reference angle for each of the given angles.
    a. 500°   **b.** $-60°$ **c.** $\frac{2\pi}{3}$ **d.** $\frac{5\pi}{6}$ **e.** 82°

19. Write each as a function of an acute angle.
    a. cos 200°   **b.** sin$(-200°)$ **c.** tan$(-192°)$
    d. cot 520°   **e.** csc 300°

20. Evaluate. Round to four decimal places.
    a. cos 2   **b.** sin 4   **c.** tan .85
    d. csc(-1.5) **e.** sec(-8)

21. From memory, fill in the blanks indicated by lowercase letters. (This problem also involves objective 19.)

Angle in degrees	60°			300°	$-210°$			225°
Angle in radians		$\frac{\pi}{2}$	$\frac{5\pi}{6}$			$-\frac{11\pi}{3}$	3π	
cosine	**a.**	**b.**	**c.**	**d.**	**e.**	**f.**	**g.**	**h.**
sine	**i.**	**j.**	**k.**	**l.**	**m.**	**n.**	**o.**	**p.**
tangent	**q.**	**r.**	**s.**	**t.**	**u.**	**v.**	**w.**	**x.**

DISCUSSION PROBLEMS

1. Discuss the measurement of angles. Include measurements such as revolutions, degrees, radians, as well as the possibility of other types of measures.

2. Discuss the concept of arc length. Why not just use a ruler to measure the length of an arc?

3. What is a trigonometric function?

4. Discuss the evaluation of a trigonometric function. Why do we sometimes obtain approximate values and sometimes exact values? Why derive any exact values?

5. Write a 500-word paper on the history of trigonometry.

MISCELLANEOUS PROBLEMS

1. If α is an angle in standard position, what are the coordinates of the point of intersection of the terminal side of α and a circle with radius 2?

2. If β is an angle in standard position, what are the coordinates of the point of intersection of the terminal side of β and a circle with radius 5?

3. If 23° is the measure of an angle in standard position, what are the coordinates of the point of intersection of the terminal side of this angle and a unit circle?

4. If 30° is the measure of an angle in standard position, what are the coordinates of the point of intersection of the terminal side of this angle and a unit circle?

Change the degree measures to radian measures (correct to the nearest hundredth).

  5. 150°   **6.** $-30°$ **7.** $-100°$ **8.** 24.5°

Change the radian measures to degree measures (correct to the nearest degree).

  9. $\frac{2\pi}{3}$ **10.** $-\frac{5\pi}{4}$

11. -2.5 **12.** .4589

Find the arc length (approximate answers correct to the nearest unit).

13. $r = 5$ in.; $\theta = 6$ **14.** $r = \sqrt{50}$ ft; $\theta = \frac{\pi}{6}$

15. $r = 6.4$ ft; $\theta = 65°$ **16.** $r = 34$ in.; $\theta = 125°$

Determine the quadrant of each angle, and find a positive angle less than one revolution that is coterminal with the given angle.

17. 375°   **18.** $-15°$

19. 815°   **20.** $-160°$

Use a calculator to find the value of the trigonometric functions correct to four decimal places.

21. sin 85.2°   **22.** tan 106.9°

23. cos 2   **24.** sec 3

25. cos(−3.5)   **26.** −sin(−18.5°)

27. cot 6   **28.** csc 158°

29. −tan 34.5°   **30.** sin .5°

31. sin .5   **32.** sec(−55.6°)

Use the definition of the trigonometric functions to find the values for an angle θ whose terminal side passes through the given point.

33. (4, 5)   **34.** (−2, 3)

35. (5, $\sqrt{11}$)   **36.** ($\sqrt{55}$, 3)

Find the exact value of the following trigonometric functions.

37. sin 60°   **38.** cos(−60°)

39. tan $\frac{\pi}{3}$ **40.** cot $\frac{5\pi}{3}$

41. cos 135°   **42.** sin(−135°)

43. −cot 45°   **44.** −tan 225°

45. sec 0°   **46.** csc 2π

47. cot $\frac{3\pi}{2}$ **48.** tan π

49. csc 240°   **50.** sec 210°

51. tan$\left(-\frac{11\pi}{3}\right)$ **52.** sin(−330°)

▌ Angular and Linear Velocity—From Windmills to Pistons

Courtesy of Sara Hunsaker

In Section 1.2 we defined radian measure for angles and derived an equation for finding the length of an arc:

$$s = r\theta$$

where s is the arc length, r is the radius of the circle, and θ is the angle measured in radians. We can use this notion to define two concepts used extensively in engineering and physics, one giving a linear equation and the other giving a quadratic equation.

Suppose a bike wheel with a 24-in. diameter is turning completely around five times every second. We say that the wheel is rotating at the rate of 5 revolutions per second (rsp), or 300 revolutions per minute (rpm). A spoke of this wheel will turn 2π radians for each rotation, so in each second the spoke will turn

$$5 \cdot 2\pi \text{ radians per second} = 10\pi \text{ radians per second or } 10\pi \text{ radians/sec}$$

We call this the **angular velocity**, denoted by ω.

ANGULAR VELOCITY

> ω = angle (in radians) per unit of time or $\omega = \dfrac{\theta}{t}$

explain ───→ Since the radius of the bike wheel is 12 in., a point on the circumference will move 12 in. for each radian. But since the wheel turns 10π radians per second, a point on the circumference moves

$$10\pi \cdot 12 \text{ in. per sec} = 120\pi \text{ in. per sec} \qquad \text{or} \qquad 10\pi \cdot 1 \text{ ft per sec} = 10\pi \text{ ft per sec}$$

We call this the **linear velocity**, denoted by V.

EXAMPLE 1 **Finding the linear velocity for a bicycle**

Calculate the linear velocity for this bike in miles per hour.

SOLUTION The angular velocity in radians per hour is

$$5\frac{\text{rev}}{\text{sec}} \cdot 60\frac{\text{sec}}{\text{min}} \cdot 60\frac{\text{min}}{\text{hr}} \cdot 2\pi\frac{\text{radians}}{\text{rev}} = 36{,}000\pi\frac{\text{radians}}{\text{hr}}$$

These convert   This converts
seconds to   revolutions to
hours.   radians.

Thus, the angular velocity (per radian) $= 36{,}000\pi\dfrac{\text{radian}}{\text{hr}} \cdot \dfrac{1}{\text{radian}} = \dfrac{36{,}000\pi}{\text{hr}}$.

Next, the radius (in miles) is

$$12\text{ in.} = 1\text{ ft} = \frac{1}{5{,}280}\text{ mi}$$

LINEAR VELOCITY $=$ (ANGULAR VELOCITY PER RADIAN)(RADIUS)

$$= \frac{36{,}000\pi}{\text{hr}}\left(\frac{1}{5{,}280}\text{ mi}\right)$$

$$= (36{,}000\pi)\left(\frac{1}{5{,}280}\right)\frac{\text{mi}}{\text{hr}}$$

$$\approx 21.42\,\frac{\text{mi}}{\text{hr}}$$

The bike is traveling at about 21.4 mph.

In general, if

$$s = r\theta \qquad \theta \text{ is measured in radians.}$$

then

$$\frac{s}{t} = r \cdot \frac{\theta}{t} \qquad t \text{ is the unit of time.}$$

Since

$$V = \frac{s}{t} \quad \text{and} \quad \omega = \frac{\theta}{t}$$

we have, by substitution,

$$V = r\omega$$

which is the number of radians per unit of time.

| **LINEAR VELOCITY** | $V = r\omega$ |

EXAMPLE 2   Finding the angular velocity of a pulley

A belt runs a pulley of radius 40 cm at 50 rpm. Find the angular velocity of the pulley and the linear velocity of the belt.

SOLUTION   For the angular velocity of the pulley:

IN RADIANS PER MINUTE

$$\omega = \frac{\theta}{t}$$

$$= \frac{50 \cdot 2\pi \text{ radians}}{1 \text{ min}}$$

$$= 100\pi \text{ radians/min}$$

IN RADIANS PER HOUR

$$\omega = \frac{100\pi \text{ radians}}{1 \text{ min}} \cdot \frac{60 \text{ min}}{1 \text{ hr}}$$

$$= \frac{6{,}000\pi \text{ radians}}{\text{hr}} \text{ or } 6{,}000\pi \text{ rad/hr}$$

For the linear velocity of the belt:

IN CENTIMETERS PER MINUTE

$$V = r\omega$$

$$= 40 \text{ cm} \cdot \frac{100\pi}{\text{min}}$$

$$= \frac{4{,}000\pi \text{ cm}}{\text{min}} \text{ or } 4{,}000\pi \text{ cm/min}$$

IN KILOMETERS PER HOUR

$$V = 40 \text{ cm} \cdot \frac{6{,}000\pi}{\text{hr}}$$

$$= \frac{240{,}000\pi \text{ cm}}{\text{hr}} \cdot \frac{1 \text{ km}}{100{,}000 \text{ cm}}$$

$$= \frac{2.4\pi \text{ km}}{\text{hr}} \approx 7.54 \text{ kph}$$ ▮

A second application is found by looking at the action of a piston, connecting rod, and crankshaft in an internal-combustion engine (see Figure 1.14).

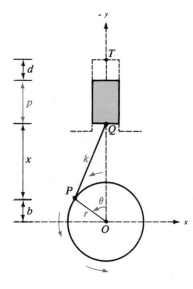

FIGURE 1.14

We wish to determine the position of the piston indicated by the distance of its top from the fully extended position at any given time of the rotation of the crankshaft (see Figure 1.14).

Fully extended position: $|OT| = p + k + r$

Partial-stroke position: $|OT| = b + x + p + d$

Solve for d: $d = |OT| - b - x - p$

Substitute the relationship for the fully extended position into the equation that was solved for d. We obtain

$$d = (p + k + r) - b - x - p$$

Now, let θ be a rotation on the crankshaft as shown in Figure 1.15, and let P be a point on the terminal side of θ with coordinates (a, b). Then

$$\cos \theta = \frac{b}{r} \quad \text{or} \quad b = r \cos \theta$$

$$\sin \theta = \frac{a}{r} \quad \text{or} \quad a = r \sin \theta$$

Also, by the Pythagorean theorem,

$$a^2 + x^2 = k^2$$
$$x^2 = k^2 - a^2$$
$$x = \sqrt{k^2 - a^2} \qquad \text{Positive value, since } x \text{ is a distance}$$
$$= \sqrt{k^2 - r^2 \sin^2\theta} \qquad \text{Since } a = r \sin \theta$$

We now have an expression for d in terms of k, r, and θ:

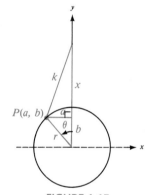

FIGURE 1.15

$$d = (p + k + r) - b - x - p$$
$$= k + r - b - x$$
$$= k + r - r \cos \theta - \sqrt{k^2 - \sin^2\theta}. \qquad \text{Substitute } b = r \cos \theta \text{ and } x = \sqrt{k^2 - r^2 \sin^2\theta}.$$

The distance d is the value you are looking for, but you will usually know the velocity of the crankshaft rather than θ, so perform one last substitution to make this a useful formula. If s is the arc length, then $s = r\theta$ and

arc length = (circumference) (number of revolutions)

arc length = (circumference) (revolutions per minute) (time in minutes)

$$s = (2\pi r)(v)(t) \qquad \text{Where } v = rpm, \quad t = \text{time in minutes}$$

$$r\theta = 2\pi rvt \qquad \text{Since } s = r\theta$$

$$\theta = 2\pi vt \qquad \text{Divide both sides by } r.$$

POSITION OF A PISTON

$$d = k + r - r\cos(2\pi vt) - \sqrt{k^2 - r^2 \sin^2 (2\pi vt)}$$

where d = depth of stroke of a piston
k = length of connecting rod
r = radius of the crankshaft
v = velocity of the crankshaft in rpm
t = time in minutes

EXAMPLE 3 **Finding the position of a piston**

Find the position of a piston after .06 second if the connecting rod is 15 cm and the radius of the crankshaft is 5 cm. Also suppose that the crankshaft is turning at 600 rpm.

SOLUTION Given $k = 15$, $r = 5$, $v = 600$, and $t = .06/60 = .001$. Thus,

$$d = 15 + 5 - 5\cos\theta - \sqrt{15^2 - 5^2 \sin^2\theta}$$

where $\theta = 2\pi vt = 1{,}200\pi (.001)$. This calculator sequence is not a particularly easy one, so we will take it slowly, one step at a time.

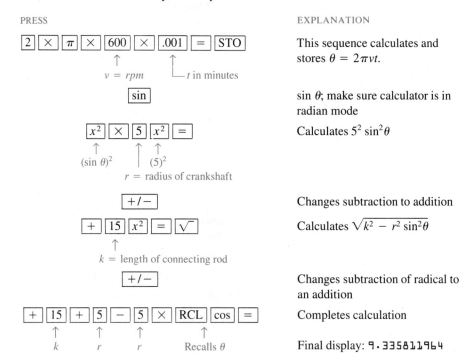

PRESS	EXPLANATION
2 × π × 600 × .001 = STO	This sequence calculates and stores $\theta = 2\pi vt$.
v = rpm t in minutes
sin	$\sin\theta$; make sure calculator is in radian mode
x^2 × 5 x^2 =	Calculates $5^2 \sin^2\theta$
(sin θ)²   (5)² r = radius of crankshaft
+/−	Changes subtraction to addition
+ 15 x^2 = √	Calculates $\sqrt{k^2 - r^2 \sin^2\theta}$
k = length of connecting rod
+/−	Changes subtraction of radical to an addition
+ 15 + 5 − 5 × RCL cos =	Completes calculation
k r r Recalls θ cos θ	Final display: 9.335811964

The piston is about 9.3 cm from the top of the stroke after .06 sec.

PROBLEMS FOR FURTHER STUDY: ANGULAR AND LINEAR VELOCITY

1. **Physics** A flywheel with radius 15 cm is spinning so that a point on the rim has a linear velocity of 120 cm/sec. Find the angular velocity.

2. **Physics** The wheel of an automobile is 60 cm in diameter. What is the angular velocity of the wheel in radians per second for this car if it is traveling at 30 kph?

3. **Physics** An airplane propeller measures 4 m from tip to tip and rotates at 1,800 rpm. Find the angular velocity and the linear velocity of a point on the tip of one of the blades. (Assume that the airplane itself is not moving.)

In Problems 4–6, find the position of the piston in Example 3 after the given times.

4. .01 second   5. 2 seconds   6. 3.852 seconds

7. **Space Science** An earth satellite travels in a circular orbit at 32,000 kph. If the radius of the orbit is 6,770 km, what is the angular velocity (in radians per hour)?

8. **Physics** The biggest windmill in the world was built in Ulfborg, Denmark, to generate electricity. Its concrete tower soars 54 meters and is topped by three giant fiberglass propeller blades, each 27 meters long. If the propeller rotates at 30 rpm, find the angular velocity and the linear velocity of a point on the tip of one of the blades.

9. **Physics** On January 28, 1978, the first federally funded commercial wind generator in the United States was dedicated in Clayton, New Mexico. The rotors span 125 feet

each ($r = 125$) and generate enough electricity for about 60 homes. If the rotor rotates at 40 rpm, find the angular velocity and the linear velocity of a point on the tip of one of the rotors.

10. **Engineering** Find the position of a piston after $\frac{1}{2}$ second if the crankshaft is turning at 60 revolutions per minute and its radius is 4 inches. Suppose that the connecting rod is 8 inches long.

11. **Engineering** Find the position of the piston in Problem 10 if the crankshaft is turning at 200 rpm.

12. **Engineering** The connecting rod in Problem 10 is twice the length of the radius of the crankshaft. Rewrite the general equation for d in terms of any radius r after t seconds, and then simplify for $\theta = 1$ to the nearest hundredth.

Hipparchus
(c. 180–125 B.C.)

Trigonometry contains the science of continually undulating magnitude: meaning magnitude which becomes alternately greater and less, without any termination to succession of increase or decrease. . . . All trigonometric functions are not undulating; but it may be stated that in common algebra nothing but infinite series undulate; in trigonometry nothing but infinite series do not undulate.

AUGUSTUS DE MORGAN
Trigonometry and Double Algebra

During the second half of the second century B.C., the astronomer Hipparchus of Nicaea compiled the first trigonometric tables in 12 books. It was he who used the 360° of the Babylonians and thus introduced trigonometry with the degree angle measure that we still use today. Hipparchus' work formed the foundation for Ptolemy's *Mathematical Syntaxis*, the most significant early work in trigonometry.

The words *tangent* and *secant* come from the relationship of the ratios to the lengths of the tangent and secant lines drawn on a circle. However, the word *sine* has a more interesting origin. According to Howard Eves in his book *In Mathematical Circles,* Aryabhata called it *jya-adga* (chord half) and abbreviated it *jya,* which the Arabs first wrote as *jiba* but later shortened to *jb.* Later writers saw *jb*, and since *jiba* was meaningless to them, they substituted *jaib* (cove or bay). The Latin equivalent for *jaib* is *sinus*, from which our present word *sine* is derived. Finally, Edmund Gunter (1581–1626) first used the prefix *co* to invent the words *cosine* and *cotangent*. In 1620 Gunter published a seven-place table of the common logarithms of the sines and tangents of angles of intervals of a minute of an arc. Gunter originally entered the ministry but later decided on astronomy as a career. He was such a poor preacher, Eves states, that Gunter left the ministry in 1619 to the "benefit of both occupations."

tan θ and sec θ are the same as the lengths of the tangent and secant lines drawn on a unit circle:

$\tan \theta = \frac{y}{x} = \frac{y}{1} = |PA| \qquad \overleftrightarrow{PA}$ is tangent

$\sec \theta = \frac{r}{x} = r = |PO| \qquad \overleftrightarrow{PO}$ is a secant.

2

GRAPHS OF TRIGONOMETRIC FUNCTIONS

Contents

2.1 Graphs of Cosine, Sine, and Tangent Functions

2.2 General Cosine, Sine, and Tangent Curves

2.3 Graphs of Secant, Cosecant, and Cotangent Functions

2.4 Addition of Ordinates

2.5 Chapter 2 Summary and Review

Application for Further Study
    Computer Graphing of Trigonometric Functions

Cumulative Review for Chapters 1 and 2

Preview

In this chapter we introduce the idea of graphing the trigonometric functions. You can understand these functions more easily if you form a mental image or "picture" of them. A quick, simple method called *framing* is introduced to allow you to graph general cosine, sine, and tangent functions. This framing technique does not require background practice working on amplitude, period, and phase displacement, but instead graphs the functions very easily and efficiently in one step. Also, using available computer and calculator technology, you can more easily "see" many relationships and variations of those functions. The optional Application for Further Study at the end of this chapter gives you some insight into how you might use a computer as you progress through the material of this section.

▌ Perspective

You would probably expect to encounter the graph of a trigonometric curve in a science or physics course, but such curves are used to describe all sorts of periodic or cyclic phenomena. For example, the accompanying news clipping is from *The Wall Street Journal*.

Nikolai Kondratieff's Wave Is Running a Bit Behind Schedule (We Hope)

In the 1920s, Soviet economist Nikolai Kondratieff identified what he believed to be a long-term cycle governing economic activity. Based on studies of such factors as wages and commodity prices in 19th century Britain, France, and America, he espoused the theory that capitalist economies expand and contract in cycles of 50 to 60 years.

The idealized wave shown here is often used as a simplified portrayal of Kondratieff's theory. The crests and troughs represent relative highs and lows in economic activity. Although the long wave uncannily pretraced the Great Depression and the post-war boom,

it is dismissed by many modern economists as a statistical absurdity. So, they aren't worried that the wave now suggests the West is heading toward another economic crisis.

As for Kondratieff, he was branded a counterrevolutionary because his waves were inconsistent with the Marxist belief that capitalism would self-destruct. The economist was imprisoned in a Siberian labor camp where, according to Aleksandr Solzhenitsyn, he "became mentally ill and died."

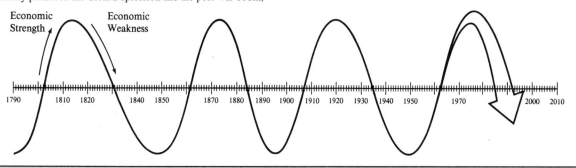

2.1 Graphs of Cosine, Sine, and Tangent Functions

In the last chapter we introduced the core of a trigonometry course—the general definition of the trigonometric functions. Our task is now to amplify, explain, discuss, and apply that definition in a variety of situations.

The nature of a function can often be more easily understood by looking at a picture or graph of it. In this chapter we will consider the graphs of these trigonometric functions. With most functions we begin by plotting points to determine the general shape. We then generalize so you can graph the curve without a lot of calculations of individual points. In order to plot points we need to relate the trigonometric functions to the Cartesian coordinate system. For example, we may wish to write $y = \tan x$, but in so doing we change the meaning of x and y as they were used in the last chapter. Recall from the *definition* of the trigonometric functions that

$$\tan \theta = \frac{y}{x}$$

where (x, y) represented a point on the terminal side of θ. *Now*, when we write

$$y = \tan x$$

x represents the *angle*, and y is the value of the tangent function of x.

Graph of Standard Sine Function

We will begin by considering the graph of $y = \sin x$. To graph $y = \sin x$, begin by plotting familiar values for the sine:

EXACT VALUES:

x = REAL NUMBER	0	$\frac{\pi}{2}$	π	$\frac{3\pi}{2}$	2π
$y = \sin x$	0	1	0	-1	0

These are exact values, and they are plotted as colored points in Figure 2.1.

CALCULATOR VALUES:

x	y
-1	$-.84$
1	$.84$
2	$.91$
3	$.14$
4	$-.76$
5	$-.96$
6	$-.28$
7	$.66$

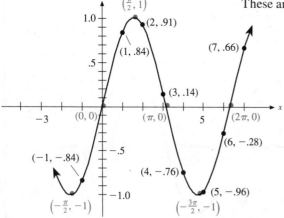

FIGURE 2.1 Graph of $y = \sin x$ by plotting points

You should plot additional points using more exact values, or a calculator (see the table with Figure 2.1; these are plotted as black points in the figure). The plotted points are then connected by a smooth curve called the **sine curve**.

The domain for x is all real numbers, so what about values other than $0 \le x \le 2\pi$? Using the reduction principle, we know that

$$\sin x = \sin(x + 2\pi) = \sin(x - 2\pi) = \sin(x + 4\pi) = \sin(x + 6\pi) = \dots$$

More generally,

$$\sin(\theta + 2n\pi) = \sin \theta$$

for any integer n. In other words, the values of the sine function repeat themselves after 2π. We describe this by saying that sine is **periodic** with period 2π. The sine curve is shown in Figure 2.2. Notice that, even though the domain of the sine function is all real numbers, the range is restricted to values between -1 and 1 (inclusive).

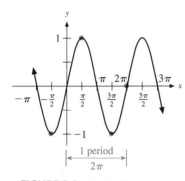

FIGURE 2.2 Graph of $y = \sin x$
Domain: all real x
Range: $-1 \le y \le 1$

FRAMING A SINE CURVE

Notice that, for the period labeled in Figure 2.2, the sine curve starts at $(0, 0)$, goes *up* to $(\frac{\pi}{2}, 1)$ and then *down* to $(\frac{3\pi}{2}, -1)$ passing through $(\pi, 0)$, and then goes back up to $(2\pi, 0)$, which completes one period. This graph shows that the **range** of the sine function is $-1 \le y \le 1$. We can summarize a technique for sketching the since curve (see Figure 2.3 on page 52) called **framing the curve**.

<table>
<tr><td>

FRAMING A SINE CURVE

⊗ The standard sine curve has period 2π. ⊗

</td><td>

The standard sine function

$$y = \sin x$$

has domain $-\infty < x < \infty$ and range $-1 \le y \le 1$, and is periodic with period 2π. One period of this curve can be sketched by framing, as follows:

1. *Start* at the origin $(0, 0)$.

2. *Height* of this frame is two units: one unit up and one unit down from the starting point $(-1 \le y \le 1)$.

3. *Length* of this frame is 2π units (about 6.28) from the starting point (the period is 2π).

4. The curve is now framed. Plot five critical points within the frame:
    a. Endpoints (along axis)
    b. Midpoint (along axis)
    c. Quarterpoints (up first, then down)

5. Draw the curve through the critical points, remembering the shape of the sine curve.

</td></tr>
</table>

FIGURE 2.3 Framing a sine curve

You can draw a standard sine curve using any point as a starting point; the procedure is always the same.

EXAMPLE 1   Drawing one period of a standard sine curve

Draw one period of a standard sine curve using (3, 1) as the starting point for building a frame.

SOLUTION   Plot the point (3, 1) as the starting point. Draw the standard frame as shown in the following figure.

Remember, when framing a curve, plot the starting point, and then *count squares on your graphing paper* to build the frame. ⊗

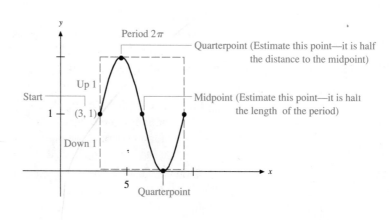

Framing a Cosine Curve

We can graph the cosine curve by plotting points, as we did with the sine curve. The details of plotting these points are left as an exercise. The cosine curve $y = \cos x$ is framed as described in Figure 2.4.

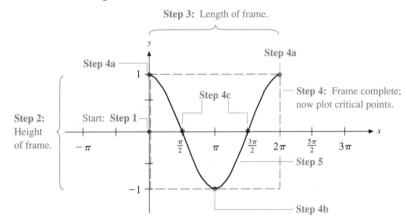

FIGURE 2.4 Framing a cosine curve

Notice that the only difference in the steps for graphing the cosine and the sine curves is in Step 4; the procedures for building the frame are identical.

FRAMING A COSINE CURVE

The standard cosine curve has period 2π. ⊗

The standard cosine function

$$y = \cos x$$

has domain $-\infty < x < \infty$ and range $-1 \leq y \leq 1$ and is periodic with period 2π. One period of this curve can be sketched by framing, as follows:

1. *Start* at the origin $(0, 0)$.
2. *Height* of the frame is two units: one unit up and one unit down from the starting point $(-1 \leq y \leq 1)$.
3. *Length* of the frame is 2π units (about 6.28) from the starting point (the period is 2π).
4. The curve is now framed. Plot five critical points within the frame:
    a. Endpoints (at the top corners of the frame)
    b. Midpoint (at the bottom of the frame)
    c. Quarterpoints (along the axis)
5. Draw the curve through the critical points, remembering the shape of the cosine curve.

Since values for x greater than 2π or less than 0 are coterminal with those already considered, we see that **the period of the cosine function is 2π**. The cosine curve is shown in Figure 2.5. Notice that the domain and range of the cosine function are the same as they were for the sine functions; D: all reals; R: $-1 \le \cos x \le 1$.

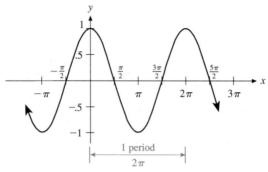

FIGURE 2.5 Graph of $y = \cos x$
Domain: all real x
Range: $-1 \le y \le 1$

EXAMPLE 2   Drawing one period of a standard cosine curve

Draw one period of a standard cosine curve using $\left(3, -\frac{1}{2}\right)$ as the starting point for building a frame.

SOLUTION   Plot the point: $\left(3, -\frac{1}{2}\right)$ as the starting point. Notice that the graph in this example is drawn using integers (compare with Example 1) so that you must approximate the distance 2π.

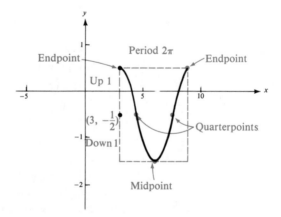

Framing a Tangent Curve

By setting up a table of values and plotting points (the details are left as an exercise), we notice that $y = \tan x$ does not exist at $\frac{\pi}{2}, \frac{3\pi}{2}$, or $\frac{\pi}{2} + \cdot n\pi$ for any integer n. The lines $x = \frac{\pi}{2}, x = \frac{3\pi}{2}, \ldots, x = \frac{\pi}{2} + n\pi$ for which the tangent is not defined are called **asymptotes**. The procedure is summarized in Figure 2.6.

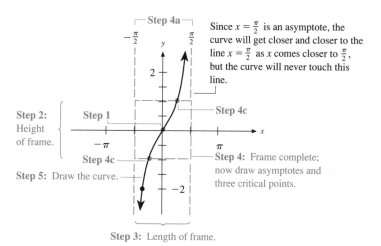

FIGURE 2.6 Framing a tangent curve

FRAMING A TANGENT
CURVE

The standard tangent curve
has period π.

The standard tangent function

$$y = \tan x$$

has as domain and range all real numbers except $x \neq \frac{\pi}{2} + n\pi$ and is periodic with period π. One period of this curve can be sketched by framing, as follows:

1. *Start* at the origin $(0, 0)$; for the tangent curve this is the *center* of the frame.

2. *Height* of this frame is two units; one unit up and one unit down from the starting point.

3. *Length* of this frame is π units (about 3.14) and is drawn so that it is $\frac{\pi}{2}$ (about 1.57) units on each side of the starting point (the period is π).

4. The curve is now framed. Draw the asymptotes and plot three critical points within the frame:
   a. Extend the vertical sides of the frame; these are the asymptotes.
   b. Midpoint (this was the starting point)
   c. Quarterpoints (down first; then up on the frame)

5. Draw the curve through the critical points, using the asymptotes as guides and remembering the shape of the tangent curve.

The tangent curve is indicated in Figure 2.7 on page 56. Even though the curve repeats for values of x greater than 2π or less than 0, it also repeats after it has passed through an interval with length π. For this reason, $\tan(\theta + n\pi) = \tan \theta$ for any integer n, and we see that **the tangent has a period of π**. The domain of the tangent function is restricted so that multiples of π added to $\frac{\pi}{2}$ are excluded; this is because the tangent is not defined for these values. The range, on the other hand, is unrestricted; it is the set of all real numbers.

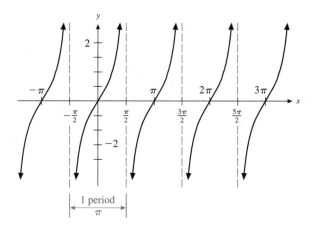

FIGURE 2.7 Graph of $y = \tan x$

Domain: all real x, $x \neq \dfrac{n\pi}{2}$ (n any integer)

Range: all real y

EXAMPLE 3   Drawing one period of a standard tangent curve

Draw one period of a standard tangent curve using $(\pi, -2)$ as a starting point for building a frame.

SOLUTION   Plot the point $(\pi, -2)$ and build the frame as shown.

⧺ Remember, the starting point is in the *center* of the frame for the tangent curve. ⊛

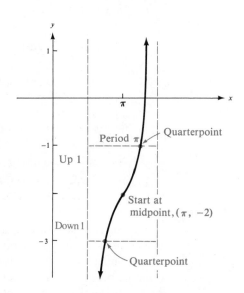

2.1 Problem Set

A *In Problems 1–5 give the functional value of each from memory.*

1. a. $\sec 270°$ **b.** $\cot \frac{3\pi}{2}$ **c.** $\sin 0$
  d. $\cos \frac{\pi}{6}$ **e.** $\csc 0$ **f.** $\tan 0°$

2. a. $\csc 30°$ **b.** $\tan \pi$ **c.** $\sec \frac{\pi}{2}$
  d. $\sin \frac{\pi}{6}$ **e.** $\cot 180°$ **f.** $\sin 45°$

3. a. $\tan 90°$ **b.** $\sec \pi$ **c.** $\cot 90°$
  d. $\cos 0$ **e.** $\csc 90°$ **f.** $\cos \frac{\pi}{3}$

4. a. $\cot \frac{\pi}{6}$ **b.** $\sin 60°$ **c.** $\cos 45°$
  d. $\sec 0°$ **e.** $\tan \frac{\pi}{6}$ **f.** $\csc \pi$

5. a. $\sin \frac{\pi}{2}$ **b.** $\csc 45°$ **c.** $\tan 60°$
  d. $\cot \frac{\pi}{4}$ **e.** $\cos \frac{\pi}{2}$ **f.** $\sin 270°$

6. Draw a quick sketch of $y = \sin x$ from memory by framing the curve.

7. Draw a quick sketch of $y = \cos x$ from memory by framing the curve.

8. Draw a quick sketch of $y = \tan x$ from memory by framing the curve.

In Problems 9–20 plot the given point; then draw a frame using the given point as a starting point to draw one period of the requested curve.

9. $(\pi, 1)$; cosine curve   **10.** $(\pi, 1)$; sine curve

11. $(\pi, 1)$; tangent curve   **12.** $\left(-\frac{\pi}{2}, 2\right)$; cosine curve

13. $\left(-\frac{\pi}{2}, 2\right)$; sine curve   **14.** $\left(-\frac{\pi}{2}, 2\right)$; tangent curve

15. $\left(-\frac{\pi}{4}, -2\right)$; cosine curve   **16.** $(-1, -2)$; cosine curve

17. $\left(-\frac{\pi}{4}, -2\right)$; tangent curve   **18.** $(-1, 2)$; sine curve

19. $\left(-\frac{\pi}{4}, -2\right)$; sine curve   **20.** $(-1, -2)$; tangent curve

B

21. Complete the table at the right by filling in values for $y = \cos x$:

22. Use the table in Problem 21, along with other values if necessary, to plot $y = \cos x$.

23. Complete a table of values like the one in Problem 21 for $y = \tan x$.

24. Use the table in Problem 23, along with other values if necessary, to plot $y = \tan x$.

25. Complete a table of values like the one in Problem 21 for $y = \sec x$.

26. Use the table in Problem 25, along with other values if necessary, to plot $y = \sec x$.

27. Complete a table of values like the one in Problem 21 for $y = \csc x$.

28. Use the table in Problem 27, along with other values if necessary, to plot $y = \csc x$.

29. Complete a table of values like the one in Problem 21 for $y = \cot x$.

30. Use the table in Problem 29, along with other values if necessary, to plot $y = \cot x$.

x	y
−1	·84
0	0
1	.84
2	.91
3	.14
4	−.76
5	−.96
6	−.28
7	.66

C

31. **CALCULATOR PROBLEM**   Explain (in detail) how you would graph the trigonometric curves on your calculator.*

32. Plot points (with or without a graphics calculator) to graph $y = 3 \cos(2x + \pi) - 3$.

33. Plot points (with or without a graphics calculator) to graph $y = 2 \sin(3x + \frac{\pi}{4}) + 1$.

34. Plot points (with or without a graphics calculator) to graph $y = \frac{\sin x}{x}$.

35. Take a piece of paper and wrap it around a candle, as shown in the illustration. Make cuts at A and B, and unroll the paper.

  a. What do you think the edges A and B will look like?

  b. Perform the experiment to see if your guess was right.

Cut here

Candle

Paper wrapped around candle

A B

---

*This problem requires a calculator with graphing capabilities.

For Problems 36–41 *label the origin of a coordinate system O. Draw a unit circle with center at* $(-1, 0)$. *Let* θ *be an angle drawn with the vertex at the center of this circle, initial side the positive x-axis, and the point P(x', y') the intersection of the terminal side of this angle and the unit circle. Let* $|PQ|$ *be the perpendicular drawn to the x'-axis and* $|PR|$ *the perpendicular drawn to the y'-axis. Finally, let S be the intersection of the line determined by the terminal side and the y-axis.*

36. Show $\sin \theta = |PQ|$.   **37.** Show $\cos \theta = |PR|$.

38. Show $\tan \theta = |SO|$.

39. By Problem 36, $\sin \theta = |PQ|$, and you can use this fact to help you sketch $y = \sin \theta$. For example, when $\theta = \frac{\pi}{4}$, draw this angle, and then measure $|PQ|$ and *plot this height* at $\theta = \frac{\pi}{4}$ on the θ-axis as shown below. As P makes one revolution, it is easy to quickly plot the points on the curve $y = \sin \theta$.

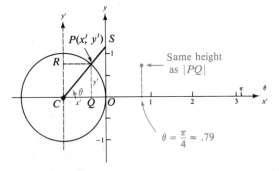

40. By Problem 38, $\tan \theta = |SO|$, and you can use this fact to help you sketch $y = \tan \theta$. For example, when $\theta = \frac{\pi}{4}$, draw this angle, and then measure $|SO|$ and *plot this height* at $\theta = \frac{\pi}{4}$ on the θ-axis as shown at the right. As P makes one revolution, notice that $|SO|$ does not exist at $\theta = \frac{\pi}{2}$ and $\frac{3\pi}{2}$.

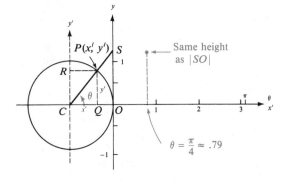

41. By Problem 37, $\cos \theta = |PR|$, and you can use this fact to help you sketch $y = \cos \theta$. To do this, rotate the unit circle described in the directions by 90°. For example, when $\theta = \frac{\pi}{4}$, draw this angle, and then measure $|PR|$ and plot this height at $\theta = \frac{\pi}{4}$ on the θ-axis as shown at the right. As P makes one revolution, it is easy to quickly plot the points on the curve $y = \cos \theta$.

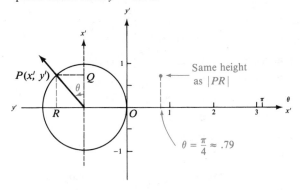

42. COMPUTER PROBLEM   There are many computer graphics programs available for every type of computer. For example, Converge, which is available on IBM format from JEMware (phone 808-523-9911), graphs trigonometric curves. Obtain a graphics program, explore graphing trigonometric functions, and then write a short paper describing the program you are using.*

---

*See also the Application for Further Study at the end of this chapter.

2.2   General Cosine, Sine, and Tangent Curves

Translations and Phase Shift

FIGURE 2.8 Graph of $y = f(x)$

Once you are given a particular function, it is possible to shift the graph of that function to other locations. For example, let $y = f(x)$ be the function shown in Figure 2.8. It is possible to shift the entire curve up, down, right, or left, as shown in Figure 2.9.

FIGURE 2.9   Shifting the graph of $y = f(x)$

FIGURE 2.10
Shifting the axes to (h, k)

Instead of considering the curve shifting relative to fixed axes, consider the effect of shifting the axes. If the coordinate axes are shifted up k units, the origin of this new coordinate system would correspond to the point $(0, k)$ on the old coordinate system. If the axes are shifted h units to the right, the origin would correspond to the point $(h, 0)$ of the old system. A horizontal shift of h units followed by a vertical shift of k units would shift the new coordinate axes so that the origin corresponds to a point (h, k) on the old axes. Suppose a *new* coordinate system with origin at (h, k) is drawn, and the new axes are labeled x' and y', as shown in Figure 2.10.

Every point on a given curve can now be denoted in two ways, as shown in Figure 2.11:

1. as (x, y) measured from the old origin, and
2. as (x', y') measured from the new origin.

To find the relationship between (x, y) and (x', y'), consider the graph shown in Figure 2.11.

$$\begin{array}{cc} x = x' + h \\ y = y' + k \end{array} \quad \text{or} \quad \begin{array}{c} x' = x - h \\ y' = y - k \end{array}$$

FIGURE 2.11   Comparison of coordinate axes

This relationship says that, if

$$y - k = f(x - h)$$

the graph of this function is the same as the graph of

$$y' = f(x')$$

where point (x', y') is measured from the new origin located at (h, k). This fact can greatly simplify the work of graphing, since $y' = f(x')$ is usually easier to graph than $y - k = f(x - h)$. This shifting is sometimes called a **phase shift**, but in this book we also call it a **translation**.

EXAMPLE 1   **Finding a translation point, (h, k)**

    a. $y - 5 = f(x - 7)$; $(h, k) = (7, 5)$

    b. $y + 6 = f(x - \pi)$; $(h, k) = (\pi, -6)$. Did you notice that $y + 6 = y - (-6)$, so that $k = -6$?

    c. $y + 1 = f\left(x + \frac{\pi}{3}\right)$; $(h, k) = \left(-\frac{\pi}{3}, -1\right)$

    d. $y = f(x)$; $(h, k) = (0, 0)$. This indicates no shift.

    e. $y - 15 = f\left(x + \frac{1}{3}\right)$; $(h, k) = \left(-\frac{1}{3}, 15\right)$

    f. $y - 6 = f(x) + 15$. Write the equation as $y - 21 = f(x)$; thus, $(h, k) = (0, 21)$. This indicates no horizontal shift.   ▌

In the case of the graphs of the circular functions, it will simply be necessary to determine (h, k) and then build a frame at (h, k) rather than at the origin.

EXAMPLE 2   **Graphing one period of a translated sine curve**

Graph one period of $y = \sin\left(x + \frac{\pi}{2}\right)$.

SOLUTION   STEP 1:   Frame the curve as shown in Figure 2.12a.

    a.   Plot $(h, k) = \left(-\frac{\pi}{2}, 0\right)$.

    b.   The period of the sine curve is 2π, and it has a high point up one unit and a low point down one unit.

STEP 2:   Plot the five critical points (two endpoints, the midpoint, and two quarterpoints). For the sine curve, plot the endpoint (h, k) and use the frame to plot the other endpoint and the midpoint. For the quarterpoints, remember that the sine curve is "up–down"; use the frame to plot the quarterpoints as shown in Figure 2.12b.

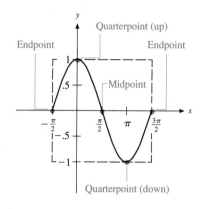

Period $= 2\pi$; this is the length of the frame.

a. Framing the curve. (This step is the same whether you are graphing a sine or a cosine function.)

FIGURE 2.12

After plotting the point $(h, k) = (-\frac{\pi}{2}, 0)$, you build the frame by counting up and down $a = 1$ unit, and to the right 2π units. It is not necessary to calculate values to find the endpoints of the frame. ⊗

STEP 3:   Remembering the shape of the sine curve, sketch the graph of one period of $y = \sin(x + \frac{\pi}{2})$ using the frame and the five critical points. If you wish to show more than one period, just repeat the same pattern.   ▌

Notice from Figure 2.12b that the graph of $y = \sin(x + \frac{\pi}{2})$ is the same as the graph of $y = \cos x$. Thus, $\sin(x + \frac{\pi}{2}) = \cos x$.

EXAMPLE 3   **Graphing one period of a translated cosine curve**

Graph one period of $y - 2 = \cos(x - \frac{\pi}{6})$.

SOLUTION   STEP 1:   Frame the curve as shown in Figure 2.13. Notice that $(h, k) = (\frac{\pi}{6}, 2)$.

STEP 2:   Plot the five critical points. For the cosine curve, the left and right endpoints are at the upper corners of the frame; the midpoint is at the bottom of the frame; the quarterpoints are on a line through the middle of the frame.

STEP 3:   Draw one period of the curve as shown in Figure 2.13.

FIGURE 2.13 Graph of one period of $y - 2 = \cos(x - \frac{\pi}{6})$ ▌

EXAMPLE 4   **Graphing one period of a translated tangent curve**

Graph one period of $y + 3 = \tan\left(x + \frac{\pi}{3}\right)$.

SOLUTION   STEP 1:   Frame the curve as shown in Figure 2.14. Notice that $(h, k) = \left(-\frac{\pi}{3}, -3\right)$, and remember that the period of the tangent is π.

STEP 2:   For the tangent curve, (h, k) is the midpoint of the frame. The endpoints, each of which is a distance of $\frac{1}{2}$ the period from the midpoint, determine the location of the asymptotes. The top and the bottom of the frame are one unit from (h, k). Locate the quarterpoints at the top and bottom of the frame, as shown in Figure 2.14.

FIGURE 2.14 Graph of one period of $y + 3 = \tan\left(x + \frac{\pi}{3}\right)$

STEP 3:   Sketch one period of the curve as shown in Figure 2.14. Remember that the tangent curve is not contained within the frame.

Changes in Amplitude and Period

We will discuss two additional changes for a function defined by $y = f(x)$. The first, $y = af(x)$, changes the scale on the y-axis; the second, $y = f(bx)$, changes the scale on the x-axis. For a function $y = af(x)$, it is clear that the y value is a times the corresponding value of $f(x)$, which means that $f(x)$ is stretched or shrunk in the y direction by the multiple of a. For example, if $y = f(x) = \cos x$, then $y = 3f(x) = 3 \cos x$ is the graph of $\cos x$ that has been stretched so that the high point is at 3 units and the low point is at -3 units. The value $2|a|$ gives the height of the frame for f. To graph $y = 3 \cos x$, frame the cosine using $a = 3$ rather than 1 (see Figure 2.15). For sine and cosine curves, $|a|$ is called the **amplitude** of the function. When $a = 1$, the amplitude is 1, so $y = \sin x$ and $y = \cos x$ are said to have amplitude 1.

EXAMPLE 5   **Graphing a trigonometric curve with an arbitrary amplitude**

Graph: $y = 3 \cos x$

SOLUTION   From the starting point $(0, 0)$, draw a frame with amplitude $a = 3$ and period 2π, as shown in Figure 2.15. *After* drawing the frame, plot the five critical points using the frame as a guide.

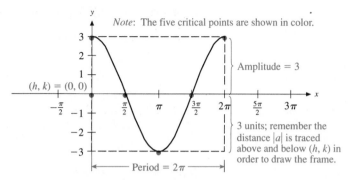

FIGURE 2.15 Graph of one period of $y = 3 \cos x$

For a function $y = f(bx)$ with $b > 0$, b affects the scale on the x-axis. Recall that $y = \sin x$ has a period of 2π ($f(x) = \sin x$, so $b = 1$). A function $y = \sin 2x$ ($f(x) = \sin x$ and $f(2x) = \sin 2x$) must complete one period as $2x$ varies from 0 to 2π. This statement means that one period is completed as x varies from 0 to π. (Remember that for each value of x, the result is doubled *before* we find the sine of that number.) In general, **the period, p, of $y = \sin bx$ is $\frac{2\pi}{b}$, and the period of $y = \cos bx$ is $\frac{2\pi}{b}$.** However, since the period of $y = \tan x$ is π, **$y = \tan bx$ has a period of $\frac{\pi}{b}$.** Therefore, when framing the curve, use $\frac{2\pi}{b}$ for the sine and cosine, and $\frac{\pi}{b}$ for the tangent.

EXAMPLE 6   **Graphing a trigonometric curve with an arbitrary period**

Graph one period of $y = \sin 2x$.

SOLUTION   The period is $p = \frac{2\pi}{2} = \pi$; thus the endpoints of the frame are $(0, 0)$ and $(\pi, 0)$, as shown in Figure 2.16.

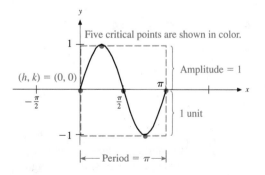

FIGURE 2.16 Graph of one period of $y = \sin 2x$

Graphs of General Trigonometric Curves

Summarizing all the preceding results, we have the *general* cosine, sine, and tangent curves:

GENERAL FORM OF COSINE, SINE, AND TANGENT CURVES

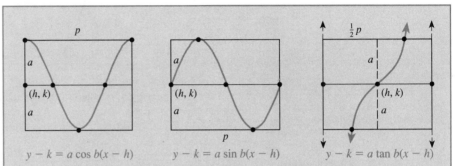

$$y - k = a \cos b(x - h) \qquad y - k = a \sin b(x - h) \qquad y - k = a \tan b(x - h)$$

1. Algebraically, put the equation into one of the general forms shown above.
2. Identify (by inspection) the following values: (h, k), a, and b. Then calculate p.
3. Draw the frame:
    a. Plot (h, k); this is the starting point.
    b. Draw a; the *height of the frame* is $2|a|$—up a units from (h, k) and down a units from (h, k).*
    c. Draw p; the *length of the frame* is p. Remember $p = \frac{2\pi}{b}$ for cosine and sine, and $p = \frac{\pi}{b}$ for tangent.
4. Locate the critical values using the frame as a guide, and then sketch the appropriate graphs. **You do not need to know the coordinates of these critical values.**

EXAMPLE 7   **Graphing a general sine curve**

Graph $y + 1 = 2 \sin \frac{2}{3}\left(x - \frac{\pi}{2}\right)$.

SOLUTION   Find (h, k), a, and p: $(h, k) = \left(\frac{\pi}{2}, -1\right)$, $a = 2$, and $b = \frac{2}{3}$; thus

$$p = \frac{2\pi}{\frac{2}{3}} = 2\pi \cdot \frac{3}{2} = 3\pi.$$

Now plot (h, k) and frame the curve. Then plot the five critical points (two endpoints, the midpoint, and two quarterpoints). Finally, after sketching one period, draw the other periods as shown in Figure 2.17.

---

*In this section we have considered only the case where a is positive. The graphs of these functions where a is negative will be considered in Section 3.4.

CALCULATOR COMMENT

Curve sketching by plotting points can be done efficiently on a calculator. Many people believe that it is not necessary to think when using a calculator, that the "machine" will do all of the work and you will no longer understand the principles involved if you begin using calculators for graphing. However, this is not true; with a calculator your attention must be focused on different matters. For example, in this section we are concerned with graphing $y = \cos x$. The first task is to decide on the domain and the range. If you know what the curve looks like, then it is easy to make an intelligent choice about the domain and range. However, if you do not know anything about the curve, you may have to make several attempts at setting the domain and the range before you obtain a "good-looking" graph. Without the work of this section prior to graphing a cosine or a sine curve on a calculator, the task would be very difficult—so difficult, in fact, that when you purchase a Casio fx-7000G calculator, many graph functions are "preprogrammed" so that you will not become discouraged when you try graphing your own functions and probably obtain very disappointing results.

For the graph shown at the top right, the following choices were made for the domain and range values:

Xmin: −1.57   Ymin: −1.1

Xmax: 6.28   Ymax: 1.1

scl: 1   scl: .2

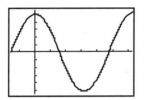

Graph of the function $y = \cos x$ on a Casio fx-7000G

Many of the newer graphing calculators are making it easier to graph functions. For example, on the TI-81 or TI-82, you simply need to press

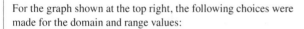

to obtain the following output:

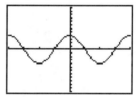

Graph of the function $y = \cos x$ on a Texas Instruments TI-81

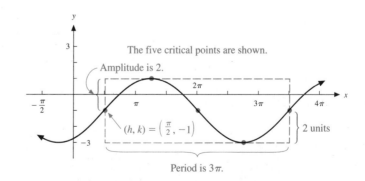

The five critical points are shown.

Amplitude is 2.

2π

$(h, k) = \left(\frac{\pi}{2}, -1\right)$

2 units

Period is 3π.

FIGURE 2.17 Graph of $y + 1 = 2 \sin \frac{2}{3}\left(x - \frac{\pi}{2}\right)$

EXAMPLE 8   Graphing a general cosine curve

Graph $y = 3 \cos\left(2x + \frac{\pi}{2}\right) - 2$.

SOLUTION   Rewrite in standard form to obtain $y + 2 = 3 \cos 2\left(x + \frac{\pi}{4}\right)$. Notice that $(h, k) = \left(-\frac{\pi}{4}, -2\right)$; $a = 3$, $b = 2$, and $p = \frac{2\pi}{2} = \pi$. Plot (h, k), and frame the curve as shown in Figure 2.18.

Do not forget to factor 2 first. ⊗

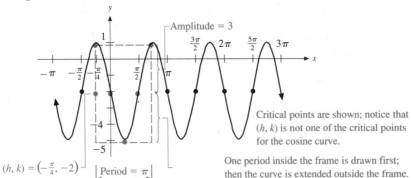

Amplitude $= 3$

Critical points are shown; notice that (h, k) is not one of the critical points for the cosine curve.

One period inside the frame is drawn first; then the curve is extended outside the frame.

$(h, k) = \left(-\frac{\pi}{4}, -2\right)$ Period $= \pi$

Graph of $y + 2 = 3 \cos 2\left(x + \frac{\pi}{4}\right)$

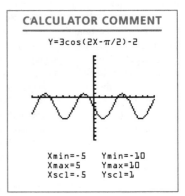

CALCULATOR COMMENT

Y=3cos(2X-π/2)-2

Xmin=-5   Ymin=-10
Xmax=5   Ymax=10
Xscl=.5   Yscl=1

EXAMPLE 9   Graphing a general tangent curve

Graph $y - 2 = 3 \tan \frac{1}{2}\left(x - \frac{\pi}{3}\right)$.

SOLUTION   Notice that $(h, k) = \left(\frac{\pi}{3}, 2\right)$; $a = 3$, $b = \frac{1}{2}$, and $p = \frac{\pi}{\frac{1}{2}} = 2\pi$. Plot (h, k), and frame the curve as shown in Figure 2.19.

CALCULATOR COMMENT

Y=3tan(.5(X-π/3))+2

Xmin=-10   Ymin=-10
Xmax=10   Ymax=10
Xscl=1   Yscl=1

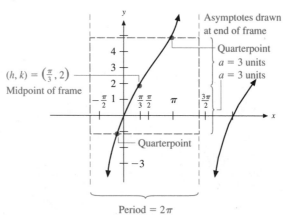

Asymptotes drawn at end of frame

Quarterpoint
$a = 3$ units
$a = 3$ units

$(h, k) = \left(\frac{\pi}{3}, 2\right)$
Midpoint of frame

Quarterpoint

Period $= 2\pi$

FIGURE 2.19 Graph of $y - 2 = 3 \tan \frac{1}{2}\left(x - \frac{\pi}{3}\right)$

Sometimes it is necessary to find an equation that fits a set of data points. The following example is from Jim Malone (Mercersburg Academy, Mercersburg, Penn. 17236).

EXAMPLE 10   **Graphing a curve from given data points**

The flywheel on a lawn mower has 16 evenly spaced cooling fins. If we rotate the engine through two complete revolutions, we can generate 32 data points, with the first coordinate representing time and the second representing the depth of the piston. Determine an equation of the curve generated by the lawn-mower-engine data shown in Table 2.1.

SOLUTION   We begin with a plot of the data points. We recognize the graph as a sine curve.

TABLE 2.1

TIME	DEPTH	TIME	DEPTH
0	0	16	0
1	.2	17	.2
2	.8	18	.8
3	1.7	19	1.7
4	2.5	20	2.5
5	3.4	21	3.4
6	4.0	22	4.0
7	4.4	23	4.4
8	4.5	24	4.5
9	4.4	25	4.4
10	4.0	26	4.0
11	3.5	27	3.5
12	2.2	28	2.2
13	1.7	29	1.7
14	.9	30	.9
15	.2	31	.2
		32	0

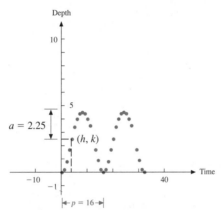

The graph has a period of 16 and an amplitude of 2.25. This means that for the general sine curve

$$y - k = a \sin b(x - h)$$

Thus, $a = 2.25 = \frac{9}{4}$; period $= \frac{2\pi}{b} = 16$ so that $b = \frac{\pi}{8}$. Finally, we need to find an appropriate starting point, (h, k): from the graph we choose $(4, 2.5)$ to find the equation

$$y - \tfrac{5}{2} = \tfrac{9}{4} \sin \tfrac{\pi}{8}(x - 4)$$

2.2   Problem Set

A *Graph one period of each of the functions given in Problems 1–21.*

1. $y = \sin(x + \pi)$

2. $y = \cos\left(x + \frac{\pi}{2}\right)$

3. $y = \sin\left(x - \frac{\pi}{3}\right)$

4. $y = \cos\left(x + \frac{3\pi}{2}\right)$

5. $y = 2 \cos x$

6. $y = 3 \sin x$

7. $y = \cos 2x$

8. $y = \sin 3x$

9. $y = \tan\left(x + \frac{\pi}{6}\right)$

10. $y = \tan\left(x - \frac{3\pi}{2}\right)$

11. $y = \frac{1}{3} \tan x$

12. $y = \frac{1}{2} \sin x$

13. $y - 2 = \sin\left(x - \frac{\pi}{2}\right)$

14. $y + 1 = \cos\left(x + \frac{\pi}{3}\right)$

15. $y - 3 = \tan\left(x + \frac{\pi}{6}\right)$

16. $y - \frac{1}{2} = \frac{1}{2} \cos x$

17. $y - 1 = 2 \cos\left(x - \frac{\pi}{4}\right)$

18. $y - 1 = \cos 2\left(x - \frac{\pi}{4}\right)$

19. $y + 2 = 3 \sin\left(x + \frac{\pi}{6}\right)$

20. $y + 2 = \sin 3\left(x + \frac{\pi}{6}\right)$

21. $y + 2 = \tan\left(x + \frac{\pi}{4}\right)$

B *Graph each of the curves given in Problems* 22–36.

22. $y = \sin(4x + \pi)$

23. $y = \sin(3x + \pi)$

24. $y = \tan\left(2x - \frac{\pi}{2}\right)$

25. $y = \tan\left(\frac{x}{2} + \frac{\pi}{3}\right)$

26. $y = \frac{1}{2}\cos\left(x + \frac{\pi}{6}\right)$

27. $y = \cos\left(\frac{1}{2}x + \frac{\pi}{12}\right)$

28. $y = 2\cos(3x + 2\pi) - 2$

29. $y = 4\sin\left(\frac{1}{2}x + 2\right)$

30. $y = \sqrt{2}\cos\left(x - \sqrt{2}\right) - 1$

31. $y = \sqrt{3}\sin\left(\frac{1}{3}x - \sqrt{\frac{1}{3}}\right)$

32. $y = 2\sin 2\pi x$

33. $y = 3\cos 3\pi x$

34. $y = 4\tan\dfrac{\pi x}{5}$

35. $y + 2 = \frac{1}{2}\cos(\pi x + 2\pi)$

36. $y - 3 = 3\cos(2\pi x + 4)$

37. ELECTRICAL ENGINEERING   The current I (in amperes) in a certain circuit (for some convenient unit of time) generates the following set of data points:

TIME	CURRENT	TIME	CURRENT
0	−60	10	30
1	−58.6889	11	40.14784
2	−54.8127	12	48.54102
3	−48.541	13	54.81273
4	−40.1478	14	58.68886
5	−30	15	60
6	−18.541	16	58.68886
7	−6.27171	17	54.81273
8	6.271708	18	48.54102
9	18.54102	19	40.14784
		20	30

Determine an equation of the curve generated by this circuit.

38. ENGINEERING   Suppose a point P on a waterwheel with a 30-ft radius is d units from the water as shown in the figure.

If it turns at 6 revolutions per minute, the height of the point P above the water is given by the following set of data points:

TIME	HEIGHT	TIME	HEIGHT
0	−1	10	−1
1	4.729499	11	4.729476
2	19.7295	12	19.72946
3	38.27052	13	38.27048
4	53.27051	14	53.27049
5	59	15	59
6	53.27051	16	53.27054
7	38.27052	17	38.27055
8	19.7295	18	19.72954
9	4.729499	19	4.729522
		20	−1

Determine an equation of the curve generated by these data points.

39. SPACE SCIENCE   The distance that a certain satellite is north or south of the equator is given by

$$y = 3{,}000 \cos\left(\frac{\pi}{60}t + \frac{\pi}{5}\right)$$

where t is the number of minutes that have elapsed since lift-off.

a. Graph the equation for $0 \le t \le 120$.

b. What is the greatest distance that the satellite ever reaches north of the equator?

c. How long does it take to complete one period?

2.3 Graphs of Secant, Cosecant, and Cotangent Functions

Graph of Cosecant Function

Procedures for graphing the reciprocal functions—namely, cosecant, secant, and cotangent—can be developed by plotting points as shown in Section 2.1. Instead, however, we will use the idea of a frame along with the idea of what it means to be a reciprocal. For example, the graph of $y = \csc x$ is related to the graph of $y = \sin x$ because

$$\csc x = \frac{1}{\sin x}$$

Remember the meaning of reciprocal:

FUNCTION	RECIPROCAL
$\sin x$	$\csc x$
1	1
$\frac{1}{2}$	2
$\frac{1}{3}$	3
$.1 = \frac{1}{10}$	10

Notice that as $\sin x$ approaches 0 (written as $\sin x \to 0$), its reciprocal $\csc x$ gets larger without limit (written as $\csc x \to \infty$). Thus, every t for which $\sin x = 0$ is an asymptote $x = t$ for the cosecant curve. The graph is shown in Figure 2.20.

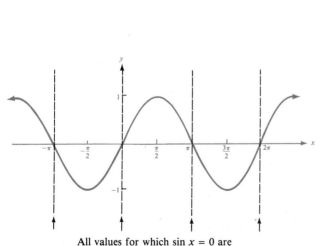

All values for which $\sin x = 0$ are asymptotes for the cosecant curve.

FIGURE 2.20

Reciprocal point, $y = 2$

$y = \sin x = \dfrac{1}{2}$

1. Draw asymptotes
2. Plot reciprocals

$y = \sin x$

FIGURE 2.21 Procedure for sketching one period of a cosecant curve

Look at one period of $y = \csc x$. Begin by drawing a frame for $y = \sin x$ (that is, the reciprocal of the curve that you want to graph). See Figure 2.21.

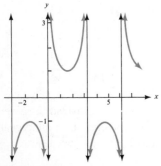

FIGURE 2.22 Graph of $y = \csc x$
Domain: all real x,
$\qquad x \neq n\pi$ (n an integer)
$\qquad$ Range: $y \geq 1$ or $y \leq -1$

Notice that the cosecant curve is not a continuous function. The domain for this curve is all real x, except those values corresponding to the asymptotes. These values are $x = n\pi$ for any integer n. The range is all values greater than (or equal to) 1 or less than (or equal to) -1. This curve is shown in Figure 2.22.

Graph of Secant Curve

The graph of the secant is also found by looking at the reciprocals, as shown in Figure 2.23.

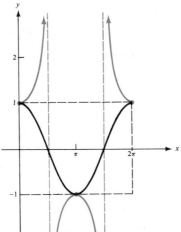

FIGURE 2.23 Procedure for graphing one period of a secant curve

PROCEDURE FOR GRAPHING $y = \sec x$

1. Draw the reciprocal curve ($y = \cos x$). One period is shown here.

2. Draw asymptotes (values for which $y = \cos x$ is 0).

3. Plot reciprocals. (Notice that you do not need to plot a lot of points—not many more than the reciprocals of the critical points for the cosine curve.)

4. Draw the curve.

The entire curve is drawn by repeating the parts drawn for the frame shown in Figure 2.23. The result is Figure 2.24.

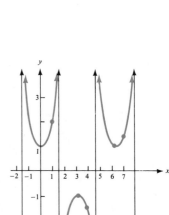

FIGURE 2.24 Graph of $y = \sec x$
Domain: all real x, $x \neq \frac{n\pi}{2}$ (n an integer)
Range: $y \leq -1$ or $y \geq 1$

Graph of Cotangent Curve

The cotangent is, of course, the reciprocal of the tangent. But remember that $y = \tan x$ itself has asymptotes. That is,

FUNCTION	RECIPROCAL
$\tan x$	$\cot x$
1	1
2	$\frac{1}{2}$
3	$\frac{1}{3}$
10	$\frac{1}{10}$
100	$\frac{1}{100}$
1,000	$\frac{1}{1,000}$

$$\tan x \to \infty \quad \text{as } x \to \frac{\pi}{2}$$

Now, if the tangent values are getting large, then the values of the reciprocals must be approaching 0. That is, $\cot x = 0$ for all values corresponding to the vertical asymptotes $x = t$ of the tangent curve. To draw one period of $y = \cot x$, look at the frame of the tangent curve.

It is difficult to "see" the cotangent curve based only on what is shown in Figure 2.25 because the cotangent curve is not continuous. However, if you repeat this period more than once and draw the entire curve as shown in Figure 2.26, it is much easier to visualize the cotangent curve.

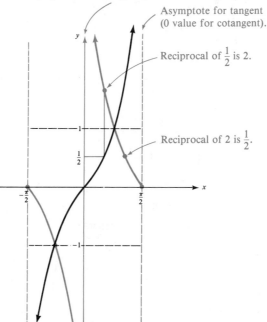

FIGURE 2.25 Detail for graphing $y = \cot x$

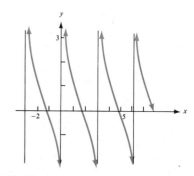

FIGURE 2.26 Graph of $y = \cot x$
Domain: all real x, $x \neq n\pi$ (n any integer)
Range: all real y

Graphs of General Trigonometric Curves

Since the graphs of the standard cosecant, secant, and tangent curves are related to the frames of the corresponding reciprocal functions, it is easy to extend the ideas of this section to the general curves as summarized in the following box.

GENERAL COSECANT, SECANT, AND COTANGENT FUNCTIONS

$y - k = a \csc b(x - h)$	Relate to the frame of the sine; starting point (h, k), height $2\lvert a\rvert$, length $\frac{2\pi}{b}$.
$y - k = a \sec b(x - h)$	Relate to the frame of the cosine; starting point (h, k), height $2\lvert a\rvert$, length $\frac{2\pi}{b}$.
$y - k = a \cot b(x - h)$	Relate to the frame of the tangent; starting point (h, k) is the center of the frame, height $2\lvert a\rvert$, length $\frac{\pi}{b}$.

EXAMPLE 1   Graph of a general cosecant curve

Graph $y - 1 = \frac{1}{2} \csc 2\left(x - \frac{\pi}{3}\right)$.

SOLUTION $(h, k) = \left(\frac{\pi}{3}, 1\right)$, $a = \frac{1}{2}$, and $p = \frac{2\pi}{2} = \pi$. Draw the frame for sine.

CALCULATOR COMMENT

Y=.5(sin(2(X-π/3)))^-1+1

EXAMPLE 2   Graph of a general secant curve

Graph $y + 2 = \frac{1}{2} \sec \frac{1}{2}\left(x + \frac{\pi}{4}\right)$.

SOLUTION $(h, k) = \left(-\frac{\pi}{4}, -2\right)$, $a = \frac{1}{2}$, $b = \frac{1}{2}$, and $p = 2\pi / \left(\frac{1}{2}\right) = 4\pi$. Relate to the frame of a cosine curve.

CALCULATOR COMMENT

Y=.5(cos(.5(X+π/4)))^-1-2

EXAMPLE 3   **Graph of a general cotangent curve**

Graph $y = 1.5 \cot \pi(x - .5)$.

SOLUTION $(h, k) = (.5, 0)$, $a = 1.5$, $b = \pi$, and $p = 1$.
Relate to the frame of a tangent curve.

CALCULATOR COMMENT

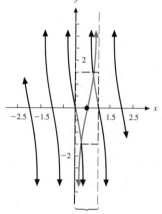

Frame and one period for tangent
Reciprocal values for cotangent

It is worthwhile to summarize the domain and range for each of the trigonometric functions. You should relate the information in Table 2.2 to the graph of each function. Remember that the domain is the set of replacements for the first component, x, and the range is the set of replacements for the second component, y. If you memorize the general shape of the trigonometric graphs, as well as the domain and range for each one, your future work will be simplified.

TABLE 2.2

FUNCTION	DOMAIN LET n BE ANY INTEGER.	RANGE
$y = \cos x$	all reals	$-1 \le y \le 1$
$y = \sin x$	all reals	$-1 \le y \le 1$
$y = \tan x$	$x \ne \dfrac{\pi}{2} + n\pi$	all reals
$y = \sec x$	$x \ne \dfrac{\pi}{2} + n\pi$	$y \le -1$ or $y \ge 1$
$y = \csc x$	$x \ne n\pi$	$y \le -1$ or $x \ge 1$
$y = \cot x$	$x \ne n\pi$	all reals

2.3 Problem Set

A *In Problems 1–6 give each functional value from memory.*

1. a. $\cos 0$ **b.** $\sec \dfrac{3\pi}{2}$ **c.** $\tan \pi$
   d. $\cot \pi$ **e.** $\csc \dfrac{3\pi}{2}$ **f.** $\sin 0$

2. a. $\cot \dfrac{\pi}{2}$ **b.** $\cos \dfrac{\pi}{6}$ **c.** $\sec \pi$
   d. $\tan \dfrac{\pi}{2}$ **e.** $\sin \dfrac{\pi}{6}$ **f.** $\csc \pi$

3. a. $\cot \dfrac{\pi}{3}$ **b.** $\csc \dfrac{\pi}{6}$ **c.** $\cos \dfrac{\pi}{4}$
   d. $\sin \dfrac{\pi}{4}$ **e.** $\tan \dfrac{\pi}{3}$ **f.** $\sec \dfrac{\pi}{2}$

4. a. $\csc \dfrac{\pi}{4}$ **b.** $\cot \dfrac{\pi}{4}$ **c.** $\sin \dfrac{\pi}{2}$
   d. $\cos \dfrac{\pi}{2}$ **e.** $\sec \dfrac{\pi}{3}$ **f.** $\tan \dfrac{\pi}{4}$

5. a. $\cot \dfrac{\pi}{6}$ **b.** $\sin \pi$ **c.** $\csc \dfrac{\pi}{3}$
   d. $\sec \dfrac{\pi}{4}$ **e.** $\cos \pi$ **f.** $\tan \dfrac{\pi}{6}$

6. a. $\sin \dfrac{3\pi}{2}$ **b.** $\csc \dfrac{\pi}{4}$ **c.** $\sec \dfrac{\pi}{6}$

　　d. $\tan 2\pi$　　**e.** $\tan 0$　　**f.** $\cos \frac{3\pi}{2}$

7. Draw a quick sketch of $y = \sec x$ from memory.

8. Draw a quick sketch of $y = \csc x$ from memory.

9. Draw a quick sketch of $y = \cot x$ from memory.

Graph the functions in each of Problems 10–30.

10. $y = 2 \sec x$

11. $y = \sec 2x$

12. $y = 2 \sec x + 1$

13. $y = 2 \sec(x + 1)$

14. $y = 2 \csc x$

15. $y = \csc 2x$

16. $y = \csc 2x + 1$

17. $y = \csc(2x + 1)$

18. $y = 2 \sec\left(x + \frac{\pi}{3}\right)$

19. $y - 2 = 2 \sec\left(x + \frac{2\pi}{3}\right)$

20. $y = \csc\left(2x - \frac{\pi}{2}\right)$

21. $y - 1 = \csc\left(2x - \frac{\pi}{2}\right)$

22. $y = 2 \cot x$

23. $y = \cot 2x$

24. $y = 2 \cot x - 1$

25. $y + 1 = 2 \cot\left(x - \frac{\pi}{6}\right)$

26. $y = 2 \cot(x - 1)$

27. $y = \sec(x - 2) + 1$

28. $y = 2 \csc(x - 1) - 2$

29. $y = 2 \cot(x + 1) - 2$

30. $y = \frac{1}{2} \csc(x - 2) - 1$

2.4　Addition of Ordinates

From radio modulation systems and sounds of musical instruments, to the quality of a singer's voice and the pattern of a human heartbeat, the periodic nature of a sine curve plays a role. For example, the following illustration shows how computers can be used to eliminate unwanted sound waves. The idea being illustrated is the idea we consider in this section—namely, that of graphing the sum of trigonometric functions.

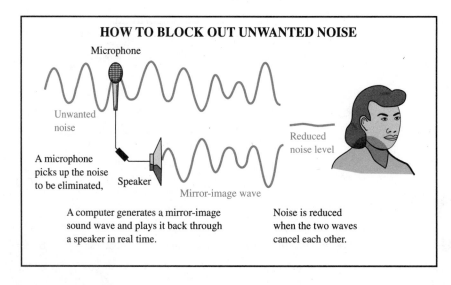

HOW TO BLOCK OUT UNWANTED NOISE

Microphone

Unwanted noise

A microphone picks up the noise to be eliminated,

Speaker

Mirror-image wave

A computer generates a mirror-image sound wave and plays it back through a speaker in real time.

Reduced noise level

Noise is reduced when the two waves cancel each other.

Adding Ordinates

We can graph the sum of two curves by **adding ordinates**. That is, we will add second components, point by point, until we have plotted enough points to see the shape of the curve that is the sum. For example, the sum of y_1 and y_2 at x_1 is found by adding the length of segment y_1 to the point (x_1, y_2) as shown in Figure 2.27.

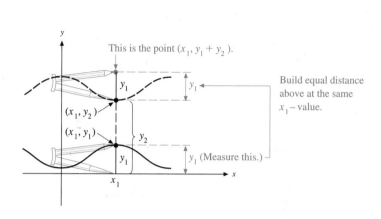

FIGURE 2.27 Finding $(x_1, y_1 + y_2)$

Sum $y_1 + y_2$ passes through points shown in color.

FIGURE 2.28 Adding ordinates

If either, or both, of y_1 or y_2 is negative, then the addition is performed by considering y_1 and y_2 as directed distances. This process is shown for several points in Figure 2.28.

EXAMPLE 1   Graphing by adding ordinates

Graph $y = \dfrac{x}{2} + \cos x$.

SOLUTION Let $y_1 = \dfrac{x}{2}$ and $y_2 = \cos x$, and write $y = y_1 + y_2$. First graph y_1 (the dotted line in Figure 2.29a). Next, graph y_2 (the solid black curve in Figure 2.29a). Choose some x value, and, using your eye or a ruler, measure y_1 and y_2 and then plot $y_1 + y_2$. Do this by counting or measuring *squares* on your graph; do *not* think in term of units. Repeat this procedure for several points, as shown by the color dots in Figure 2.29a. Finally, connect the dots for $y = y_1 + y_2$, as shown in Figure 2.29b.

(a)   (b)

FIGURE 2.29 Graph of $y = \frac{x}{2} + \cos x$

EXAMPLE 2   Graphing by adding ordinates

Graph $y = 2 \cos 3x + 3 \sin 2x$ by adding ordinates.

SOLUTION   Let $y = y_1 + y_2$, where $y_1 = 2 \cos 3x$ and $y_2 = 3 \sin 2x$. First graph $y_1 = 2 \cos 3x$ and $y_2 = 3 \sin 2x$ on the same axes, as shown in Figure 2.30. Next, select a particular x value—say, x_1—and graphically find $(x_1, y_1 + y_2)$ by adding ordinates. Do this for several points, as shown, and draw a smooth curve (in color) to represent $y = y_1 + y_2$.

CALCULATOR COMMENT

Y=2cos(3X)+3sin(2X)

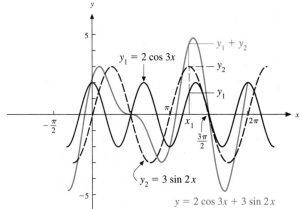

FIGURE 2.30 Graph of $y = 2 \cos 3x + 3 \sin 2x$ by adding ordinates

Graphs of Other Trigonometric Curves

EXAMPLE 3   Graphing a product

Graph $y = \dfrac{x}{2} \sin x$.

SOLUTION   This function is a product where $y_1 = \dfrac{x}{2}$ and $y_2 = \sin x$.

The procedure for finding a product is similar to that for finding sums: using your eye or a ruler, count squares to find y_1 and y_2 and then plot the *product* $y_1 y_2$. If you focus on values of x for which either y_1 or y_2 is 0 or 1, you will find this process to be particularly easy, as shown in Figure 2.31.

CALCULATOR COMMENT

Y=(X/2)sinX

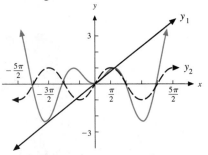

FIGURE 2.31 Graph of $y = \frac{x}{2} \sin x$

If you have two engines or motors that are running side by side at almost the same speed, then you can hear the noise as a series of beats. These beats are the result of the peaks of the addition of ordinates of the tones of the separate engines. For example, if $y_1 = \sin 7x$ and $y_2 = \sin 8x$, then the result of $y = y_1 + y_2$ is a series of beats as shown in Figure 2.32.

$y_1 = \sin 7x$ $y_2 = \sin 8x$ $y = y_1 + y_2$

FIGURE 2.32 Graphs of $y_1 = \sin 7x$, $y_2 = \sin 8x$, and $y = \sin 7x + \sin 8x$ showing the beats in noise

In radio, modulation is the modification of the amplitude, period (frequency), and starting point (phase) of an electrical wave in accordance with a modulating signal wave. The common systems are amplitude modulation (as in AM radio) and frequency modulation (as in FM radio). Example 4 illustrates a function that can occur in radio transmission that involves multiplication of ordinates.

EXAMPLE 4   **Graphing a product**

Graph $y = (3 \sin x)(\cos 6x)$ from 0 to 2π.

SOLUTION   Graph this curve by plotting points as shown in Figure 2.33.

CALCULATOR COMMENT

Y=(3sinX)(cos(6X))

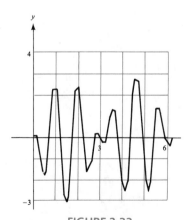

FIGURE 2.33

Graph of $y = (3 \sin x)(\cos 6x)$ by plotting points

2.4 Problem Set

A *Graph the functions given in Problems 1–24 by adding ordinates.*

1. $y = \frac{x}{2} + \sin x$

2. $y = \frac{x}{3} + \sin x$

3. $y = \frac{x}{3} + \cos x$

4. $y = \frac{2x}{3} + \cos x$

5. $y = \frac{2x}{3} + \sin x$

6. $y = x + \sin x$

7. $y = x + \cos x$

8. $y = x + \cos 2x$

9. $y = 2x + \cos 2x$

B

10. $y = \sin x + \cos x$

11. $y = 2 \sin x + \cos x$

12. $y = \sin x + 2 \cos x$

13. $y = \sin 2x + \cos 2x$

14. $y = 2 \sin 2x + \cos 2x$

15. $y = \sin 2x + 2 \cos 2x$

16. $y = 2 \sin \frac{x}{2} + \cos x$

17. $y = 2 \sin x + \cos \frac{x}{2}$

18. $y = 3 \sin \frac{x}{3} + \cos \frac{x}{3}$

19. $y = x \sin x$

20. $y = x \cos x$

21. $y = \frac{x}{2} \cos x$

22. $y = -\frac{x}{2} \cos x$

23. $y = -\frac{x}{2} \sin x$

24. $y = -x \cos x$

25. BUSINESS The sales of rototillers are very seasonal. A new business can describe its sales by $y = .02x + \left| \sin \frac{x\pi}{6} \right|$, where x is the number of months since going into business on January 1, 1992. Plot the sales graph for the first three years.

26. BUSINESS FORECASTING Using the information from Problem 25, plot the graph for rototiller sales for the year 1995.

27. PHYSICS Two light waves are said to be *destructive* if the amplitude of the sum of the waves is zero at every point. Show that $y = \sin x$ and $y = \cos\left(x + \frac{\pi}{2}\right)$ combine to form destructive interference.

28. OCEANOGRAPHY The sun–moon tide curves are shown here.* During a new moon, the sun and moon tidal bulges are centered at the same longitude, so their effects are added to produce maximum high tides and minimum low tides. This produces maximum tidal range.

NEW MOON (Spring tide)

Write possible equations for the sun curve, moon curve, and for the combined curve. Assume the tidal range is 10 ft and the period is 12.

29. OCEANOGRAPHY The sun–moon tide curves are shown here.* During the third quarter (neap tide) the sun and moon tidal bulges are at right angles to each other. Thus, the sun tide reduces the effect of the moon tide.

THIRD QUARTER (Neap tide)

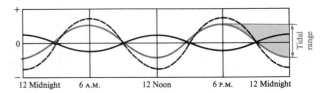

Write possible equations for the sun curve, moon curve, and for the combined curve. Assume the tidal range is 4 ft and the period is 12.

*From *Introductory Oceanography*, 5th edition, by H. V. Thurman, p. 253. Reprinted with permission of Merrill, an imprint of Macmillan Publishing Company. Copyright © 1988 Merrill Publishing Company, Columbus, Ohio.

30. The pattern of CO_2 in the atmosphere can be illustrated by a graph whose equation is $y = x + \sin x$. Graph this curve for $0 \le x \le 20$.

[C] *Sketch the curves in Problems 31–39 from 0 to 2π by plotting points or by using a graphics calculator.*

31. $y = x \sin x$

32. $y = x^2 \cos x$

33. $y = \dfrac{1}{x} \cos x$

34. $y = \dfrac{\sin x}{1 + x^2}$

35. $y = (2 \sin x)(\cos 6x)$

36. $y = (3 \cos x)(\sin 6x)$

37. $y = (4 \sin x)(\cos 8x)$

38. $y = \cos^2 x$

39. $y^3 = \sin x$

40. PHYSICS   Graph the sound wave given by
$$y = .04 \sin 200\pi x + .06 \sin 400\pi x \quad \text{for } 0 \le x \le \frac{1}{100}$$

41. Graph $y = \sin 7x + \sin 9x$, and notice the effect of beats.

42. ELECTRICAL ENGINEERING   The current I from two power sources is given by
$$I = 3 \sin 240\pi t + 6 \cos 120\pi t$$
where t is time in seconds. Sketch the graph of the current for $0 \le t \le \dfrac{1}{30}$.

CALCULATOR COMMENT

The French mathematician Joseph Fourier (1768–1830) was studying problems of heat flow and used trigonometry to describe what he saw. In more advanced mathematics, a series is defined as a sum of terms of a sequence and if these terms are trigonometric, then the series is called a **Fourier series**. The following is an example of a Fourier series:

$$y = \cos x + \frac{1}{3} \cos 3x + \frac{1}{5} \cos 5x + \cdots$$

Addition of ordinates can be used with Fourier series. The main characteristic of a Fourier series is that it can be used to represent functions that ordinarily are represented by different expressions in different parts of some interval. Fourier series are used today in studying heat flow, electrical fields, and sound waves.

Computers and calculators have greatly simplified the graphing and analysis of Fourier series. In the Application for Further Study at the end of this chapter, we use computer software to graph trigonometric curves.

2.5 Summary and Review

| *Objectives* | *Pages / Examples / Problems* |

2.1 Graphs of Cosine, Sine, and Tangent Functions

1. Graph the cosine, sine, and tangent curves from memory by using a frame.   pp. 50–56; Problems 6–8.

2. Graph the standard cosine, sine, and tangent curves by building a frame from a given starting point.   pp. 52–56; Examples 1–3; Problems 9–20.

3. Graph trigonometric curves by plotting points.   p. 50; Problems 21–34.

2.2 General Cosine, Sine, and Tangent Curves

4. Know the standard forms of the cosine, sine, and tangent curves. p. 64

5. Graph one period of the general cosine, sine, and tangent curves by using a frame. pp. 60–62; Examples 2–4; Problems 1–21.

6. Graph the general cosine, sine, and tangent curves. pp. 62–66; Examples 5–9; Problems 22–36.

2.3 Graphs of Secant, Cosecant, and Cotangent Functions

7. Graph the secant, cosecant, and cotangent curves from memory. pp. 69–71; Problems 7–9.

8. Graph the general secant, cosecant, and cotangent curves. pp. 71–73; Examples 1–3; Problems 10–30.

2.4 Addition of Ordinates (Optional)

9. Graph trigonometric curves by adding ordinates. pp. 74–76; Examples 1–2; Problems 1–18.

10. Graph trigonometric curves by multiplying ordinates. pp. 76–77; Examples 3–4; Problems 19–24.

TERMS

Adding ordinates [2.4] Cosine curve [2.1] Period [2.1, 2.2] Sine curve [2.1]
Amplitude [2.2] Cotangent curve [2.3] Phase shift [2.2] Tangent curve [2.1]
Asymptote [2.1, 2.2] Frame [2.1, 2.2] Secant curve [2.3] Translation [2.2]
Cosecant curve [2.3]

CHAPTER 2 REVIEW OF OBJECTIVES

The problem numbers correspond to the objectives listed in Section 2.5.

1. From memory draw a quick sketch of the following curves.

   a. $y = \cos x$ **b.** $y = \sin x$ **c.** $y = \tan x$

2. Plot the given point, and draw a frame using the given point as a starting point. Draw one period of the requested curve.

   a. $\left(\frac{\pi}{6}, -\frac{3}{2}\right)$; sine curve   **b.** $\left(-\frac{3\pi}{2}, 1\right)$; cosine curve

   c. $\left(-\frac{\pi}{3}, 2\right)$; tangent curve   **d.** $(-1, 1)$; sine curve

3. Graph the given curves by plotting points.

   a. $y = \sin x + \cos x$ **b.** $y = \dfrac{\tan x}{x}$

4. Fill in the blanks.

   Standard forms: $y - k = a \cos b(x - h)$
   $y - k = a \sin b(x - h)$
   $y - k = a \tan b(x - h)$

   The origin is translated to _____**a.**_____.
   The height of the frame is _____**b.**_____.

The length of the frame is _____**c.**_____ for cosine, _____**d.**_____ for sine, and _____**e.**_____ for tangent.
The a value for the sine curve is called its _____**f.**_____, and the length of the frame is called its _____**g.**_____.

5. Graph one period of the given curves.

   a. $y - 2 = \sin\left(x - \frac{\pi}{6}\right)$ **b.** $y = \cos\left(x + \frac{\pi}{4}\right)$

   c. $y = \frac{1}{2}\tan\frac{1}{2}x$ **d.** $y = \tan\left(x - \frac{2\pi}{3}\right) - 2$

   e. $y = 2\cos\frac{2}{3}x$

6. Graph the given curves.

   a. $y = 2\cos\left(2x + \frac{\pi}{3}\right)$ **b.** $y = 3\sin\frac{\pi x}{4}$

   c. $y + 3 = 3\tan\frac{1}{2}\left(x - \frac{\pi}{3}\right)$ **d.** $y + 1 = \cos\left(x - \frac{\pi}{3}\right)$

   e. $y = \frac{3}{2}\tan\frac{3}{2}x$

7. From memory draw a quick sketch of the following curves.

   a. $y = \sec x$ **b.** $y = \csc x$ **c.** $y = \cot x$

8. Graph the given curves.

   a. $y - 1 = \sec(\tfrac{1}{2}x - \tfrac{1}{4})$ **b.** $y + 2 = \cot(x - \tfrac{\pi}{3})$

   c. $y + 1 = \csc 2(x - \tfrac{\pi}{4})$ **d.** $y - \tfrac{3}{2} = \cot(x - 1)$

9. Graph the given curves by adding ordinates.

   a. $y = \dfrac{3x}{4} + \cos x$ **b.** $y = \sin 3x + 2 \cos x$

10. Graph the given curves by multiplying ordinates.

    a. $y = 2x \sin x$ **b.** $y = \dfrac{2x}{3} \cos 2x$

DISCUSSION PROBLEMS

1. In your own words, explain the "framing procedure" for graphing.

2. What is a sinusoid curve?

3. Find as many examples of trigonometric curves as possible from magazines, newspapers, or books (other than mathematics books).

4. Write a paper on the impact of the calculator on trigonometry.

5. Write a 500-word paper on the relationship between music and sine curves.

MISCELLANEOUS PROBLEMS

Graph each of the curves given in Problems 1–30.

1. $y = 5 \cos x$
2. $y = 5 \sin x$
3. $y = 5 \tan x$
4. $y = \sin 5x$
5. $y = \tan 5x$
6. $y = \cos 5x$
7. $y = \tan 5x - 3$
8. $y = \cos 5x - 1$
9. $y = \sin 5x + 1$
10. $y = 5 \cos x - 3$
11. $y = 5 \sin x - 3$
12. $y = 5 \tan x - 3$
13. $y = 5 \sin(x - 3)$
14. $y = 5 \tan \tfrac{1}{2}(x - 3)$
15. $y = 5 \cos(x - 3)$
16. $y = \tan(5x - 3)$

17. $y = \cos(5x - 3)$
18. $y = \sin(5x - 3)$
19. $y - 2 = \cos(3x - 1)$
20. $y + 1 = \sin(2x + 1)$
21. $y + \tfrac{1}{2} = \tan(4x - 1)$
22. $y + 1 = \sin(2x - \pi)$
23. $y - \tfrac{\pi}{3} = \cos(3x + 2)$
24. $y - 2 = \tan(2x + \tfrac{\pi}{2})$
25. $y = 2 \sec 3x$
26. $y = \sqrt{2} \csc 2x$
27. $y = \pi \cot x$
28. $y - 1 = \sec(x - \tfrac{\pi}{6})$
29. $y + 1 = \csc(x - \tfrac{\pi}{4})$
30. $y - 1 = \cot(x + \tfrac{\pi}{6})$

CALCULATOR COMMENT

There are many commercially available computer programs to help you do trigonometric graphing. One such package is called Converge from JEMware (phone 808-523-9911). It requires an IBM-compatible computer. This program can be used to graph complicated curves; it can also serve as an instructional aid in teaching you how to graph the curves in the first place. If you use Converge, you will need to make some menu choices before you start. Choose the Graphing menu. This is a general menu that allows you to graph one of many different types of functions. Choose the form $y = f(x)$ and then simply type in the trigonometric function or functions you wish to graph. To input the real number π, you should type PI. You can choose to have the scale on the x-axis in terms of real numbers 1, 2, 3, . . . or in terms of real numbers that are multiples of $\pi/4$.

Finally, you have the option of adding, subtracting, or multiplying functions by using the Alg/Trig method with the choice called Operations of Functions.

Computer Graphing of Trigonometric Curves

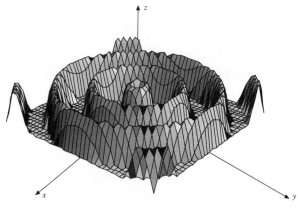

Graph of $z = \sin\sqrt{x^2 + y^2}$

Getting Started

Begin by graphing the standard cosine, sine, and tangent curves. Most programs require parentheses following the function you are graphing; the input will look like this:

$$Y = COS(X); \qquad Y = SIN(X); \qquad Y = TAN(X)$$

Remember, a computer graphs a function simply by plotting at a very high speed a large number of points, giving the illusion that it is drawing a smooth curve. This effect is analogous to the well-known illusion of motion produced by the individual frames of a moving picture.

Changes in Amplitude

Experiment 1 Graph $y = \sin x$. Without erasing this curve, graph each of the following curves:

a. $y = 3 \sin x$ **b.** $y = \frac{1}{2} \sin x$ **c.** $y = \pi \sin x$

What is the effect of a numerical coefficient on a sine curve? What will $y = a \sin x$ look like?

SOLUTION **a.** $y = 3 \sin x$. Many programs require an asterisk for multiplication between the numerical coefficient and the trigonometric function, as shown here:

$$Y = 3*SIN(X)$$

This graph is shown in Figure 2.34a.

b. $y = \frac{1}{2} \sin x$. *Input:* $Y = (1/2)*SIN(X)$

This graph is shown in Figure 2.34b.

c. $y = \pi \sin x$. *Input:* $Y = PI*SIN(X)$
Notice that an approximate value of π must be input. The graph is shown in Figure 2.34c.

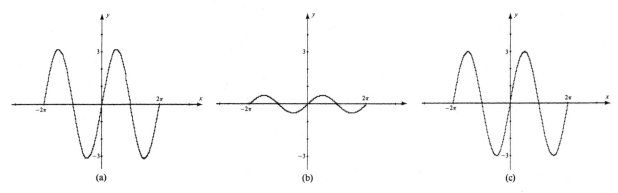

(a)   (b)   (c)

FIGURE 2.34 Graphs showing changes in amplitude

The numerical coefficient of the trigonometric function gives the amplitude; the graph of $y = a \sin x$ is a standard sine curve with an amplitude of $|a|$ rather than of 1.

Changes in Period

Experiment 2 Graph $y = \cos x$. Without erasing this curve, graph

a. $y = \cos 2x$ **b.** $y = \cos \frac{1}{2}x$ **c.** $y = \cos \pi x$

What is the effect of a numerical coefficient on the angle of a cosine curve? What will $y = \cos bx$ look like?

SOLUTION **a.** $y = 2x$. *Input:* $Y = COS(2*X)$
The graph is shown in Figure 2.35a.

b. $y = \cos \frac{1}{2}x$. *Input:* $Y = COS((1/2)*X)$
The graph is shown in Figure 2.35b.

c. $y = \cos \pi x$. *Input:* $Y = COS(PI*X)$
The graph is shown in Figure 2.35c.

(a)

(b)

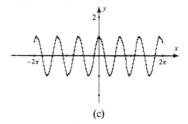
(c)

FIGURE 2.35 Graphs showing changes in period

The numerical coefficient of the angle of a cosine curve changes the period of the cosine curve. The graph of $y = \cos bx$ is a standard cosine curve with a period of $\frac{2\pi}{b}$ instead of 2π.

Phase Displacement

Experiment 3 Graph $y = \tan x$. Without erasing this curve, graph

    a. $y = \tan\left(x - \frac{\pi}{2}\right)$ **b.** $y = \tan\left(x + \frac{\pi}{2}\right)$

    c. $y = \tan(x + 1)$ **d.** $y = \tan(x - 2)$

What is the effect of a number subtracted from the angle of a tangent curve? What will $y = \tan(x - h)$ look like?

SOLUTION   **a.** $y = \tan\left(x - \frac{\pi}{2}\right)$. *Input:*   `Y=TAN(X-PI/2)`

        The graph is shown in Figure 2.36a.

    b. $y = \tan\left(x + \frac{\pi}{2}\right)$. *Input:*   `Y=TAN(X+PI/2)`

        The graph is shown in Figure 2.36b.

    c. $y = \tan(x + 1)$. *Input:*   `Y=TAN(X+1)`

        The graph is shown in Figure 2.36c.

    d. $y = \tan(x - 2)$. *Input:*   `Y=TAN(X-2)`

        The graph is shown in Figure 2.36d.

(a) and (b)

(c)

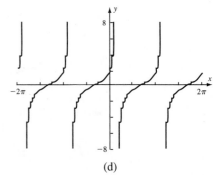
(d)

FIGURE 2.36 Graphs showing changes in phase displacement

The number subtracted from the angle is the amount that the graph is displaced from the standard tangent curve.

Translation

Experiment 4   Graph $y = \sin x$. Without erasing this curve, graph

  a. $y + 1 = \sin\left(x - \frac{\pi}{2}\right)$ **b.** $y - 1 = \sin(x - 2)$

  c. $y - 2 = \sin(x + 1)$
  What will $y - k = \sin(x - h)$ look like?

SOLUTION   **a.** $y + 1 = \sin\left(x - \frac{\pi}{2}\right)$.   *Input:*   `Y=SIN(X-PI/2)-1`
     The graph is shown in Figure 2.37a.

   b. $y - 1 = \sin(x - 2)$.   *Input:*   `Y=SIN(X-2)+1`
     The graph is shown in Figure 2.37b.

   c. $y - 2 = \sin(x + 1)$.   *Input:*   `Y=SIN(X+1)+2`
     The graph is shown in Figure 2.37c.

(a)

(b)

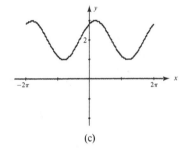
(c)

FIGURE 2.37   Graphs showing a translation

     The curve $y - k = \sin(x - h)$ is a standard sine curve that has been translated to
     the point (h, k).

Graphing General Trigonometric Curves

After using a computer to learn the shapes of the standard trigonometric curves and the
effects of amplitude, period, displacement changes, and translations, you can then use
it to help you graph some very complicated trigonometric curves.

EXAMPLE 1   **Graphing a quotient**

     Graph $y = \dfrac{10 \sin x}{x}$.

SOLUTION   *Input:*   `Y=10*SIN(X)/X`
     The resulting graph is shown in
     Figure 2.38. What happens at $x = 0$?

FIGURE 2.38   Graph of $y = \dfrac{10 \sin x}{x}$

EXAMPLE 2   Graphing a sum of squares

Graph $y = \sin^2 x$, $y = \cos^2 x$, and $y = \sin^2 x + \cos^2 x$ on the same grid.

SOLUTION   *Input:*   Y=(SIN(X))^2, Y=(COS(X))^2,   and

Y=(SIN(X))^2,+(COS(X))^2

Graph these curves as shown in Figure 2.39.

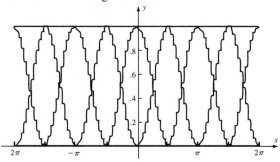

FIGURE 2.39

EXAMPLE 3   Graphing a Fourier series

Graph $y = \sin x$, $y = \sin x + \dfrac{\sin 3x}{3}$, and $y = \sin x + \dfrac{\sin 3x}{3} + \dfrac{\sin 5x}{5}$.

SOLUTION   *Input:*   Y=SIN(X), Y=SIN(X)+SIN(3X)/3,   and

Y=SIN(X)+SIN(3X)/3+SIN(5X)/5

Graph these curves as shown in Figure 2.40.

FIGURE 2.40

Joseph Fourier used combinations of the trigonometric functions, which today are used in advanced mathematics in the study of electrical circuits, sound, and heat flow, to mention just a few applications. Example 3 shows the first three terms of a *Fourier series*:

$$\sin x + \frac{\sin 3x}{3} + \frac{\sin 5x}{5} + \frac{\sin 7x}{7} + \cdots$$

The wave displayed on an oscilloscope for the sound of a tuning fork vibrating at f cycles per second with amplitude a is given by

$$y = a \sin 2\pi f x$$

Now, for a vibrating string (as in a piano) of length L, tension T, and mass m per unit length, the value for f is

$$f = \frac{1}{2L} \sqrt{\frac{T}{m}}$$

For example, the frequency of middle C on a properly tuned piano is 261.626. Using this value for f and an amplitude of 8, we obtain the equation $y = 8 \sin 2\pi fx$. The first overtone for this note is $y = 4 \sin 4\pi fx$, producing the sound wave described by the equation

$$y = 8 \sin 2\pi fx + 4 \sin 4\pi fx$$

where $f = 261.626$. This sound wave can be graphed by inputting

```
Y=8*SIN(2*PI*261.626*X)+4*SIN(4*PI*261.626*X)
```

The graph is shown in Figure 2.41.

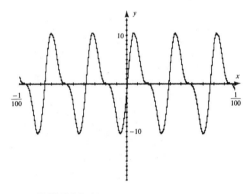

FIGURE 2.41   Graph of a sound wave

Problems for Further Study: Computer Graphing of Trigonometric Curves

1. Repeat Experiment 1 for $y = \cos x$.

2. Repeat Experiment 1 for $y = \tan x$.

3. Repeat Experiment 2 for $y = \sin x$.

4. Repeat Experiment 2 for $y = \tan x$.

5. Repeat Experiment 3 for $y = \sin x$.

6. Repeat Experiment 3 for $y = \cos x$.

7. Repeat Experiment 4 for $y = \tan x$.

8. Repeat Experiment 4 for $y = \cos x$.

Graph the curves in Problems 9–20 by using a computer. You can either use a printout of the graph or copy what is on the screen onto your own graph paper.

9. $y = \dfrac{10 \cos x}{x}$ **10.** $y = \dfrac{\tan x}{5x}$

11. $y = \dfrac{x}{\cos x}$ **12.** $y = \dfrac{x}{\sin x}$

13. $y = \dfrac{x}{\tan x}$ **14.** $y = \dfrac{3 \sin x}{\cos x}$

15. $y = 2 \sin 3x + 3 \sin 2x$ **16.** $y = \cos 9x + \sin 11x$

17. $y = 3 \sin x - 2 \cos \pi x$ **18.** $y = \sin x + \cos 2x - \sin 3x$

19. $y = \cot^2 x - \csc^2 x$ **20.** $y = \sec^2 x - \tan^2 x$

21. Graph the following curves on the same grid.

   a. $y = \cos x$ **b.** $y = \cos x + \dfrac{\cos 3x}{3}$

   c. $y = \cos x + \dfrac{\cos 3x}{3} + \dfrac{\cos 5x}{5}$

22. Graph the following curves on the same grid.

   a. $y = \sin x$ **b.** $y = \sin x + \dfrac{\sin 2x}{2}$

   c. $y = \sin x + \dfrac{\sin 2x}{2} + \dfrac{\sin 3x}{3}$

23. Music   Graph the sound wave for $f = 294$.

24. Music   Graph the sound wave for $f = 294$, along with its first overtone.

25. Electrical   The current l for a two-power-source circuit is
$$y = 2 \sin 100\pi t + 4 \cos 120\pi t$$
Graph this current from $-.2$ to $.2$.

26. Graph $y = 20 \sin(328\pi x)$.

▌ Chapters 1–2

Sample Test 1: Objective Form

1. State the generalized definition of the trigonometric functions.

2. Fill in the blanks.

FUNCTION	COFUNCTION	RECIPROCAL
cosine	a. _____	b. _____
tangent	c. _____	d. _____
secant	e. _____	f. _____
cosecant	g. _____	h. _____
sine	i. _____	j. _____

3. Using the definition of the trigonometric functions, find their values for an angle α whose terminal side passes through $(\sqrt{7}, 3)$.

4. The westernmost city in the continental United States is Ferndale, California, and it is located at 124°W longitude, 41°N latitude. How far (to the nearest 100 miles) is Ferndale from the equator if you assume that the earth's radius is approximately 3,950 miles?

5. Complete the following table of exact values.

angle in degrees	$-45°$	a.	b.	210°	90°
angle in radians	c.	π	$\frac{-4\pi}{3}$	d.	e.
sine	f.	g.	h.	i.	j.
cosine	k.	l.	m.	n.	o.
tangent	p.	q.	r.	s.	t.

6. From memory, draw a quick sketch of each of the following. Label the axes and scale.
   a. $y = \sin x$ **b.** $y = \cot x$ **c.** $y = \csc x$

7. Graph one period of $y = 2 \cos 3x$.

8. Graph one period of $y = \frac{1}{2} \tan \frac{3}{2}x$.

9. Graph $y - 3 = \sin\left(x - \frac{\pi}{3}\right)$.

10. Graph $y + 2 = \cos(3x + 2\pi)$.

Sample Test 2: Fill In and Multiple Choice

Fill in the blanks in Problems 1–5 with the word or words that make the statements complete and correct.

1. Let θ be an angle in standard position with a point $P(x, y)$ on the terminal side a distance of r from the origin ($r \neq 0$). Then
$$\sin \theta = \text{_____} ; \qquad \sec \theta = \text{_____} ; \qquad \cot \theta = \text{_____} .$$

2. The angle with measure 2 is in Quadrant(s) _____ .

3. An angle rotated in a clockwise direction is _____ .

4. In Quadrant II, the _____ and _____ are negative.

5. The period of the standard sine curve is _____ .

*Choose the **best** answer from the choices given.*

6. The reciprocal of secant is
   A. cosecant   B. sine
   C. cosine   D. cotangent
   E. none of these

7. csc 15.8° correct to four decimal places is
   A. 3.6727   B. .9622
   C. 1.0393   D. -10.8806
   E. none of these

8. If θ is an angle in standard position and P_θ is a point on a unit circle on the terminal side of θ, then the coordinates of θ are
   A. impossible to find   B. $(0, 1)$
   C. $(\sin \theta, \cos \theta)$ D. $(\cos \theta, \sin \theta)$
   E. none of these

9. Calculate 25 cot 38° using a calculator.
   A. 19.53214066   B. 7.757741525
   C. 80.56468471   D. 31.9985408
   E. none of these

10. If $\theta = -3$, then its reference angle is
   A. 3   B. -3
   C. $\pi - 3$ D. $3 - \pi$
   E. none of these

11. The length of an arc with a central angle of 40° and radius of 1 is
   A. less than 1   B. equal to 1
   C. between 1 and 40   D. equal to 40
   E. none of these

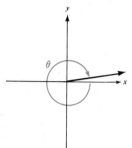

12. The size of θ (see angle in margin) is about

    A. $\frac{\pi}{6}$ B. $\frac{1}{2}$ C. -6 D. $-\frac{5\pi}{4}$ E. none of these

13. An angle with measure equal to $\frac{3\pi}{4}$ is

    A. $45°$ B. $-45°$ C. $135°$ D. all are equal   E. none of these

14. $\cos \frac{5\pi}{6}$ is

    A. $\frac{1}{2}$ B. $\frac{\sqrt{3}}{2}$

    C. $\frac{\sqrt{2}}{2}$ D. $\frac{\sqrt{3}}{3}$

    E. none of these

15. $\tan\left(-\frac{3\pi}{4}\right)$ is

    A. $\frac{\sqrt{3}}{3}$ B. -1

    C. $-\sqrt{3}$ D. $\sqrt{3}$

    E. none of these

16. The length of the frame for $y = 3 \tan 2x$ is

    A. $\frac{\pi}{2}$ B. π

    C. $\frac{2\pi}{3}$ D. 2π

    E. none of these

17. The graph of the cosecant curve passes through the point

    A. $(0, 0)$ B. $\left(\frac{\pi}{2}, 1\right)$

    C. $\left(\frac{3\pi}{2}, 1\right)$ D. all of the above

    E. none of these

18. The center of the frame for $y - 1 = 2 \tan\left(x - \frac{\pi}{3}\right)$ is

    A. $(1, 2)$ B. $\left(1, \frac{\pi}{3}\right)$

    C. $\left(\frac{\pi}{3}, 1\right)$ D. $\left(\frac{\pi}{3}, 2\right)$

    E. none of these

19. When graphing the curve $y + 1 = 3 \sin\left(x + \frac{\pi}{3}\right)$ the point (h, k) is

    A. $\left(-1, \frac{\pi}{3}\right)$ B. $\left(\frac{\pi}{3}, -1\right)$

    C. $(0, 3)$ D. $\left(-\frac{\pi}{3}, 1\right)$

    E. none of these

20. The top right-hand corner of the frame for $y = 2 \tan\left(x + \frac{3\pi}{4}\right)$ is

    A. $\left(-\frac{\pi}{4}, 0\right)$ B. $\left(\frac{\pi}{4}, 0\right)$

    C. $\left(\frac{\pi}{4}, 2\right)$ D. $\left(-\frac{\pi}{4}, 2\right)$

    E. none of these

▌ Karl Gauss
(1777–1855)

▌ Sin²φ is odious to me, even though Laplace made use of it; shoult it be feared that sin ϕ^2 might become ambiguous, which would perhaps never occur, or at most very rarely when speaking of sin(ϕ^2), well then, let us write (sin ϕ)², but not sin²φ, which by analogy shoult signify sin(sin ϕ).

KARL GAUSS
Gauss-Schumacher Briefweches, Bd. 3, p. 292; Bd. 4, p. 63

We might add that the notation we use today does not conform to Gauss' suggestion, because sin²φ is used today to mean (sin ϕ)².

Ptolemy lived in Alexandria and worked at the museum there. Along with Hipparchus and Menelaus, he invented trigonometry to build a quantitative astronomy that could be used to predict the paths and positions of the heavenly bodies and to aid in telling time, calendar-reckoning, navigating, and studying geography. In his book *Syntaxis Mathematica*, he derived many of the identities introduced in this chapter. Although his purposes were more related to solving triangles (see Chapter 5), today a major focus of trigonometry is on the relationships of the functions themselves, which is what we emphasize in this chapter.

There is ample evidence in the history of trigonometry about the construction of the trigonometric tables. Ptolemy's calculations of the trigonometric table for the values of sine follow from the following theorem, known as

Ptolemy's theorem: In a cyclic quadrilateral, the product of the diagonals is equal to the sum of the products of the two pairs of opposite sides. Ptolemy's procedure is outlined in Howard Eves' *Great Moments in Mathematics Before 1650*, pages 99–105. Others who have been prominent in constructing tables of values are Abû'l-Wefâ (940–998), Georg von Peurbach (1423–1461), Regiomontanus (1436–1476), and Rhaeticus (1514–1576). It is interesting to note that such tables are rarely used today and have even been deleted from this text. The Mathematics Advisory Board of the *CRC Handbook of Tables for Mathematics* almost voted to delete the trigonometric tables because calculators have made these tables rather superfluous. Today, the focus in trigonometry is on the functions themselves, and on the identities presented in this chapter.

3

FUNDAMENTAL TRIGONOMETRIC IDENTITIES

▌Contents

3.1 Proving Identities

3.2 Proving Identities (Continued)

3.3 Disproving Identities

3.4 Cofunction and Opposite-Angle Identities

3.5 Addition Laws

3.6 Double-Angle and Half-Angle Identities

3.7 Product and Sum Identities

3.8 Chapter 3 Summary and Review

Application for Further Study
   Sunrise, Sunset, and Linear Interpolation

From page 446 of *Calculus and Analytic Geometry*, Third Edition, by Robert Ellis and Denny Gulick, copyright © 1986 by Harcourt Brace Jovanovich, Inc., reprinted by permission of the publisher.

▌Preview

In this chapter we focus on building a new skill, one that will be needed in subsequent courses. It is called *proving identities*, but it means showing that a given open equation is true for all values in the domain. The first three sections of this chapter are designed to given you confidence in proving identities and then the remaining sections develop those trigonometric identities you will need for your subsequent work in mathematics. The identities you need are listed on the inside back cover of the book for your easy reference. You should be familiar with these identities, but, at the same time, you should keep them in proper perspective. Do not become overwhelmed with the long list of derived identities; instead, approach each one as just a new identity to prove. After each is derived, you can add it to your list for future reference.

▌Perspective

If you plan on taking more advanced mathematics courses, you will need to be able to work with trigonometric identities. For example, trigonometric identities are frequently used in a variety of calculus applications. In the example shown on this page, the application is finding the length of a part of a curve. Notice that first the fundamental identity $\cos^2 t + \sin^2 t$ is used (without a remark or warning), and then some half-angle identities are used. We introduced the fundamental identities in Chapter 1, and we review them in Section 3.1; the half-angle identities are introduced in Section 3.6.

Solution   First we notice that

$$\frac{dx}{dt} = r - r \cos t \quad \text{and} \quad \frac{dy}{dt} = r \sin t$$

Therefore by (4) we have

$$\mathscr{L} = \int_0^{2\pi} \sqrt{(r - r \cos t)^2 + (r \sin t)^2} \; dt$$

$$= \int_0^{2\pi} \sqrt{r^2 - 2r^2 \cos t + r^2 \cos^2 t + r^2 \sin^2 t} \; dt$$

$$= \int_0^{2\pi} \sqrt{2r^2 - 2r^2 \cos t} \; dt$$

$$= r \int_0^{2\pi} \sqrt{2(1 - \cos t)} \; dt$$

By the half-angle formula for $\sin t/2$.

$$\frac{t - \cos t}{2} = \sin^2 \frac{t}{2}, \quad \text{so that} \quad \sqrt{2(1 - \cos t)} = \sqrt{4 \sin^2 \frac{t}{2}}$$

Since $\sin t/2 \geq 0$ for $0 \leq t \leq 2\pi$, we conclude that

---

3.1 Proving Identities

Fundamental Identities

In Chapter 1 we considered eight fundamental identities that are used to simplify and change the form of a great variety of trigonometric expressions. We will repeat them here for your easy reference.

FUNDAMENTAL IDENTITIES

Reciprocal Identities

1. $\sec \theta = \dfrac{1}{\cos \theta}$ 2. $\csc \theta = \dfrac{1}{\sin \theta}$ 3. $\cot \theta = \dfrac{1}{\tan \theta}$

Ratio Identities

4. $\tan \theta = \dfrac{\sin \theta}{\cos \theta}$ 5. $\cot \theta = \dfrac{\cos \theta}{\sin \theta}$

Pythagorean Identities

6. $\cos^2\theta + \sin^2\theta = 1$ 7. $1 + \tan^2\theta = \sec^2\theta$ 8. $\cot^2\theta + 1 = \csc^2\theta$

In addition to these fundamental identities, you will need to remember from arithmetic and algebra the proper procedure for simplifying expressions, especially the concepts of common denominators and complex fractions. You will combine this algebraic simplification with a knowledge of the fundamental identities, as shown by the next example.

EXAMPLE 1   **Algebraically simplifying an expression with trigonometric functions**

Simplify $\dfrac{\sin \theta + \dfrac{\cos^2\theta}{\sin \theta}}{\dfrac{\cos \theta}{\sin \theta}}$ and leave your answer in terms of sine and cosine functions only.

SOLUTION

$$\frac{\sin \theta + \dfrac{\cos^2\theta}{\sin \theta}}{\dfrac{\cos \theta}{\sin \theta}} = \frac{\dfrac{\sin^2\theta + \cos^2\theta}{\sin \theta}}{\dfrac{\cos \theta}{\sin \theta}} \qquad \text{Common denominator}$$

$$= \frac{\sin^2\theta + \cos^2\theta}{\sin \theta} \cdot \frac{\sin \theta}{\cos \theta} \qquad \text{Dividing rational expressions}$$

$$= \frac{1}{\cos \theta} \qquad \text{Substitution}$$

CALCULATOR COMMENT

You can verify each of these identities on a graphing calculator. For example, for identity 6 graph Y1 = (SIN X)^2 + (COS X)^2. The graph is a constant function:

You can similarly check the other Pythagorean identities or the reciprocal identities. For the ratio identities, graph Y1 = SIN X/COS X. The result is recognized as a tangent curve:

Procedure for Proving That Trigonometric Equations Are Identities

⊗ If you are trying to prove that an equation is an identity you CANNOT perform operations on both sides of the equation because you do not know both sides are equal (indeed, that is what you are proving). ⊗

In this section you will learn how to prove that a given expression is an identity. Remember, an identity is an open equation that is true for all replacements for the variable that are in the domain. The procedure for proving identities is not the same as the procedure for solving an equation. Recall that in Chapter 1 we wrote

$$\sec \theta = \frac{1}{\cos \theta} \quad \text{as} \quad \sec \theta \cos \theta = 1$$

by multiplying both sides by cos θ. We could do this because *we had previously proved that it was an identity.*

Suppose you are given a trigonometric equation such as

$$\tan \theta + \cot \theta = \sec \theta \csc \theta$$

and are asked to show that it is an identity. This equation has a left side, L, and a right side, R, and you must prove that $L = R$. There are three ways to do this:

1. Reduce the left-hand side to the right-hand side by using algebra and the fundamental identities.
2. Reduce the right-hand side to the left-hand side.
3. Reduce both sides **independently** to the same expression.

When you are asked to prove an identity, we mean prove that the given equation is an identity. We will illustrate these techniques with several examples.

EXAMPLE 2   Comparing methods for proving an identity

Prove that $\tan \theta + \cot \theta = \sec \theta \csc \theta$ is an identity.

SOLUTION   METHOD I: Reduce the left-hand side to the right-hand side.

$$\tan \theta + \cot \theta = \frac{\sin \theta}{\cos \theta} + \frac{\cos \theta}{\sin \theta}$$

$$= \frac{\sin^2\theta}{\cos \theta \sin \theta} + \frac{\cos^2\theta}{\cos \theta \sin \theta}$$

$$= \frac{\sin^2\theta + \cos^2\theta}{\cos \theta \sin \theta}$$

$$= \frac{1}{\cos \theta \sin \theta}$$

$$= \frac{1}{\cos \theta} \cdot \frac{1}{\sin \theta}$$

$$= \sec \theta \csc \theta$$

In proving identities, it is often advantageous to change the trigonometric functions to cosines and sines.

When you obtain an expression that is the same as the other side of the equation, you are finished with the proof. Thus, $\tan \theta + \cot \theta = \sec \theta \csc \theta$.

METHOD II: Reduce the right-hand side to the left-hand side.

$$\sec \theta + \csc \theta = \frac{1}{\cos \theta} \cdot \frac{1}{\sin \theta}$$

$$= \frac{1}{\cos \theta \sin \theta}$$

$$= \frac{\sin^2\theta + \cos^2\theta}{\cos \theta \sin \theta}$$

$$= \frac{\sin^2\theta}{\cos \theta \sin \theta} + \frac{\cos^2\theta}{\cos \theta \sin \theta}$$

$$= \frac{\sin \theta}{\cos \theta} + \frac{\cos \theta}{\sin \theta}$$

$$= \tan \theta \cot \theta$$

You would, of course, choose only one of these three methods.

Thus, $\tan \theta + \cot \theta = \sec \theta \csc \theta$.

METHOD III: Reduce both sides to a form that is the same. Be sure to work with each side separately; do not treat this as an equation and add expressions to both sides or multiply both sides by the same expression.

$$\tan \theta + \cot \theta = \frac{\sin \theta}{\cos \theta} + \frac{\cos \theta}{\sin \theta} \qquad \sec \theta \csc \theta = \frac{1}{\cos \theta} \cdot \frac{1}{\sin \theta}$$

$$= \frac{\sin^2\theta + \cos^2\theta}{\cos \theta \sin \theta} \qquad\qquad = \frac{1}{\cos \theta \sin \theta}$$

$$= \frac{1}{\cos \theta \sin \theta}$$

These are identical.

Thus, $\tan \theta + \cot \theta = \sec \theta \csc \theta$.

Of course, we would pick only one of these methods when proving a particular identity. Usually it is easier to begin with the more complicated side and try to reduce it to the simpler side. If both sides seem equally complicated, you might change all the functions to sines and cosines and then simplify.

CALCULATOR COMMENT

You can use a graphing calculator to verify that the given equations are identities. Let Y1 be the function on the left of the equation, and Y2 be the function on the right of the equation. Graph both Y1 and Y2, and if the graphs are identical, then you can assume that the given equation is an identity. For Example 1, we found both graphs looked like the following:

$$Y1 = TAN(X) + COT(X) \qquad Y2 = SEC(X)CSC(X)$$

EXAMPLE 3   Proving an identity by combining fractions

Prove that $\dfrac{1}{1 + \cos \theta} + \dfrac{1}{1 - \cos \theta} = 2 \csc^2 \theta$ is an identity.

SOLUTION
$$\frac{1}{1 + \cos \theta} + \frac{1}{1 - \cos \theta} = \frac{(1 - \cos \theta) + (1 + \cos \theta)}{(1 + \cos \theta)(1 - \cos \theta)}$$

Begin with the more complicated side.

$$= \frac{2}{1 - \cos^2 \theta}$$

$$= \frac{2}{\sin^2 \theta}$$

$$= 2 \csc^2 \theta$$

CALCULATOR COMMENT

Y1=LeftSide Y2=RightSide

EXAMPLE 4   Proving an identity by changing to sines and cosines

Prove that $\sec 4\theta + \cos 4\theta = 2 \sec 4\theta - \tan 4\theta \sin 4\theta$ is an identity.

SOLUTION
$$\sec 4\theta + \cos 4\theta = \frac{1}{\cos 4\theta} + \cos 4\theta$$

When both sides seem equally complicated, a good procedure to follow is to change everything to cosines and sines and then simplify.

$$= \frac{1 + \cos^2 4\theta}{\cos 4\theta}$$

$$= \frac{1 + (1 - \sin^2 4\theta)}{\cos 4\theta}$$

Notice that the angle here is 4θ. The identity is proved in terms of 4θ.

$$= \frac{2 - \sin^2 4\theta}{\cos 4\theta}$$

$$= \frac{2}{\cos 4\theta} - \frac{\sin^2 4\theta}{\cos 4\theta}$$

$$= 2 \cdot \frac{1}{\cos 4\theta} - \frac{\sin 4\theta}{\cos 4\theta} \cdot \frac{\sin 4\theta}{1}$$

$$= 2 \sec 4\theta - \tan 4\theta \sin 4\theta$$

EXAMPLE 5   Proving an identity by separating a fraction into a sum

Prove that $\dfrac{\sec 2\lambda + \cot 2\lambda}{\sec 2\lambda} = 1 + \csc 2\lambda - \sin 2\lambda$ is an identity.

SOLUTION   Begin with the left-hand side. When working with a fraction consisting of a single term as a denominator, it is often helpful to separate the fraction into the sum of several fractions:

$$\frac{\sec 2\lambda + \cot 2\lambda}{\sec 2\lambda} = \frac{\sec 2\lambda}{\sec 2\lambda} + \frac{\cot 2\lambda}{\sec 2\lambda}$$

$$= 1 + \cot 2\lambda \cdot \frac{1}{\sec 2\lambda}$$

$$= 1 + \frac{\cos 2\lambda}{\sin 2\lambda} \cdot \cos 2\lambda$$

$$= 1 + \frac{\cos^2 2\lambda}{\sin 2\lambda}$$

$$= 1 + \frac{1 - \sin^2 2\lambda}{\sin 2\lambda}$$

$$= 1 + \frac{1}{\sin 2\lambda} - \frac{\sin^2 2\lambda}{\sin 2\lambda}$$

$$= 1 + \csc 2\lambda - \sin 2\lambda$$

CALCULATOR COMMENT

We will consider some additional simplifications of trigonometric identities in the next section.

3.1   Problem Set

A *Simplify each of the expressions in Problems 1–6 using only cosines, sines, and the fundamental identities.*

1. $\dfrac{1 - \cos^2\theta}{\sin \theta}$

2. $\dfrac{1 - \sin^2\theta}{\cos \theta}$

3. $\dfrac{\dfrac{\sin \theta}{\cos \theta} + \dfrac{\cos \theta}{\sin \theta}}{\dfrac{1}{\sin \theta \cos \theta}}$

4. $\sin \theta \dfrac{\cos^2\theta}{\sin \theta}$

5. $\dfrac{\cos \theta + \dfrac{\sin^2\theta}{\cos \theta}}{\sin \theta}$

6. $\dfrac{\dfrac{\cos^4\theta}{\sin^2\theta} + \cos^2\theta}{\dfrac{\cos^2\theta}{\sin^2\theta}}$

Reduce each of the expressions in Problems 7–12 to involve only sines and cosines, and then simplify.

7. $\sin \theta + \cot \theta$

8. $\sec \theta + \tan \theta$

9. $\dfrac{\tan \theta + \cot \theta}{\sec \theta \csc \theta}$

10. $\dfrac{\sec \theta + \csc \theta}{\tan \theta \cot \theta}$

11. $\sec^2\theta + \tan^2\theta$

12. $\csc^2\theta + \cot^2\theta$

Prove that the equations in Problems 13–24 are identities.

13. $\tan \theta = \sin \theta \sec \theta$

14. $\cot \theta = \cos \theta \csc \theta$

15. $\sec \theta \dfrac{\cos \theta}{1 - \sin^2\theta}$

16. $\csc \theta = \dfrac{\sin \theta}{1 - \cos^2\theta}$

17. $\tan 2\theta \csc 2\theta = \sec 2\theta$

18. $\cos 2\theta = \sin 2\theta \cot 2\theta$

19. $\tan \theta \cos \theta = \sin \theta$ **20.** $\cot \theta \sin \theta = \cos \theta$

21. $\dfrac{\sin \theta \csc \theta}{\tan \theta} = \cot \theta$ **22.** $\dfrac{\cot \theta \tan \theta}{\sin \theta} = \csc \theta$

23. $\dfrac{\cot t}{\sin^2 t + \cos^2 t} = \dfrac{1}{\tan t}$ **24.** $\dfrac{\sec^2 t}{1 + \tan^2 t} = 1$

B *Prove that the equations in Problems 25–31 are identities.*

25. $\csc \theta + \sin \theta = 2 \csc \theta - \cot \theta \cos \theta$
26. $\csc \theta - \sin \theta = \cos^2 \theta \csc \theta$
27. $(\csc 2\theta - \sin 2\theta)(\csc 2\theta + \sin 2\theta) = \csc^2 2\theta(1 - \sin^4 2\theta)$

28. $\dfrac{\tan 2\theta + \cot 2\theta}{\sec 2\theta} = \csc 2\theta$

29. $\dfrac{\tan 3\theta + \cot 3\theta}{\csc 3\theta} = \sec 3\theta$

30. $\dfrac{\sec \lambda + \tan^2 \lambda}{\sec \lambda} = 1 + \sec \lambda - \cos \lambda$

31. $\dfrac{\sin 2\lambda}{\tan 2\lambda} + \dfrac{\cos 2\lambda}{\cot 2\lambda} = \cos 2\lambda + \sin 2\lambda$

C *Prove that the equations in Problems 32–41 are identities.*

32. $(\tan 2\beta - \cot 2\beta)^2 = \tan^2 2\beta(2 - \csc^2 2\beta)^2$
33. $(1 - \sin^2\alpha)(1 + \tan^2\alpha) = 1$
34. $(1 - \cos^2 2\alpha)(1 + \cot^2 2\alpha) = 1$

35. $\dfrac{\cos^2\gamma + \tan^2\gamma - 1}{\sin^2\gamma} = \tan^2\gamma$

36. $\cot^2\gamma = \dfrac{\sin^2\gamma + \cot^2\gamma - 1}{\cos^2\gamma}$

37. $\dfrac{\sec^2\gamma + \tan^2\gamma - 1}{\csc^2\gamma} = 2\tan^2\gamma - 2\sin^2\gamma$

38. $(\cos \alpha - \cos \beta)^2 + (\sin \alpha - \sin \beta)^2$
   $= 2 - 2(\cos \alpha \cos \beta + \sin \alpha \sin \beta)$

39. $(\sec \alpha + \sec \beta)^2 - (\tan \alpha - \tan \beta)^2$
   $= 2 + 2(\sec \alpha \sec \beta + \tan \alpha \tan \beta)$

40. $\tan A + \cot B = (\sin A \sin B + \cos A \cos B) \sec A \csc B$

41. $\sec A + \csc B = (\cos A \sin^2 B + \sin B \cos^2 A) \sec^2 A \csc^2 B$

3.2 Proving Identities (Continued)

There are many "tricks of the trade" that can be used in proving identities. Some of these "tricks" are developed in the following examples. Since $a + b$ is the **conjugate** of $a - b$, the procedure illustrated in Example 1 is sometimes called **multiplying by the conjugate**.

EXAMPLE 1   Proving an identity by using a conjugate

Prove that $\dfrac{\cos \theta}{1 - \sin \theta} = \dfrac{1 + \sin \theta}{\cos \theta}$ is an identity.

SOLUTION   Sometimes, when there is a binomial in the numerator or denominator, the identity can be proved by multiplying the numerator and denominator by the conjugate of the binomial. That is, multiply both the numerator and the denominator of the left side by $1 + \sin \theta$. (This does not change the value of the expression because you are multiplying by 1.)

$$\dfrac{\cos \theta}{1 - \sin \theta} = \dfrac{\cos \theta}{1 - \sin \theta} \cdot \dfrac{1 + \sin \theta}{1 + \sin \theta}$$

$$= \dfrac{\cos \theta(1 + \sin \theta)}{1 - \sin^2\theta}$$

$$= \dfrac{\cos \theta(1 + \sin \theta)}{\cos^2\theta}$$

$$= \dfrac{1 + \sin \theta}{\cos \theta}$$

CALCULATOR COMMENT

You could also have proved this identity by multiplying the numerator and denominator of the right-hand side by $(1 - \sin \theta)$.

EXAMPLE 2   Proving an identity by factoring and reducing

Prove that $\dfrac{\sec^2 2\theta - \tan^2 2\theta}{\sec 2\theta + \tan 2\theta} = \dfrac{\cos 2\theta}{1 + \sin 2\theta}$ is an identity.

SOLUTION   Sometimes the identity can be proved by factoring and reducing.

$$\frac{\sec^2 2\theta - \tan^2 2\theta}{\sec 2\theta + \tan 2\theta} = \frac{(\sec 2\theta + \tan 2\theta)(\sec 2\theta - \tan 2\theta)}{\sec 2\theta + \tan 2\theta}$$

Notice that the angle here is 2θ. The identity is proved in terms of 2θ.

$$= \sec 2\theta - \tan 2\theta$$

$$= \frac{1}{\cos 2\theta} - \frac{\sin 2\theta}{\cos 2\theta}$$

$$= \frac{1 - \sin 2\theta}{\cos 2\theta}$$

$$= \frac{1 - \sin 2\theta}{\cos 2\theta} \cdot \frac{1 + \sin 2\theta}{1 + \sin 2\theta}$$

$$= \frac{1 - \sin^2 2\theta}{\cos 2\theta(1 + \sin 2\theta)}$$

$$= \frac{\cos^2 2\theta}{\cos 2\theta(1 + \sin 2\theta)}$$

$$= \frac{\cos 2\theta}{1 + \sin 2\theta}$$

CALCULATOR COMMENT

EXAMPLE 3   Proving an identity by multiplying by 1

Prove that $\dfrac{-2 \sin \theta \cos \theta}{1 - \sin \theta - \cos \theta} = 1 + \sin \theta + \cos \theta$ is an identity.

SOLUTION   Sometimes, when there is a fraction on one side, the identity can be proved by multiplying the other side by 1, where 1 is written so that the desired denominator is obtained. Thus, for this example,

CALCULATOR COMMENT

$$1 + \sin \theta + \cos \theta = (1 + \sin \theta + \cos \theta) \cdot \frac{1 - \sin \theta - \cos \theta}{1 - \sin \theta - \cos \theta}$$

$$= \frac{(1 + \sin \theta + \cos \theta)(1 - \sin \theta - \cos \theta)}{1 - \sin \theta - \cos \theta}$$

$$= \frac{1 - \sin \theta - \cos \theta + \sin \theta - \sin^2 \theta - \sin \theta \cos \theta + \cos \theta - \cos \theta \sin \theta - \cos^2 \theta}{1 - \sin \theta - \cos \theta}$$

$$= \frac{1 - (\sin^2 \theta + \cos^2 \theta) - 2 \sin \theta \cos \theta}{1 - \sin \theta - \cos \theta} = \frac{-2 \sin \theta \cos \theta}{1 - \sin \theta - \cos \theta}$$

In summary, there is no one best way to proceed in proving identities. However, the following hints should help.

HINTS FOR PROVING
TRIGONOMETRIC
IDENTITIES

Tricks of the Trade

1. If one side contains one function only, write all the trigonometric functions on the other side in terms of that function.

2. If the denominator of a fraction consists of only one term, break up the fraction into the sum of several fractions.

3. Simplify by combining fractions.

4. Factoring is sometimes helpful.

5. Change all trigonometric functions to sines and cosines and simplify.

6. Multiply numerator and denominator by the conjugate of either the numerator or the denominator.

7. If there are squares of functions, look for alternate forms of the Pythagorean identities.

8. Avoid the introduction of radicals.

3.2 Problem Set

A *Prove that the equations in Problems 1–16 are identities.*

1. $\sin \theta = \sin^3\theta + \cos^2\theta \sin \theta$

2. $\sec \theta = \sec \theta \sin^2\theta + \cos \theta$

3. $\tan \theta = \cot \theta \tan^2\theta$

4. $\dfrac{\sin \theta \cos \theta + \sin^2\theta}{\sin \theta} = \cos \theta + \sin \theta$

5. $\tan^2\theta - \sin^2\theta = \tan^2\theta \sin^2\theta$

6. $\cot^2\theta \cos^2\theta = \cot^2\theta - \cos^2\theta$

7. $\tan A + \cot A = \sec A \csc A$

8. $\cot A = \csc A \sec A - \tan A$

9. $\sin x + \cos x = \dfrac{\sec x + \csc x}{\csc x \sec x}$

10. $\dfrac{\cos \gamma + \tan \gamma \sin \gamma}{\sec \gamma} = 1$

11. $\dfrac{1 - \sec^2 t}{\sec^2 t} = -\sin^2 t$

12. $\dfrac{1 + \cot^2 t}{\cot^2 t} = \sec^2 t$

13. $(\sec \theta - \cos \theta)^2 = \tan^2\theta - \sin^2\theta$

14. $\dfrac{\sin \theta}{\csc \theta} + \dfrac{\cos \theta}{\sec \theta} = 1$

15. $1 - \sin 2\theta = \dfrac{1 - \sin^2 2\theta}{1 + \sin 2\theta}$

16. $\dfrac{1 - \tan^2 3\theta}{1 - \tan 3\theta} = 1 + \tan 3\theta$

B *Prove that the equations in Problems 17–31 are identities.*

17. $\csc 3\beta - \cos 3\beta \cot 3\beta = \sin 3\beta$

18. $\dfrac{1 + \cot^2 A}{1 + \tan^2 A} = \cot^2 A$

19. $\dfrac{\sin^2\beta - \cos^2\beta}{\sin \beta + \cos \beta} = \sin \beta - \cos \beta$

20. $\dfrac{\tan^2\gamma - \cot^2\gamma}{\tan \gamma + \cot \gamma} = \tan \gamma - \cot \gamma$

21. $\tan^2 2\gamma + \sin^2 2\gamma + \cos^2 2\gamma = \sec^2 2\gamma$

22. $\cot^2 C + \cos^2 C + \sin^2 C = \csc^2 C$

23. $\dfrac{\tan \theta + \cot \theta}{\sec \theta \csc \theta} = 1$

24. $\dfrac{\tan \theta - \cot \theta}{\sec \theta \csc \theta} = \sin^2\theta - \cos^2\theta$

25. $\dfrac{1}{\sin \theta + \cos \theta} + \dfrac{1}{\sin \theta - \cos \theta} = \dfrac{\sin \theta}{\sin^2\theta - \frac{1}{2}}$

26. $\dfrac{1}{\sec \theta + \tan \theta} + \dfrac{1}{\sec \theta - \tan \theta} = 2 \sec \theta$

27. $\dfrac{1 + \tan B}{1 - \tan B} = \dfrac{\sec^2 B + 2 \tan B}{2 - \sec^2 B}$

28. $(\cot x + \csc x)^2 = \dfrac{\sec x + 1}{\sec x - 1}$

29. $\dfrac{\sin^3 x - \cos^3 x}{\sin x - \cos x} = 1 + \sin x \cos x$

30. $\dfrac{\tan^3 t \cot^3 t}{\tan t - \cot t} = \sec^2 t + \cot^2 t$

31. $\sqrt{(3 \cos\theta - 4 \sin\theta)^2 + (3 \sin\theta + 4 \cos\theta)^2} = 5$

33. $\dfrac{2 \tan^2\theta + 2 \tan \theta \sec \theta}{\tan \theta + \sec \theta - 1} = \tan \theta + \sec \theta + 1$

34. $\dfrac{\csc \theta + 1}{\csc \theta - 1} - \dfrac{\sec \theta - \tan \theta}{\sec \theta + \tan \theta} = 4 \tan \theta \sec \theta$

35. $\dfrac{\cos \theta + \sin \theta}{\cos \theta - \sin \theta} + \dfrac{\cot \theta - 1}{\cot \theta + 1} = \dfrac{-2}{\sin^2\theta - \cos^2\theta}$

36. $\dfrac{\cos \theta + 1}{\cos \theta - 1} + \dfrac{1 - \sec \theta}{1 + \sec \theta} = -2 \cot^2\theta - 2 \csc^2\theta$

C *Prove that the equations in Problems 32–37 are identities.*

32. $\sin \theta + \cos \theta + 1 = \dfrac{2 \sin \theta \cos \theta}{\sin \theta + \cos \theta - 1}$

37. $\dfrac{\sin \theta}{1 - \cos \theta} + \dfrac{\cos \theta}{1 - \sin \theta} = (1 + \sin \theta + \cos \theta)(\sec \theta \csc \theta)$

3.3   Disproving Identities

As you worked the problems in the last two sections, you may have come upon an equation that you could not prove to be an identity and in desperation said "This can't be done!" There is a very simple way to prove that a given equation is not an identity. If you can find *one replacement* of the variable for which the functions are defined that will make the equation false, then you have proved that the equation is not an identity. When you prove that an equation is not an identity, we say you have **disproved an identity**.

EXAMPLE 1   Finding a counterexample

A common mistake for trigonometry students is to write $\sin 2\theta$ as $2 \sin \theta$. Show that

$$\sin 2\theta = 2 \sin \theta$$

is not an identity.

SOLUTION Let $\theta = 30°$.

LEFT SIDE	RIGHT SIDE
$\sin 2\,\theta = \sin 2\,(30°)$	$2 \sin 30° = 2\left(\dfrac{1}{2}\right)$
$= \sin 60°$	$= 1$
$= \dfrac{\sqrt{3}}{2}$	

CALCULATOR COMMENT

Graph: Y1=sin(2X) Y2=sinX

Thus, $\sin 2\theta = 2 \sin \theta$ is not an identity; it is disproved by a counterexample.

EXAMPLE 2 **Deciding whether an equation is an identity**

Is $\cos^2 x - \sin^2 x = \sin x$ an identity?

SOLUTION Begin by assuming it is an identity, and try to prove it as shown in the last section.

$$\cos^2 x - \sin^2 x = (1 - \sin^2 x) - \sin^2 x$$
$$= 1 - 2\sin^2 x$$

CALCULATOR COMMENT

Graph:
Y1=(cosX)^2-(sinX)^2
Y2=sinX

As you can see, graphs are not the same so Y1 ≠ Y2. The equation is not an identity.

If you suspect that it might not be an identity, you can try to prove that it is not an identity by counterexample. Choose some value for the variable, say, $x = 30°$.

LEFT SIDE

$$\cos^2 30° - \sin^2 30° = \left(\frac{\sqrt{3}}{2}\right)^2 - \left(\frac{1}{2}\right)^2$$
$$= \frac{3}{4} - \frac{1}{4} = \frac{1}{2}$$

RIGHT SIDE

$$\sin 30° = \frac{1}{2}$$

For $x = 30°$, $\cos^2 x - \sin^2 x = \sin x$, so you have neither proved nor disproved that the equation is an identity. Choose some other value for the variable, say, $x = 60°$.

LEFT SIDE

$$\cos^2 60° - \sin^2 60° = \left(\frac{1}{2}\right)^2 - \left(\frac{\sqrt{3}}{2}\right)^2$$
$$= \frac{1}{4} - \frac{3}{4}$$
$$= -\frac{1}{2}$$

RIGHT SIDE

$$\sin 60° = \frac{\sqrt{3}}{2}$$

For $x = 60°$, $\cos^2 x - \sin^2 x \neq \sin x$, so the statement is not an identity; it is disproved by a counterexample. ▮

As you can see from Example 2, if you happen to pick a value that satisfies the equation, you have proved nothing. For this reason, it is often worthwhile to use an unusual value when checking for a counterexample, which is easy if you use a calculator. For Example 2, suppose we pick $x = 17°$. Evaluate both the left and right sides of the equation using a calculator.

For the left side, evaluate $\cos^2 17° - \sin^2 17°$ by calculator:

DISPLAY

ALGEBRAIC: 17 cos x² − 17 sin x² = .8290375726

RPN: 17 cos ENTER × 17 sin ENTER × − .8290375724

GRAPHING: (cos 17) x² − (sin 17) x² ENTER .8290375726

For the right side, $\sin 17° = 0.2923717047$.

Therefore, $\cos^2 x - \sin^2 x \neq \sin x$

"The exception proves the rule" is a saying that is, of course, completely false. In fact, *any* exception disproves the rule, and this is what counterexamples are all about.

However, you should not pick values for counterexamples that cause any of the functions to be undefined. For example, when working with a tangent function, do not pick the angle equal to $\frac{\pi}{2}$ or to $\frac{3\pi}{2}$. Even the fundamental identities do not have meaning for undefined values.

3.3   Problem Set

A *Disprove each identity in Problems 1–10 by finding a counterexample. Be sure you do not choose a value that makes any of the functions undefined.*

1. $\sec^2 x - 1 = \sqrt{3}\tan x$ **2.** $\sec^2 x - 1 = \tan x$

3. $2\cos 2x \sin 2x = \sin 2x$ **4.** $2\cos 2x \sin 2x = \cos 2x$

5. $\sin x = \cos x$ **6.** $\tan x = \cot x$

7. $\cos^2 x - 3\sin x + 3 = 0$ **8.** $\sin^2 x + \cos x = 0$

9. $\tan^2 x = \sec x$ **10.** $2\sin^2 x - 2\cos^2 x = 1$

B *Prove or disprove that each of the equations in Problems 11–20 is an identity.*

11. $\sec^2\theta \cos\theta + 2\tan\theta \cos\theta = \sec^2\theta \cos\theta + 2\sin\theta$

12. $\sin^2\theta \cos\theta + 2\tan\theta \sin\theta = \sin^2\theta \cos\theta + 2\cos\theta$

13. $\sin\theta = \cot\theta \sec\theta - \cot\theta \cos\theta$

14. $\cos\lambda = \tan\lambda \csc\lambda - \tan\lambda \sin\lambda$

15. $\tan\lambda + \cot\lambda = \sec\lambda \csc\lambda$

16. $\dfrac{\sin\alpha}{\tan\alpha} + \dfrac{\cos\alpha}{\cot\alpha} = \cos\alpha + \sin\alpha$

17. $\dfrac{\cot\alpha + \sin^2\alpha \sec\alpha}{\sin\alpha} = \sec\alpha \sin\alpha$

18. $\dfrac{\tan\beta + \cot\beta}{\sec\beta \csc\beta} = 1$

19. $\sin\beta + \cot\beta \csc\beta = \sec\beta$

20. $\dfrac{\cos^4\gamma + \cos^2\gamma \sin^2\gamma}{\sin^2\gamma \cot^2\gamma} = \cos^2\gamma$

C *Prove or disprove each of the identities in Problems 21–31.*

21. $\dfrac{1 + \tan\gamma}{1 - \tan\gamma} = \dfrac{\sec^2\gamma \cos\gamma + 2\sin\gamma}{2\cos\gamma - \sec\gamma}$

22 $\csc\theta + \cot\theta = \dfrac{1}{\csc\theta - \cot\theta}$

23. $\sec^2\lambda - \csc^2\lambda = (2\sin^2\lambda - 1)(\sec^2\lambda + \csc^2\lambda)$

24. $2\csc A = 2\csc A - \cot A \cos A + \cos^2 A \csc A$

25. $\sec^2 2\lambda + \csc^2 2\lambda = \csc^2 2\lambda \sec^2 2\lambda$

26. $\dfrac{\tan\theta}{\cot\theta} - \dfrac{\cot\theta}{\tan\theta} = \sec^2\theta - \csc^2\theta$

27. $\dfrac{\cos^4\theta - \sin^4\theta}{(\cos^2\theta - \sin^2\theta)^2} = \dfrac{\cos\theta}{\cos\theta + \sin\theta} + \dfrac{\sin\theta}{\cos\theta - \sin\theta}$

28. $\dfrac{\cos^2\theta - \cos\theta \csc\theta}{\cos^2\theta \csc\theta - \cos\theta \csc^2\theta} = \sin\theta$

29. $\dfrac{\tan^2\theta - 2\tan\theta}{2\tan\theta - 4} = \tfrac{1}{2}\tan\theta$

30. $\dfrac{\cos\theta + \cos^2\theta}{\cos\theta + 1} = \dfrac{\cos\theta \sin\theta + \cos^2\theta}{\sin\theta + \cos\theta}$

31. $\dfrac{\csc\theta + 1}{\cot^2\theta + \csc\theta + 1} = \dfrac{\sin^2\theta + \sin\theta \cos\theta}{\sin\theta + \cos\theta}$

3.4 Cofunction and Opposite-Angle Identities

When proving identities, it is sometimes necessary to simplify the functional value of the sum or difference of two angles. That is, if α and β represent any two angles, we know that in general

$$\cos(\alpha - \beta) \neq \cos \alpha - \cos \beta$$

For example, if $\alpha = 60°$ and $\beta = 30°$,

$$\cos(60° - 30°) = \cos 30° \quad \text{while} \quad \cos 60° - \cos 30° = \frac{1}{2} - \frac{\sqrt{3}}{2}$$

$$= \frac{\sqrt{3}}{2} \quad \leftarrow \text{These are not the same} \rightarrow \quad = \frac{1 - \sqrt{3}}{2}$$

Thus,

$$\cos(60° - 30°) \neq \cos 60° - \sin 30°$$

In fact, we will now show that

$$\cos(\alpha - \beta) = \cos \alpha \cos \beta + \sin \alpha \sin \beta$$

This result not only will enable us to find the cosine of the difference of two angles, but also will provide the cornerstone upon which we will build a great many additional identities.

Distance Formula

To prove $\cos(\alpha - \beta) = \cos \alpha \cos \beta + \sin \alpha \sin \beta$ we will need to use the **distance formula** from algebra.

DISTANCE FORMULA

> If $P_1(x_1, y_1)$ and $P_2(x_2, y_2)$ are any two points, then the distance d between P_1 and P_2 is
>
> $$d = \sqrt{(x_2 - x_1)^2 + (y_2 - y_1)^2}$$

The proof of this formula follows directly from the Pythagorean theorem and is found in most algebra books.

EXAMPLE 1   **Using the distance formula**

Find the distance between $(1, -2)$ and $(-3, 4)$.

SOLUTION

$$d = \sqrt{(-3 - 1)^2 + (4 + 2)^2}$$
$$= \sqrt{16 + 36}$$
$$= \sqrt{52}$$
$$= 2\sqrt{13}$$

EXAMPLE 2   Using the distance formula with trigonometric functions

Find the distance between $(\cos \theta, \sin \theta)$ and $(1, 0)$.

SOLUTION

$$d = \sqrt{(1 - \cos \theta)^2 + (0 - \sin \theta)^2}$$
$$= \sqrt{1 - 2 \cos \theta + \cos^2\theta + \sin^2\theta}$$
$$= \sqrt{1 - 2 \cos \theta + 1}$$
$$= \sqrt{2 - 2 \cos \theta}$$

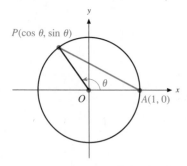

FIGURE 3.1 Length of a chord subtended by an angle θ

Chord Length

Example 2 is used to find the length of any chord (in a unit circle) whose corresponding arc subtends or is intercepted by a central angle θ, where θ is in standard position. Let A be the point $(1, 0)$ and P the point on the intersection of the terminal side of angle θ and the unit circle (see Figure 3.1). This means (from the definition of cosine and sine) that the coordinates of P are $(\cos \theta, \sin \theta)$. The length of the chord AP, denoted by $|AP|$, is the number found in Example 2.

CHORD LENGTH

The length of chord AP is

$$|AP| = \sqrt{2 - 2 \cos \theta}$$

where θ is the central angle between OA and OP.

EXAMPLE 3   Finding the length of a chord of a circle

Find $|P_\alpha P_\beta|$, where P_α and P_β are points on a unit circle determined by the angles α and β, respectively.

SOLUTION   Since $P_\alpha P_\beta$ forms a chord in a unit circle and since the central angle θ is $\alpha - \beta$ (see Figure 3.2),

$$|P_\alpha P_\beta| = \sqrt{2 - 2 \cos(\alpha - \beta)}$$

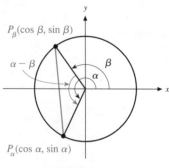

FIGURE 3.2
Distance between P_α and P_β

Suppose we rework Example 3 using the distance formula instead of the chord length formula:

$$|P_\alpha P_\beta| = \sqrt{(\cos \beta - \cos \alpha)^2 + (\sin \beta - \sin \alpha)^2}$$
$$= \sqrt{\cos^2\beta - 2 \cos \alpha \cos \beta + \cos^2\alpha + \sin^2\beta - 2 \sin \alpha \sin \beta + \sin^2\alpha}$$
$$= \sqrt{(\cos^2\beta + \sin^2\beta) + (\cos^2\alpha + \sin^2\alpha) - 2(\cos \alpha \cos \beta + \sin \alpha \sin \beta)}$$
$$= \sqrt{2 - 2(\cos \alpha \cos \beta + \sin \alpha \sin \beta)}$$

This result and the result from Example 3 look very different, yet they both represent the same distance. If we equate these expressions (since they are equivalent), we obtain a very useful result (after simplifying):

$$\sqrt{2 - 2\cos(\alpha - \beta)} = \sqrt{2 - 2(\cos\alpha\cos\beta + \sin\alpha\sin\beta)}$$

$$2 - 2\cos(\alpha - \beta) = 2 - 2(\cos\alpha\cos\beta + \sin\alpha\sin\beta)$$

$$-2\cos(\alpha - \beta) = -2(\cos\alpha\cos\beta + \sin\alpha\sin\beta)$$

$$\cos(\alpha - \beta) = \cos\alpha\cos\beta + \sin\alpha\sin\beta$$

We can now use this identity in a variety of different contexts and as a basis for deriving other identities in this chapter.

Cofunction Identities

We will begin numbering the identities with 9 to remind you that the first eight identities are the fundamental identities. These identities 9–11 are called the **cofunction identities** because they relate the sine and cosine as well as the tangent and cotangent.

COFUNCTION IDENTITIES

For any real number (or angle) θ,

9. $\cos\left(\frac{\pi}{2} - \theta\right) = \sin\theta$ 10. $\sin\left(\frac{\pi}{2} - \theta\right) = \cos\theta$ 11. $\tan\left(\frac{\pi}{2} - \theta\right) = \cot\theta$

PROOF OF IDENTITY 9   Let $\alpha = \frac{\pi}{2}$ and $\beta = \theta$ in the identity

$$\cos(\alpha - \beta) = \cos\alpha\cos\beta + \sin\alpha\sin\beta$$

$$\cos\left(\frac{\pi}{2} - \theta\right) = \cos\frac{\pi}{2}\cos\theta + \sin\frac{\pi}{2}\sin\theta$$

$$= 0 \cdot \cos\theta + 1 \cdot \sin\theta$$

$$\therefore \cos\left(\frac{\pi}{2} - \theta\right) = \sin\theta$$

Note: This is the same as the graph of $y = \sin x$.

PROOF OF IDENTITY 10   The proof depends on Identity 9. For any θ,

$$\cos\theta = \cos\left[\frac{\pi}{2} - \left(\frac{\pi}{2} - \theta\right)\right]$$

$$= \sin\left(\frac{\pi}{2} - \theta\right) \qquad \text{This is Identity 9.}$$

$$\therefore \sin\left(\frac{\pi}{2} - \theta\right) = \cos\theta$$

Identities involving the tangent are usually proved after proving similar identities for cosine and sine. The fundamental identity

$$\tan \theta = \frac{\sin \theta}{\cos \theta}$$

is applied first, allowing you then to use the appropriate identities for cosine and sine.

PROOF OF IDENTITY 11   This process is illustrated below:

$$\tan\left(\frac{\pi}{2} - \theta\right) = \frac{\sin\left(\frac{\pi}{2} - \theta\right)}{\cos\left(\frac{\pi}{2} - \theta\right)}$$

$$= \frac{\cos \theta}{\sin \theta}$$

$$= \cot \theta$$

The cofunction identities allow us to change a trigonometric function to the cofunction of its complement.

EXAMPLE 4   **Writing a trigonometric function in terms of its cofunction**

a. $\sin 28°$ **b.** $\cos 43°$ **c.** $\cot 9°$ **d.** $\sin \frac{\pi}{6}$

SOLUTION   **a.** $\sin 28° = \cos(90° - 28°)$ **b.** $\cos 43° = \sin(90° - 43°)$
    $= \cos 62°$ $= \sin 47°$

c. $\cot 9° = \tan(90° - 9°)$ **d.** $\sin \frac{\pi}{6} = \cos\left(\frac{\pi}{2} - \frac{\pi}{6}\right)$
    $= \tan 81°$ $= \cos \frac{2\pi}{6}$

                                          $= \cos \frac{\pi}{3}$

Opposite-Angle Identities

Suppose the given angle in Example 4 is larger than 90°; for example,

$$\cos 125° = \sin(90° - 125°)$$
$$= \sin(-35°)$$

The result can be further simplified using the following **opposite-angle identities**.

OPPOSITE-ANGLE IDENTITIES

For any real number (or angle) θ,

12. $\cos(-\theta) = \cos \theta$ 13. $\sin(-\theta) = -\sin \theta$ 14. $\tan(-\theta) = -\tan \theta$

PROOF OF IDENTITY 12   Let $\alpha = 0$ and $\beta = \theta$ in the identity $\cos(\alpha - \beta) = \cos\alpha\cos\beta + \sin\alpha\sin\beta$.

$$\cos(0 - \theta) = \cos 0 \cos \theta + \sin 0 \sin \theta$$
$$= 1 \cdot \cos \theta + 0 \cdot \sin \theta$$
$$= \cos \theta$$

CALCULATOR COMMENT

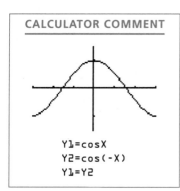

Y1=cosX
Y2=cos(-X)
Y1=Y2

But, if you simplify directly,

$$\cos(0 - \theta) = \cos(-\theta)$$
$$\therefore \cos(-\theta) = \cos\theta$$

PROOF OF IDENTITY 13   Let $\alpha = \frac{\pi}{2}$ and $\beta = -\theta$.

$$\cos\left(\frac{\pi}{2} + \theta\right) = \cos\left[\frac{\pi}{2} - (-\theta)\right]$$
$$= \cos\frac{\pi}{2}\cos(-\theta) + \sin\frac{\pi}{2}\sin(-0)$$
$$= 0 \cdot \cos(-\theta) + 1 \cdot \sin(-\theta)$$
$$= \sin(-\theta)$$

Also,

$$\cos\left(\frac{\pi}{2} + \theta\right) = \cos\left(\theta + \frac{\pi}{2}\right)$$
$$= \cos\left[\theta - \left(-\frac{\pi}{2}\right)\right]$$
$$= \cos\theta\cos\left(-\frac{\pi}{2}\right) + \sin\theta\sin\left(-\frac{\pi}{2}\right)$$
$$= \cos\theta \cdot \ \ 0 \ \ \ \ + \sin\theta \cdot (-1)$$
$$= -\sin\theta$$
$$\therefore \sin(-\theta) = -\sin\theta$$

PROOF OF IDENTITY 14

$$\tan(-\theta) = \frac{\sin(-\theta)}{\cos(-\theta)}$$
$$= \frac{-\sin\theta}{\cos\theta}$$
$$= -\tan\theta$$

EXAMPLE 5   **Writing a trigonometric function as a function of a positive angle**

a. $\cos(-19°)$ **b.** $\sin(-19°)$ **c.** $\tan(-2)$

SOLUTION   **a.** $\cos(-19°) = \cos 19°$ **b.** $\sin(-19°) = -\sin 19°$

c. $\tan(-2) = -\tan 2$

EXAMPLE 6   **Writing a trigonometric function in terms of a cofunction of a positive angle**

a. $\cos 125°$ **b.** $\sin 102°$ **c.** $\cot 2.5$

SOLUTION   **a.** $\cos 125° = \sin(90° - 125°)$ You might recall we worked problems like this in

$= \sin(-35°)$ Chapter 1 using reference angles and the reduction

$= -\sin 35°$ principle. Note that 35° is the reference angle for

$(-35°)$ and sine is negative in Quadrant IV

b. $\sin 102° = \cos(90° - 102°)$ Can you verify this using the reduction principle

$= \cos(-12°)$ from Chapter 1?

$= \cos 12°$

c. $\cot 2.5 = \tan\left(\frac{\pi}{2} - 2.5\right)$

$\approx \tan(-.9292)$

$= -\tan .9292$

You may also need to use opposite-angle identities together with other identities. For example, you know from algebra that

$$a - b \quad \text{and} \quad b - a$$

are opposites. This means that $a - b = -(b - a)$. In trigonometry you often see angles like $\frac{\pi}{2} - \theta$ and want to write $\theta - \frac{\pi}{2}$. This means that $\frac{\pi}{2} - \theta = -\left(\theta - \frac{\pi}{2}\right)$. In particular,

$$\cos\left(\frac{\pi}{2} - \theta\right) = \cos\left[-\left(\theta - \frac{\pi}{2}\right)\right]$$

$$= \cos\left(\theta - \frac{\pi}{2}\right) \qquad \text{By Identity 12}$$

EXAMPLE 7   **Using an opposite-angle identity**

a. $\sin\left(\frac{\pi}{2} - \theta\right)$ **b.** $\tan\left(\frac{\pi}{2} - \theta\right)$ **c.** $\cos(\pi - \theta)$

SOLUTION   **a.** $\sin\left(\frac{\pi}{2} - \theta\right) = \sin\left[-\left(\theta - \frac{\pi}{2}\right)\right]$

$= -\sin\left(\theta - \frac{\pi}{2}\right)$

b. $\tan\left(\frac{\pi}{2} - \theta\right) = \tan\left[-\left(\theta - \frac{\pi}{2}\right)\right]$

$= -\tan\left(\theta - \frac{\pi}{2}\right)$

c. $\cos(\pi - \theta) = \cos[-(\theta - \pi)]$

$= \cos(\theta - \pi)$

Graphing with Opposites

The procedure for graphing $y = -\cos x$ is identical to the procedure for graphing $y = \cos x$, except that, after the frame is drawn, the endpoints are at the bottom of the frame instead of at the top of the frame, and the midpoint is at the top, as shown in Figure 3.3.

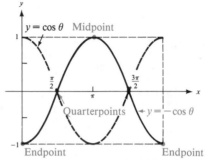

FIGURE 3.3 One period of $y = -\cos x$

EXAMPLE 8   **Graphing a trigonometric function with a negative angle or negative number**

Graph $y = \sin(-x)$.

SOLUTION   First use an opposite-angle identity, if necessary: $\sin(-x) = -\sin x$. To graph $y = -\sin x$, build the frame as before; the endpoints and midpoints are the same, but the quarterpoints are reversed, as shown in Figure 3.4.

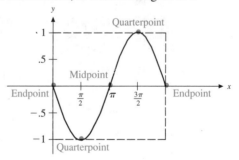

FIGURE 3.4 One period of $y = -\sin x$

3.4   Problem Set

A *Use the distance formula to find the distance between the points given in Problems 1–6.*

Write each function in Problems 7–12 in terms of its cofunction by using the cofunction identities.

1. $(6, 2)$ and $(1, 10)$

2. $(-2, 3)$ and $(5, 10)$

3. $(1, 6)$ and $(5, -8)$

4. (α, θ) and (β, ω)

5. $(\cos \alpha, \sin \alpha)$ and $(0, 1)$

6. $(\cos \theta, \sin \theta)$ and $(\cos \lambda, \sin \lambda)$

7. $\cos 15°$

8. $\sin 38°$

9. $\cot 41°$

10. $\sin \frac{\pi}{6}$

11. $\cos \frac{5\pi}{6}$

12. $\cot \frac{2\pi}{3}$

Write the functions in Problem 13–18 *as functions of positive angles.*

13. $\cos(-18°)$

14. $\tan(-49°)$

15. $\sin(-41°)$

16. $\sin(-31°)$

17. $\cos(-39°)$

18. $\tan(-24°)$

B

19. Write each as a function of $\theta - \frac{\pi}{3}$.

  a. $\cos\left(\frac{\pi}{3} - \theta\right)$ **b.** $\sin\left(\frac{\pi}{3} - \theta\right)$ **c.** $\tan\left(\frac{\pi}{3} - \theta\right)$

20. Write each as a function of $\left(\theta - \frac{2\pi}{3}\right)$.

  a. $\cos\left(\frac{2\pi}{3} - \theta\right)$ **b.** $\sin\left(\frac{2\pi}{3} - \theta\right)$ **c.** $\tan\left(\frac{2\pi}{3} - \theta\right)$

21. Write each as a function of $\alpha - \beta$.

  a. $\cos(\beta - \alpha)$ **b.** $\sin(\beta - \alpha)$ **c.** $\tan(\beta - \alpha)$

Use the opposite-angle identities, if necessary, to graph the functions in Problems 22–30.

22. $y = -2\cos x$

23. $y = -3\sin x$

24. $y = \tan(-x)$

25. $y = \cos(-2x)$

26. $y = \sin(-3x)$

27. $y - 1 = \sin(2 - x)$

28. $y - 2 = \cos(\pi - x)$

29. $y - 3 = \sin(\pi - x)$

30. $y + 1 = \tan\left(\frac{\pi}{6} - x\right)$

C *Prove that the equations in Problems* 31–32 *are identities.*

31. $\cot(-\theta) = -\cot\theta$

32. $\sec(-\theta) = \sec\theta$

3.5 Addition Laws

The identities of this section are based on the cosine identity, derived at the beginning of the last section, and are called the **addition laws**. Since subtraction can easily be written as a sum, the designation *addition laws* refers to both addition and subtraction.

ADDITION LAWS

15. $\cos(\alpha + \beta) = \cos\alpha\cos\beta - \sin\alpha\sin\beta$

16. $\cos(\alpha - \beta) = \cos\alpha\cos\beta + \sin\alpha\sin\beta$

17. $\sin(\alpha + \beta) = \sin\alpha\cos\beta + \cos\alpha\sin\beta$

18. $\sin(\alpha - \beta) = \sin\alpha\cos\beta - \cos\alpha\sin\beta$

19. $\tan(\alpha + \beta) = \dfrac{\tan\alpha + \tan\beta}{1 - \tan\alpha\tan\beta}$

20. $\tan(\alpha - \beta) = \dfrac{\tan\alpha - \tan\beta}{1 + \tan\alpha\tan\beta}$

EXAMPLE 1   Using an addition law

Write $\cos\left(\frac{2\pi}{3} + \theta\right)$ as a function of θ only.

SOLUTION   Use Identity 15:

$$\cos\left(\tfrac{2\pi}{3} + \theta\right) = \cos\tfrac{2\pi}{3}\cos\theta - \sin\tfrac{2\pi}{3}\sin\theta$$

$$= \left(-\tfrac{1}{2}\right)\cos\theta - \left(\tfrac{\sqrt{3}}{2}\right)\sin\theta$$

$$= -\tfrac{1}{2}(\cos\theta + \sqrt{3}\sin\theta)$$

Substitute exact values where possible. Simplify.

EXAMPLE 2   Evaluation of a trigonometric expression by using an addition law

Evaluate

$$\frac{\tan 18° - \tan 40°}{1 + \tan 18° \tan 40°}$$

using a calculator.

SOLUTION   You can do a lot of arithmetic, or you can use Identity 20.

$$\frac{\tan 18° - \tan 40°}{1 + \tan 18° \tan 40°} = \tan(18° - 40°)$$

$$= \tan(-22°)$$

$$= -\tan 22° \qquad \text{Identity 14 from Section 3.4}$$

$$\approx -.4040 \qquad \text{Use a calculator.}$$

EXAMPLE 3   Finding an exact value of a multiple of 15°

Find the exact value of $\cos 345°$.

SOLUTION $\cos 345° = \cos 15°$ First use the reduction principle, if appropriate.

$$= \cos(45° - 30°) \qquad \text{Next, write the angle in terms of a sum or difference of angles whose exact values you know; in this case, } 15° = 45° - 30°.$$

$$= \cos 45° \cos 30° + \sin 45° \sin 30° \qquad \text{Apply Identity 16.}$$

$$= \frac{\sqrt{2}}{2} \cdot \frac{\sqrt{3}}{2} + \frac{\sqrt{2}}{2} \cdot \frac{1}{2} \qquad \text{Use exact values.}$$

$$= \frac{\sqrt{6}}{4} + \frac{\sqrt{2}}{4} \qquad \text{Simplify.}$$

$$= \frac{\sqrt{6} + \sqrt{2}}{4}$$

We will conclude this section by proving some of the addition laws; others are left as exercises.

PROOF OF IDENTITY 15 $\cos(\alpha + \beta) = \cos[\alpha - (-\beta)]$

$$= \cos \alpha \cos(-\beta) + \sin \alpha \sin(-\beta)$$

$$= \cos \alpha \cos \beta - \sin \alpha \sin \beta \qquad \text{Use Identities 12 and 13.}$$

PROOF OF IDENTITY 16   Identity 16 is the main cosine identity proved at the beginning of Section 3.4. It is numbered and included here for the sake of completeness.

PROOF OF IDENTITY 17 $$\sin(\alpha + \beta) = \cos\left[\frac{\pi}{2} - (\alpha + \beta)\right]$$

$$= \cos\left[\left(\frac{\pi}{2} - \alpha\right) - \beta\right]$$

$$= \cos\left[\left(\frac{\pi}{2} - \alpha\right) \cos \beta + \sin\left(\frac{\pi}{2} - \alpha\right) \sin \beta\right.$$

$$= \sin \alpha \cos \beta + \cos \alpha \sin \beta$$

PROOF OF IDENTITY 18   Replace β by $-\beta$ in Identity 17; the details are left as an exercise.

PROOF OF IDENTITY 19

$$\tan(\alpha + \beta) = \frac{\sin(\alpha + \beta)}{\cos(\alpha + \beta)}$$

$$= \frac{\sin \alpha \cos \beta + \cos \alpha \sin \beta}{\cos \alpha \cos \beta - \sin \alpha \sin \beta}$$

$$= \frac{\sin \alpha \cos \beta + \cos \alpha \sin \beta}{\cos \alpha \cos \beta - \sin \alpha \sin \beta} \cdot \frac{\dfrac{1}{\cos \alpha \cos \beta}}{\dfrac{1}{\cos \alpha \cos \beta}} \qquad \text{Multiply by 1.}$$

$$= \frac{\dfrac{\sin \alpha \cos \beta}{\cos \alpha \cos \beta} + \dfrac{\cos \alpha \sin \beta}{\cos \alpha \cos \beta}}{\dfrac{\cos \alpha \cos \beta}{\cos \alpha \cos \beta} - \dfrac{\sin \alpha \sin \beta}{\cos \alpha \cos \beta'}}$$

$$= \frac{\tan \alpha + \tan \beta}{1 - \tan \alpha \tan \beta}$$

PROOF OF IDENTITY 20   The proof of Identity 20 is left as an exercise.

3.5 Problem Set

A *Change each of the expressions in Problems 1–6 to functions of θ only.*

1. a. $\cos(30° + \theta)$ **b.** $\sin(\theta - 45°)$
2. a. $\tan(45° + \theta)$ **b.** $\tan(\theta - 45°)$
3. a. $\cos\left(\frac{\pi}{3} - \theta\right)$ **b.** $\cos\left(\theta - \frac{\pi}{4}\right)$
4. a. $\sin\left(\frac{2\pi}{3} + \theta\right)$ **b.** $\sin\left(\theta - \frac{\pi}{6}\right)$
5. a. $\cos(\theta + \theta)$ **b.** $\sin(\theta + \theta)$
6. a. $\tan(\theta + \theta)$ **b.** $\cot(\theta + \theta)$

Evaluate each of the expressions in Problems 7–16.

7. $\sin 158° \cos 92° - \cos 158° \sin 92°$
8. $\cos 114° \cos 85° + \sin 114° \sin 85°$
9. $\cos 30° \cos 48° - \sin 30° \sin 48°$
10. $\sin 18° \cos 23° + \cos 18° \sin 23°$
11. $\sin 125° \cos 85° - \cos 125° \sin 85°$
12. $\cos 215° \cos 145° - \sin 215° \sin 145°$
13. $\dfrac{\tan 32° + \tan 18°}{1 - \tan 32° \tan 18°}$
14. $\dfrac{\tan 59° - \tan 25°}{1 + \tan 59° \tan 25°}$
15. $\dfrac{\tan 120° - \tan 89°}{1 + \tan 120° \tan 89°}$
16. $\dfrac{\tan 245° + \tan 210°}{1 - \tan 245° \tan 210°}$

B *Using the identities of this section, find the exact values of the sine, cosine, and tangent of each of the angles given in Problems 17–22.*

17. $-15°$ **18.** $195°$ **19.** $75°$ **20.** $165°$
21. $345°$ **22.** $105°$

Prove the identities in Problems 23–30.

23. $\sin(\alpha - \beta) = \sin \alpha \cos \beta - \cos \alpha \sin \beta$
24. $\tan(\alpha - \beta) = \dfrac{\tan \alpha - \tan \beta}{1 + \tan \alpha \tan \beta}$
25. $\cot(\alpha + \beta) = \dfrac{\cot \alpha \cot \beta - 1}{\cot \beta + \cot \alpha}$
26. $\cot(\alpha - \beta) = \dfrac{\cot \alpha \cot \beta + 1}{\cot \beta - \cot \alpha}$
27. $\dfrac{\cos 5\theta}{\sin \theta} - \dfrac{\sin 5\theta}{\cos \theta} = \dfrac{\cos 6\theta}{\sin \theta \cos \theta}$
28. $\dfrac{\sin 6\theta}{\sin 3\theta} - \dfrac{\cos 6\theta}{\cos 3\theta} = \sec 3\theta$
29. $\sin(\alpha + \beta) \cos \beta - \cos(\alpha + \beta) \sin \beta = \sin \alpha$
30. $\cos(\alpha - \beta) \cos \beta - \sin(\alpha - \beta) \sin \beta = \cos \alpha$

Prove the identities in Problems 31–47.

31. $\sin\left(\frac{3\pi}{2} - \theta\right) = -\cos\theta$

32. $\cos\left(\frac{3\pi}{2} - \theta\right) = -\sin\theta$

33. $\sin\left(\frac{3\pi}{2} + \theta\right) = -\cos\theta$

34. $\cos\left(\frac{3\pi}{2} + \theta\right) = \sin\theta$

35. $\sin(\pi - \theta) = \sin\theta$

36. $\cos(\pi - \theta) = -\cos\theta$

37. $\tan(\pi - \theta) = -\tan\theta$

38. $\sin(\pi + \theta) = -\sin\theta$

39. $\cos(\pi + \theta) = -\cos\theta$

40. $\tan(\pi + \theta) = \tan\theta$

41. $\sec\left(\frac{\pi}{2} - \theta\right) = \csc\theta$

42. $\cot\left(\frac{\pi}{2} - \theta\right) = \tan\theta$

43. $\csc\left(\frac{\pi}{2} - \theta\right) = \sec\theta$

44. $\dfrac{\sin(\theta + h) - \sin\theta}{h} = \cos\theta\left(\dfrac{\sin h}{h}\right) - \sin\theta\left(\dfrac{1 - \cos h}{h}\right)$

45. $\dfrac{\cos(\theta + h) - \cos\theta}{h} = \sin\theta\left(\dfrac{\sin h}{h}\right) - \cos\theta\left(\dfrac{1 - \cos h}{h}\right)$

46. $\sin(\alpha + \beta + \gamma) = \sin\alpha\cos\beta\cos\gamma + \cos\alpha\sin\beta\cos\gamma + \cos\alpha\cos\beta\sin\gamma - \sin\alpha\sin\beta\sin\gamma$

47. $\cos(\alpha + \beta + \gamma) = \cos\alpha\cos\beta\cos\gamma - \cos\alpha\sin\beta\sin\gamma - \sin\alpha\cos\beta\sin\gamma - \sin\alpha\sin\beta\cos\gamma$

C

48. Assume that $\tan\alpha = \frac{a}{b}$ and $\tan\beta = \frac{c}{d}$ where a and c are any real numbers and b and d are nonzero real numbers. Prove

$$\tan(\alpha + \beta) = \frac{\tan\alpha + \tan\beta}{1 - \tan\alpha\tan\beta}$$

49. Use Figure 3.5 to prove the addition law

$$\tan(\alpha + \beta) = \frac{\tan\alpha + \tan\beta}{1 - \tan\alpha\tan\beta}$$

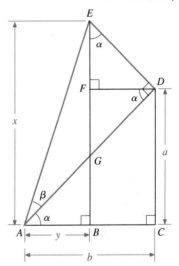

FIGURE 3.5 Geometric proof of an addition law

(*Hint:*   See the May 1989 issue of *The Mathematics Teacher*, p. 384)

3.6 Double-Angle and Half-Angle Identities

Double-Angle Identities

Two additional, special cases of the addition laws of Section 3.5 are now considered. The first is that of the **double-angle identities**.

DOUBLE-ANGLE
IDENTITIES

21. $\cos 2\theta = \cos^2\theta - \sin^2\theta$ 22. $\sin 2\theta = 2 \sin \theta \cos \theta$

$ = 2 \cos^2\theta - 1$ 23. $\tan 2\theta = \dfrac{2 \tan \theta}{1 - \tan^2\theta}$

$ = 1 - 2 \sin \theta$

To prove these identities, use the addition laws where $\alpha = \theta$ and $\beta = \theta$:

$$\cos 2\theta = \cos(\theta + \theta) = \cos \theta \cos \theta - \sin \theta \sin \theta$$
$$= \cos^2\theta - \sin^2\theta$$

$$\sin 2\theta = \sin(\theta + \theta) = \sin \theta \cos \theta + \cos \theta \sin \theta$$
$$= 2 \sin \theta \cos \theta$$

$$\tan 2\theta = \tan(\theta + \theta) = \frac{\tan \theta + \tan \theta}{1 - \tan \theta \tan \theta}$$
$$= \frac{2 \tan \theta}{1 - \tan^2\theta}$$

(You did these in Problems 5–6 in Problem Set 3.5.)

EXAMPLE 1   Using double-angle identities

a. $\cos 100x$ **b.** $\sin 120°$

SOLUTION **a.** $\cos 100x = \cos(2 \cdot 50x) = \cos^2 50x - \sin^2 50x$

b. $\sin 120° = \sin 2(60°) = 2 \sin 60° \cos 60°$

EXAMPLE 2   Using a double-angle identity for a multiple greater than 2

$$\cos 3\theta = \cos(2\theta + \theta) = \cos 2\theta \cos \theta - \sin 2\theta \sin \theta$$
$$= (\cos^2\theta - \sin^2\theta) \cos \theta - (2 \sin \theta \cos \theta) \sin \theta$$
$$= \cos^3\theta - \sin^2\theta \cos \theta - 2 \sin^2\theta \cos \theta$$
$$= \cos^3\theta - 3 \sin^2\theta \cos \theta$$
$$= \cos^3\theta - 3(1 - \cos^2\theta) \cos \theta$$
$$= \cos^3\theta - 3 \cos \theta + 3 \cos^3\theta$$
$$= 4 \cos^3\theta - 3 \cos \theta$$

EXAMPLE 3   **Evaluation of a trigonometric expression by using a double-angle identity**

Evaluate $\dfrac{2 \tan \frac{\pi}{16}}{1 - \tan^2 \frac{\pi}{16}}$.

SOLUTION   Notice that this is the right-hand side of Identity 23, so it is the same as $\tan\left(2 \cdot \frac{\pi}{16}\right) = \tan \frac{\pi}{8}$. Now use a calculator to find $\tan \frac{\pi}{8} \approx .4142$.

EXAMPLE 4   **Graphing by using a double-angle identity**

Graph $y = \sin^2 x$.

SOLUTION   Many students think of the sine curve and then *incorrectly* draw the graph of $y = |\sin x|$ in order to obtain positive values only, as shown in Figure 3.6. The graph of $y = \sin^2 x$ does not have corners (or cusps) as shown in Figure 3.6.

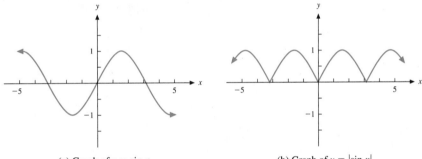

(a) Graph of $y = \sin x$ (b) Graph of $y = |\sin x|$

FIGURE 3.6 Comparison of the graphs of $y = \sin x$ and $y = |\sin x|$

Y=(sinX)^2

In order to correctly graph $y = \sin^2 x$ (other than by using a calculator, as shown in the margin), consider the double-angle identity

$$\cos 2x = 1 - 2\sin^2 x \qquad \text{or} \qquad \sin^2 x = \tfrac{1}{2} - \tfrac{1}{2}\cos 2x$$

We see that the curve $y = \sin^2 x$ is a cosine curve with $a = -\tfrac{1}{2}$, $b = 2$, and $(h, k) = \left(0, \tfrac{1}{2}\right)$ as shown in Figure 3.7.

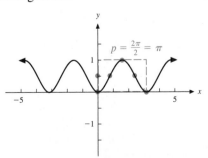

FIGURE 3.7 Graph of $y = \sin^2 x$ is

the same as the graph of $y - \tfrac{1}{2} = -\tfrac{1}{2}\cos 2x$

EXAMPLE 5   **Finding trigonometric functions of double angles, given a single angle**

If $\cos \theta = \frac{3}{5}$ and θ is in Quadrant IV, find $\cos 2\theta$, $\sin 2\theta$, and $\tan 2\theta$.

SOLUTION   Since $\cos 2\theta = 2 \cos^2 \theta - 1$,

$$\cos 2\theta = 2\left(\frac{3}{5}\right)^2 - 1$$

$$= 2\left(\frac{9}{25}\right) - 1$$

$$= -\frac{7}{25}$$

For the other functions of 2θ, you need to know $\sin \theta$. Begin with the fundamental identity relating cosine and sine.

$$\sin^2 \theta = 1 - \cos^2 \theta$$

$$\sin \theta = -\sqrt{1 - \cos^2 \theta} \qquad \text{Negative since the sine is negative in Quadrant IV}$$

$$= -\sqrt{1 - \left(\frac{3}{5}\right)^2}$$

$$= -\sqrt{1 - \frac{9}{25}}$$

$$= -\sqrt{\frac{16}{25}}$$

$$= -\frac{4}{5}$$

Now, $\sin 2\theta = 2 \sin \theta \cos \theta$

$$= 2\left(-\frac{4}{5}\right)\left(\frac{3}{5}\right)$$

$$= -\frac{24}{25}$$

Finally, $\tan 2\theta = \dfrac{\sin 2\theta}{\cos 2\theta}$

$$= \frac{-\frac{24}{25}}{-\frac{7}{25}} = \frac{24}{7}$$

Half-Angle Identities

Sometimes, as in Example 5, we want to write $\cos 2\theta$ in terms of cosines, and at other times we want to write it in terms of sines. Thus,

$$
\begin{aligned}
\cos 2\theta &= \cos^2\theta - \sin^2\theta \\
&= \cos^2\theta - (1 - \cos^2\theta) \\
&= 2\cos^2\theta - 1
\end{aligned}
\qquad \text{and} \qquad
\begin{aligned}
\cos 2\theta &= \cos^2\theta - \sin^2\theta \\
&= (1 - \sin^2\theta) - \sin^2\theta \\
&= 1 - 2\sin^2\theta
\end{aligned}
$$

These last two identities lead us to the second important special case of the addition laws, called the **half-angle identities**. We wish to solve $\cos 2\alpha = 2\cos^2\alpha - 1$ for $\cos^2\alpha$.

$$2\cos^2\alpha - 1 = \cos 2\alpha$$

$$2\cos^2\alpha = 1 + \cos 2\alpha$$

$$\cos^2\alpha = \frac{1 + \cos 2\alpha}{2}$$

Now, if $\alpha = \frac{1}{2}\theta$, then $2\alpha = \theta$ and $\cos^2\frac{1}{2}\theta = \dfrac{1 + \cos\theta}{2}$.

If $\frac{1}{2}\theta$ is in Quadrant I or IV, then $\cos\frac{1}{2}\theta = \sqrt{\dfrac{1 + \cos\theta}{2}}$.

If $\frac{1}{2}\theta$ is in Quadrant II or III, then $\cos\frac{1}{2}\theta = -\sqrt{\dfrac{1 + \cos\theta}{2}}$.

These results are summarized by writing

$$\cos\tfrac{1}{2}\theta = \pm\sqrt{\frac{1 + \cos\theta}{2}}$$

However, *you must be careful.* The sign $+$ or $-$ is chosen according to the quadrant in which $\frac{1}{2}\theta$ is located. The formula requires either $+$ or $-$, but not both. This use of $\pm$ is different from the use of $\pm$ in algebra. For example, when using $\pm$ in the quadratic formula, we are indicating *two* possible correct roots—one plus, one minus. In this trigonometric identity we will obtain *one* correct sign, plus or minus, depending on the quadrant of $\frac{1}{2}\theta$.*

For the sine, solve $\cos 2\alpha = 1 - 2\sin^2\alpha$ for $\sin^2\alpha$:

$$\cos 2\alpha = 1 - 2\sin^2\alpha$$

$$2\sin^2\alpha = 1 - \cos 2\alpha$$

$$\sin^2\alpha = \frac{1 - \cos 2\alpha}{2}$$

---

*You should also remember from algebra that $\frac{1}{2}\theta$ and $\dfrac{\theta}{2}$ mean the same thing, so $\cos\frac{1}{2}\theta$ and $\cos\dfrac{\theta}{2}$ are the same thing, but $\dfrac{\cos\theta}{2}$ is different.

Replace $\alpha = \frac{1}{2}\theta$, and $\sin^2 \frac{1}{2}\theta = \dfrac{1 - \cos\theta}{2}$ or $\sin \frac{1}{2}\theta = \pm \sqrt{\dfrac{1 - \cos\theta}{2}}$, where the sign depends on the quadrant of $\frac{1}{2}\theta$. If $\frac{1}{2}\theta$ is in Quadrant I or II, you use $+$; if it is in Quadrant III or IV, you use $-$.

Finally, to find the half-angle identity for the tangent, write

$$\tan \frac{1}{2}\theta = \dfrac{\sin \frac{1}{2}\theta}{\cos \frac{1}{2}\theta}$$

$$= \dfrac{\pm \sqrt{\dfrac{1 - \cos\theta}{2}}}{\pm \sqrt{\dfrac{1 + \cos\theta}{2}}}$$

$$= \pm \sqrt{\dfrac{1 - \cos\theta}{1 + \cos\theta}}$$

$$= \pm \sqrt{\dfrac{1 - \cos\theta}{1 + \cos\theta} \cdot \dfrac{1 - \cos\theta}{1 - \cos\theta}}$$

$$= \pm \sqrt{\dfrac{(1 - \cos\theta)^2}{\sin^2\theta}} \qquad \text{Remember that } 1 - \cos^2\theta = \sin^2\theta.$$

$$= \dfrac{1 - \cos\theta}{\sin\theta} \qquad \begin{array}{l}\text{Notice that } 1 - \cos\theta \text{ is positive. Also, since}\\ \tan \frac{1}{2}\theta \text{ and } \sin\theta \text{ have the same sign regardless of}\\ \text{the quadrant of } \theta, \text{ the desired result follows.}\end{array}$$

You can also show that $\tan \frac{1}{2}\theta = \dfrac{\sin\theta}{1 + \cos\theta}$.

HALF-ANGLE IDENTITIES

24. $\cos \frac{1}{2}\theta = \pm \sqrt{\dfrac{1 + \cos\theta}{2}}$ 26. $\tan \frac{1}{2}\theta = \dfrac{1 - \cos\theta}{\sin\theta}$

25. $\sin \frac{1}{2}\theta = \pm \sqrt{\dfrac{1 - \cos\theta}{2}}$ $= \dfrac{\sin\theta}{1 + \cos\theta}$

EXAMPLE 6   Using a half-angle identity

Find the exact value of $\cos \frac{9\pi}{8}$.

SOLUTION $\cos \dfrac{9\pi}{8} = \cos \left(\dfrac{1}{2} \cdot \dfrac{9\pi}{4} \right) = -\sqrt{\dfrac{1 + \cos \frac{9\pi}{4}}{2}}$

$= -\sqrt{\dfrac{1 + \cos \frac{\pi}{4}}{2}}$ Choose a negative sign, since $\frac{9\pi}{8}$ is in Quadrant III and the cosine is negative in this quadrant.

$= -\sqrt{\dfrac{1 + \frac{\sqrt{2}}{2}}{2}}$

$= -\sqrt{\dfrac{2 + \sqrt{2}}{4}} = -\dfrac{1}{2}\sqrt{2 + \sqrt{2}}$

EXAMPLE 7   **Finding trigonometric functions of single angles, given a double angle**

If $\cot 2\theta = \frac{3}{4}$, find $\cos \theta$, $\sin \theta$, and $\tan \theta$, where both θ and 2θ are in Quadrant I.

SOLUTION   You need to find $\cos 2\theta$ so that you can use it in the half-angle identities. To do this, first find $\tan 2\theta$:

$$\tan 2\theta = \frac{1}{\cot 2\theta} = \frac{4}{3}$$

Next, find $\sec 2\theta$:

$$\sec 2\theta = \pm\sqrt{1 + \tan^2 2\theta}$$

$$= \sqrt{1 + \frac{16}{9}}$$ It is positive because 2θ is in Quadrant I.

$$= \frac{5}{3}$$

Finally, $\cos 2\theta$ is the reciprocal of $\sec 2\theta$: $\cos 2\theta = \dfrac{3}{5}$.

Next use the half-angle identities:

$$\cos \theta = \pm\sqrt{\frac{1 + \cos 2\theta}{2}} \quad \text{and} \quad \sin \theta = \pm\sqrt{\frac{1 - \cos 2\theta}{2}}$$

Do you see that θ is one-half of 2θ in these formulas?

$$\cos \theta = +\sqrt{\frac{1 + \frac{3}{5}}{2}} \qquad\qquad \sin \theta = +\sqrt{\frac{1 - \frac{3}{5}}{2}}$$

Positive value chosen because θ is in Quadrant I

$$\cos \theta = \frac{2}{\sqrt{5}} \qquad\qquad\qquad \sin \theta = \frac{1}{\sqrt{5}}$$

$$\tan \theta = \frac{\sin \theta}{\cos \theta} = \frac{1/\sqrt{5}}{2/\sqrt{5}} = \frac{1}{2}$$

EXAMPLE 8   Proving an identity using a double-angle identity

Prove that $\sin \theta = \dfrac{2 \tan \frac{1}{2}\theta}{1 + \tan^2 \frac{1}{2}\theta}$.

SOLUTION When proving identities involving functions of different angles, you should write all the trigonometric functions in the problems as functions of a single angle:

$$\frac{2 \tan \frac{1}{2}\theta}{1 + \tan^2 \frac{1}{2}\theta} = \frac{2 \dfrac{\sin \frac{1}{2}\theta}{\cos \frac{1}{2}\theta}}{\sec^2 \frac{1}{2}\theta}$$

$$= 2 \frac{\sin \frac{1}{2}\theta}{\cos \frac{1}{2}\theta} \cdot \cos^2 \frac{1}{2}\theta$$

$$= 2 \sin \frac{1}{2}\theta \cos \frac{1}{2}\theta$$

$$= \sin \theta \qquad \text{From Identity 22 (double-angle identity)} \qquad \blacksquare$$

3.6 Problem Set

A *Use the double-angle or half-angle identities to evaluate each of Problems 1–9 using exact values.*

1. $2 \cos^2 22.5° - 1$

2. $\dfrac{2 \tan \frac{\pi}{8}}{1 - \tan^2 \frac{\pi}{8}}$

3. $\sqrt{\dfrac{1 - \cos 60°}{2}}$

4. $\cos^2 15° - \sin^2 15°$

5. $1 - 2 \sin^2 90°$

6. $-\sqrt{\dfrac{1 - \cos 420°}{2}}$

7. $\sin 22.5°$ **8.** $\cos \frac{\pi}{8}$ **9.** $\tan 22.5°$

In each of Problems 10–15, find the exact values of cosine, sine, and tangent of 2θ where $0 \le \theta < 2\pi$.

10. $\sin \theta = \frac{3}{5}$; θ in Quadrant I

11. $\sin \theta = \frac{5}{13}$; θ in Quadrant II

12. $\tan \theta = -\frac{5}{12}$; θ in Quadrant IV

13. $\tan \theta = -\frac{3}{4}$; θ in Quadrant II

14. $\cos \theta = \frac{5}{9}$; θ in Quadrant I

15. $\cos \theta = -\frac{5}{13}$; θ in Quadrant III

In each of Problems 16–21, find the exact values of cosine, sine, and tangent of $\frac{1}{2}\theta$ where $0 \le \theta < 2\pi$.

16. $\sin \theta = \frac{3}{5}$; θ in Quadrant I

17. $\sin \theta = \frac{5}{13}$; θ in Quadrant II

18. $\tan \theta = -\frac{5}{12}$; θ in Quadrant IV

19. $\tan \theta = -\frac{3}{4}$; θ in Quadrant II

20. $\cos \theta = \frac{5}{9}$; θ in Quadrant I

21. $\cos \theta = -\frac{5}{13}$; θ in Quadrant III

B *In each of Problems 22–27, find $\cos \theta$, $\sin \theta$, and $\tan \theta$ when θ is in Quadrant I ($0° < \theta < 90°$) and $\cot 2\theta$ is given.*

22. $\cot 2\theta = -\frac{3}{4}$ **23.** $\cot 2\theta = 0$

24. $\cot 2\theta = \frac{1}{\sqrt{3}}$ **25.** $\cot 2\theta = -\frac{1}{\sqrt{3}}$

26. $\cot 2\theta = -\frac{4}{3}$ **27.** $\cot 2\theta = \frac{4}{3}$

28. AVIATION An airplane flying faster than the speed of sound is said to have a speed greater than Mach 1. The Mach number is the ratio of the speed of the plane to the speed of sound and is denoted by M. When a plane flies faster than the speed of sound, a sonic boom is heard, created by sound waves that form a cone with a vertex angle θ, as shown in Figure 3.8.

FIGURE 3.8 Pattern of sound waves creating a sonic boom
It can be shown that, if $M > 1$, then

$$\sin\frac{\theta}{2} = \frac{1}{M}$$

a. If $\theta = \frac{\pi}{6}$, find the Mach number to the nearest tenth.

b. Find the exact Mach number for part **a.**

29. NAVIGATION If a boat is moving at a constant rate that is faster than the water waves it produces, then the boat sends out waves in the shape of a cone with a vertex angle θ, as shown in Figure 3.9.

FIGURE 3.9 Pattern of bow waves
If r represents the ratio of the speed of the boat to the speed of the wave and if $r > 1$, then

$$\sin\frac{\theta}{2} = \frac{1}{r}$$

a. If $\theta = \frac{\pi}{4}$, find r to the nearest tenth.

b. Find the exact value of r for part **a.**

Graph the functions given in Problems 30–35.

30. $y = \cos^2 x$

31. $y = |\cos x|$

32. $y = \dfrac{\sin x}{1 + \cos x}$

33. $y = \pm\sqrt{\dfrac{1 + \cos x}{2}}$

34. $y = \pm\sqrt{\dfrac{1 - \cos x}{2}}$

35. $y = \dfrac{1 - \cos x}{\sin x}$

Prove each of the identities in Problems 36–44.

36. $\sin\alpha = 2\sin\frac{\alpha}{2}\cos\frac{\alpha}{2}$

37. $\cos 4\theta = \cos^2 2\theta - \sin^2 2\theta$

38. $\sin 2\theta = \dfrac{2\tan\theta}{1 + \tan^2\theta}$

39. $\tan\frac{3}{2}\beta = \dfrac{2\tan\frac{3\beta}{4}}{1 - \tan^2\frac{3\beta}{4}}$

40. $\tan\frac{1}{2}\theta = \dfrac{1 - \cos\theta}{\sin\theta}$

41. $\tan\frac{1}{2}\theta = \dfrac{\sin\theta}{1 + \cos\theta}$

42. $\sin 2\theta \sec\theta = 2\sin\theta$

43. $\sin 2\theta \tan\theta = 2\sin^2\theta$

44. $\cos^4\theta - \sin^4\theta = \cos 2\theta$

C *Prove each of the identities in Problems 45–50.*

45. $\sec 2\theta - \tan 2\theta = \dfrac{1 - \tan\theta}{1 + \tan\theta}$

46. $\sin 3\theta = 3\sin\theta - 4\sin^3\theta$

47. $\tan\frac{B}{2} = \csc B - \cot B$

48. $\sin 4\theta = 4\sin\theta\cos\theta - 8\sin^3\theta\cos\theta$

49. $\frac{1}{2}\cot x - \frac{1}{2}\tan x = \cot 2x$

50. $\cos^4\theta = \frac{1}{8}(3 + 4\cos 2\theta + \cos 4\theta)$

51. Use Figure 3.10 to prove $\sin 2\theta = 2\sin\theta\cos\theta.$*

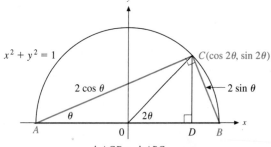

$\triangle ACD \sim \triangle ABC$
FIGURE 3.10 Geometric proof of a double-angle identity

52. Use Figure 3.10 to prove $\cos 2\theta = 2\cos^2\theta - 1.$*

*Figure 3.10 is from "Classroom Capsules" problem by Roger B. Nelsen, Lewis and Clark College, Portland, Ore. *College Mathematics Journal*, January 1989, p. 51.

3.7   Product and Sum Identities

It is sometimes convenient, or even necessary, to write a trigonometric sum as a product or a product as a sum. To do so, we again turn to the identities for the sum and difference of two angles.

Product-to-Sum Identities

Add and subtract the following pair of identities.

$$\cos \alpha \cos \beta + \sin \alpha \sin \beta = \cos(\alpha - \beta)$$
$$\cos \alpha \cos \beta - \sin \alpha \sin \beta = \cos(\alpha + \beta)$$

Adding: $2 \cos \alpha \cos \beta = \cos(\alpha - \beta) + \cos(\alpha + \beta)$
Subtracting: $2 \sin \alpha \sin \beta = \cos(\alpha - \beta) - \cos(\alpha + \beta)$

Also:

$$\sin \alpha \cos \beta + \cos \alpha \sin \beta = \sin(\alpha + \beta)$$
$$\sin \alpha \cos \beta - \cos \alpha \sin \beta = \sin(\alpha - \beta)$$

Adding: $2 \sin \alpha \cos \beta = \sin(\alpha + \beta) + \sin(\alpha - \beta)$
Subtracting: $2 \cos \alpha \sin \beta = \sin(\alpha + \beta) - \sin(\alpha - \beta)$

These identities are called the **product-to-sum identities**.

PRODUCT-TO-SUM IDENTITIES

> 27. $2 \cos \alpha \cos \beta = \cos(\alpha - \beta) + \cos(\alpha + \beta)$
> 28. $2 \sin \alpha \sin \beta = \cos(\alpha - \beta) - \cos(\alpha + \beta)$
> 29. $2 \sin \alpha \cos \beta = \sin(\alpha + \beta) + \sin(\alpha - \beta)$
> 30. $2 \cos \alpha \sin \beta = \sin(\alpha + \beta) - \sin(\alpha - \beta)$

EXAMPLE 1   Using product-to-sum identities

Write each product as the sum of two functions.

a. $2 \sin 3 \sin 1$ **b.** $\sin 40° \cos 12°$

SOLUTION   a. Use Identity 28, where $\alpha = 3$ and $\beta = 1$.

$$2 \sin 3 \sin 1 = \cos(3 - 1) - \cos(3 + 1)$$
$$= \cos 2 - \cos 4$$

b. Use Identity 29, where $\alpha = 40°$ and $\beta = 12°$.

$$2 \sin 40° \cos 12° = \sin(40° + 12°) + \sin(40° - 12°)$$
$$= \sin 52° + \sin 28°$$

But what about the coefficient 2? Since you know that the preceding is an *equation* that is true, you can divide both sides by 2 to obtain

$$\sin 40° \cos 12° = \tfrac{1}{2}(\sin 52° + \sin 28°)$$

Sum-to-Product Identities

Identities 27–30 will change a product to a sum. They can also be used to change a sum to a product. To do this, put them in a more useful form for this purpose by letting $x = \alpha + \beta$ and $y = \alpha - \beta$. To rewrite Identity 27 in sum form, substitute x for $\alpha + \beta$ and y for $\alpha - \beta$:

$$
\begin{aligned}
2 \cos \alpha \cos \beta &= \cos(\alpha - \beta) + \cos(\alpha + \beta) \\
&= \cos y + \cos x \\
&= \cos x + \cos y
\end{aligned}
$$

Now, you need to eliminate α and β:

$$
\begin{cases} x = \alpha + \beta \\ y = \alpha - \beta \end{cases}
\qquad\qquad
\begin{cases} x = \alpha + \beta \\ y = \alpha - \beta \end{cases}
$$

Add: $x + y = 2\alpha$ and Subtract: $x - y = 2\beta$

$$
\frac{x + y}{2} = \alpha
\qquad\qquad
\frac{x - y}{2} = \beta
$$

Thus,

$$
2 \cos\left(\frac{x + y}{2}\right) \cos\left(\frac{x - y}{2}\right) = \cos x + \cos y
$$

This is called a **sum-to-product identity**. If you make the same substitutions into Identities 28, 29, and 30, you will obtain the other sum identities.

**SUM-TO-PRODUCT
IDENTITIES**

$$
31.\ \cos x + \cos y = 2 \cos\left(\frac{x + y}{2}\right) \cos\left(\frac{x - y}{2}\right)
$$

$$
32.\ \cos x - \cos y = -2 \sin\left(\frac{x + y}{2}\right) \sin\left(\frac{x - y}{2}\right)
$$

$$
33.\ \sin x + \sin y = 2 \sin\left(\frac{x + y}{2}\right) \cos\left(\frac{x - y}{2}\right)
$$

$$
34.\ \sin x - \sin y = 2 \sin\left(\frac{x - y}{2}\right) \cos\left(\frac{x + y}{2}\right)
$$

EXAMPLE 2   Using a sum-to-product identity

Write $\sin 35° + \sin 27°$ as a product.

SOLUTION $x = 35°, y = 27°$, and

$$
\frac{x + y}{2} = \frac{35° + 27°}{2} = 31°
$$

$$
\frac{x - y}{2} = 4°
$$

Therefore, $\sin 35° + \sin 27° = 2 \sin 31° \cos 4°$.

You sometimes will use these product and sum identities to prove other identities.

EXAMPLE 3   Proving an identity using a sum-to-product identity

Prove $\dfrac{\sin 7\gamma + \sin 5\gamma}{\cos 7\gamma - \cos 7\gamma} = -\cot \gamma$.

SOLUTION

$$\frac{\sin 7\gamma + \sin 5\gamma}{\cos 7\gamma - \cos 5\gamma} = \frac{2 \sin\left(\dfrac{7\gamma + 5\gamma}{2}\right) \cos\left(\dfrac{7\gamma - 5\gamma}{2}\right)}{-2 \sin\left(\dfrac{7\gamma + 5\gamma}{2}\right) \sin\left(\dfrac{7\gamma - 5\gamma}{2}\right)}$$

$$= \frac{2 \sin 6\gamma \cos \gamma}{-2 \sin 6\gamma \sin \gamma}$$

$$= -\frac{\cos \gamma}{\sin \gamma}$$

$$= -\cot \gamma$$

3.7 Problem Set

A *Write each of the expressions in Problems 1–12 as the sum or difference of two functions.*

1. $2 \cos 75° \cos 35°$

2. $2 \cos 46° \cos 18°$

3. $2 \sin 35° \sin 24°$

4. $2 \sin 53° \cos 24°$

5. $\sin 70° \sin 88°$

6. $\cos 53° \cos 70°$

7. $\sin 41° \cos 19°$

8. $\cos 115° \sin 200°$

9. $\sin 225° \sin 300°$

10. $\sin 2\theta \sin 5\theta$

11. $\cos \theta \cos 3\theta$

12. $\cos 3\theta \sin 2\theta$

Write each of the expressions in Problems 13–24 as a product of two functions.

13. $\sin 43° + \sin 63°$

14. $\sin 22° - \sin 6°$

15. $\cos 81° - \cos 79°$

16. $\cos 78° + \cos 25°$

17. $\sin 215° + \sin 300°$

18. $\cos 25° - \cos 100°$

19. $\sin x - \sin 2x$

20. $\cos 5x - \cos 3x$

21. $\sin x + \sin 2x$

22. $\cos 3\theta + \cos 2\theta$

23. $\cos 5y + \cos 9y$

24. $\sin 6z - \sin 9z$

B *Prove each of the identities given in Problems 25–32.*

25. $\cos x - \cos y = -2 \sin\left(\dfrac{x + y}{2}\right) \sin\left(\dfrac{x - y}{2}\right)$

26. $\sin x + \sin y = 2 \sin\left(\dfrac{x + y}{2}\right) \cos\left(\dfrac{x - y}{2}\right)$

27. $\sin x - \sin y = 2 \sin\left(\dfrac{x - y}{2}\right) \cos\left(\dfrac{x + y}{2}\right)$

28. $\left(\sin \dfrac{\theta}{2} + \cos \dfrac{\theta}{2}\right)^2 = 1 + \sin\theta$

29. $\cos^2 \dfrac{\theta}{2} - \sin^2 \dfrac{\theta}{2} = \cos \theta$

30. $\dfrac{\sin 5\theta + \sin 3\theta}{\cos 5\theta + \cos 3\theta} = \tan 4\theta$

31. $\dfrac{\cos 3\theta - \cos \theta}{\sin \theta - \sin 3\theta} = \tan 2\theta$

32. $\dfrac{\cos 5\omega + \cos \omega}{\cos \omega - \cos 5\omega} = \dfrac{\cot 2\omega}{\tan 3\omega}$

C

33. Here is an alternate proof for the formula $\cos(\alpha + \beta) = \cos \alpha \cos \beta - \sin \alpha \sin \beta$.

   a. *Case i:*   Let $\alpha = 0$ and $\beta = 0$. Prove the given identity.

   b. *Case ii:*   Let $\alpha \neq 0$ and $\beta \neq 0$. Let α be in standard position and β be any angle drawn so that its initial side is along the terminal side of α (see Figure 3.11).

FIGURE 3.11 Figure for proof of sum formula

Let P_α be an arbitrary point on the terminal side of α. Let P_β be the point on the terminal side of β so that $P_\alpha P_\beta$ is perpendicular to OP_α. Draw perpendiculars $P_\alpha S$ and $P_\beta T$ to the x-axis. Draw $P_\alpha Q$ perpendicular to $P_\beta T$. Thus,

$$\cos(\alpha + \beta) = \frac{OT}{OP_\beta}$$

Now show

$$\cos(\alpha + \beta) = \frac{OS}{OP_\alpha} \cdot \frac{OP_\alpha}{OP_\beta} - \frac{QP_\alpha}{P_\alpha P_\beta} \cdot \frac{P_\alpha P_\beta}{OP_\beta}$$

and therefore

$$\cos(\alpha + \beta) = \cos \alpha \cos \beta - \sin \alpha \sin \beta$$

3.8 Summary and Review

Objectives	*Pages / Examples / Problems*
3.1 Proving Identities	
1. Simplify trigonometric expressions.	p. 94; Example 1; Problems 1–12.
2. Prove identities using the fundamental identities.	pp. 96–98; Examples 2–5; Problems 13–41.
3.2 Proving Identities (Continued)	
3. Prove identities by combining fractions.	p. 97; Example 3 (Section 3.1); Problems 14, 25, 26.
4. Prove identities by writing all the trigonometric functions in terms of sines and cosines.	p. 97; Example 4 (Section 3.1); Problems 2, 4, 7, 8, 10, 14, 17, 23, 24.
5. Prove identities by breaking up the fraction into a sum of fractions.	p. 98; Example 5 (Section 3.1); Problems 4, 9–12.
6. Prove identities by using the conjugate.	p. 99; Example 1; Problems 27, 28.
7. Prove identities by factoring.	p. 100; Example 2; Problems 1, 15, 16, 19, 20, 29, 30
8. Prove identities by using a variety of techniques.	p. 101; Example 3; Problems 3, 5, 6, 13, 18, 21, 22, 31.
3.3 Disproving Identities	
9. Prove that an equation is not an identity by finding a counterexample.	p. 102; Example 1; Problems 1–10.
10. Prove or disprove that a given equation is an identity.	p. 103; Example 2; Problems 11–20.
3.4 Cofunction and Opposite-Angle Identities	
11. Know and use the distance formula.	pp. 105–106; Examples 1 and 2; Problems 1–6.
12. Find the chord length given the central angle.	pp. 106–107; Example 3.

13. Use the cofunction identities.   pp. 107–108; Example 4; Problems 7–12.

14. Use the opposite-angle identities.   pp. 108–110; Examples 5–7; Problems 13–21.

15. Use the opposite-angle identities as an aid to graph certain trigonometric functions.   p. 111; Example 8; Problems 22–30.

3.5 Addition Laws

16. Use the addition laws.   p. 112; Example 1; Problems 1–16.

17. Find exact values by using the addition laws.   p. 113; Examples 2–3; Problems 17–30.

18. Prove identities using the addition laws.   p. 114; Problems 23–49.

3.6 Double-Angle and Half-Angle Identities

19. Use the double-angle identities.   pp. 116–117; Examples 1–3; Problems 1–2, 4–5.

20. Given $\cos \theta$, $\sin \theta$, or $\tan \theta$ and the quadrant, find $\cos 2\theta$, $\sin 2\theta$, and $\tan 2\theta$.   p. 118; Example 5; Problems 10–15.

21. Use the half-angle identities.   pp. 119–120; Example 6; Problems 3, 6–9, 16–21, 28, 29.

22. Given $\cot 2\theta$, find $\cos \theta$, $\sin \theta$, and $\tan \theta$.   p. 121; Example 7; Problems 22–27.

23. Graph trigonometric functions by using identities.   p. 117; Example 4; Problems 30–35.

24. Prove identities using the half-angle and double-angle identities.   p. 122; Example 8; Problems 36–50.

3.7 Product and Sum Identities

25. Use the product identities to write a product as a sum.   p. 124; Example 1; Problems 1–12.

26. Use the sum identities to write a sum as a product.   pp. 125–126; Examples 2 and 3; Problems 13–24.

27. Prove identities using the product and sum.   p. 126; Problems 25–32.

TERMS

Addition laws [3.5]	Conjugate [3.2]	Double-angle identities [3.6]	Opposite-angle identities [3.4]
Chord length [3.4]	Counterexample [3.3]	Half-angle identities [3.6]	Product-to-sum identities [3.7]
Cofunction identities [3.4]	Distance formula [3.4]	Identity [3.1]	Sum-to-product identities [3.7]

CHAPTER 3 REVIEW OF OBJECTIVES

The problem numbers correspond to the objectives listed in Section 3.8.

1. Simplify $\dfrac{\sec^2 x + \tan^2 x + 1}{\sec x}$.

Prove the identities in Problems 2–8.

2. $\dfrac{\csc^2 \alpha}{1 + \cot^2 \alpha} = 1$

3. $\dfrac{1}{\sin \theta + \cos \theta} + \dfrac{1}{\sin \theta - \cos \theta} = \dfrac{2 \sin \theta}{\sin^4 \theta - \cos^4 \theta}$

4. $\dfrac{\cos \theta}{\sec \theta} - \dfrac{\sin \theta}{\cot \theta} = \dfrac{\cos \theta \cot \theta - \tan \theta}{\csc \theta}$

5. $\tan \beta + \sec \beta = \dfrac{1 + \csc \beta}{\cos \beta \csc \beta}$

6. $\dfrac{1 - \cos x}{1 + \cos x} = \dfrac{2 - 2 \cos x - \sin^2 x}{\sin^2 x}$

7. $\dfrac{\sin^2 \theta - \cos^2 \theta}{\sin \theta + \cos \theta} = \sin \theta - \cos \theta$

8. $\dfrac{1 + \tan^2 \theta}{\csc \theta} = \sec \theta \tan \theta$

9. Prove that the given equation is not an identity:
$\tan \theta + \csc \theta = \cot \theta$.

10. Prove or disprove that the given equations are identities.

   a. $(\sin t + \cos t)^2 = 1$ **b.** $\cos t \tan t \csc t \sec t = 1$

11. Suppose the circle given in Figure 3.12 is a unit circle. Find the distance between P_α and P_β.

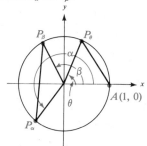

FIGURE 3.12

12. Suppose the circle given in Figure 3.12 is a unit circle. Find the length of chord AP_θ. This is the length of any chord in a unit circle subtended by a central angle θ.

13. Write each function in terms of its cofunction.

   a. $\sin 38°$ **b.** $\tan \frac{\pi}{8}$ **c.** $\cos 2$

14. Write each as a function of $\left(\theta - \frac{\pi}{6}\right)$.

   a. $\cos\left(\frac{\pi}{6} - \theta\right)$ **b.** $\sin\left(\frac{\pi}{6} - \theta\right)$ **c.** $\tan\left(\frac{\pi}{6} - \theta\right)$

15. Graph $y - 1 = 2 \sin\left(\frac{\pi}{6} - x\right)$.

16. Use exact values and the addition laws to write $\cos(\theta - 30°)$ as a function of $\cos \theta$ and $\sin \theta$.

17. Find the exact value of $\sin 105°$.

18. Prove $\cos(\alpha - \beta) = \cos \alpha \cos \beta + \sin \alpha \sin \beta$.

19. Evaluate

$$\frac{2 \tan \frac{\pi}{6}}{1 - \tan^2 \frac{\pi}{6}}$$

using exact values.

20. If $\cos \theta = \frac{4}{5}$ and θ is in Quadrant IV, find the exact values of $\cos 2\theta$, $\sin 2\theta$, and $\tan 2\theta$.

21. Evaluate

$$-\sqrt{\frac{1 + \cos 240°}{2}}$$

using exact values.

22. If $\cot 2\theta = -\frac{4}{3}$, find the exact values of $\cos \theta$, $\sin \theta$, and $\tan \theta$ where θ is in Quadrant I.

23. Graph $y = \cos^2 x$.

24. If $\cos 2\alpha = 2 \cos^2 \alpha - 1$, then show

$$\cos \tfrac{1}{2}\theta = \pm \sqrt{\frac{1 + \cos \theta}{2}}$$

where $\theta = 2\alpha$.

25. Write $\sin 3\theta \cos \theta$ as a sum.

26. Write $\sin(x + h) - \sin x$ as a product.

27. Prove $\dfrac{\sin 5\theta + \sin 3\theta}{\cos 5\theta - \cos 3\theta} = -\cot \theta$.

DISCUSSION PROBLEMS

1. What are the "fundamental identities" and why do you think these are called "fundamental"?

2. Outline your "method of attack" for proving a trigonometric identity.

3. Derive (prove) each of the identities 1–20 found on the page facing the inside back cover.

4. Derive (prove) each of the identities 21–34 found on the page facing the inside back cover.

5. Write a 500-word paper on cartography (mapmaking) and its relationship to trigonometry.

MISCELLANEOUS PROBLEMS

Prove that each of the equations in Problems 1–38 is an identity.

1. $\dfrac{\sec^2 u}{\tan^2 u} = \csc^2 u$ **2.** $\dfrac{\sec u}{\tan^2 u} = \cot u \csc u$

3. $\sec^2 \gamma + 2 \tan \gamma = \dfrac{\sec^2 \gamma \cos \gamma + 2 \sin \gamma}{\cos \gamma}$

4. $\sec \gamma + \tan \gamma = \dfrac{\csc \gamma + 1}{\csc \gamma \cos \gamma}$

5. $\tan 3\theta \sin 3\theta = \sec 3\theta - \cos 3\theta$

6. $\csc 3\theta - \sin 3\theta = \cot 3\theta \cos 3\theta$

7. $1 + \sin^2 \lambda = 2 - \cos^2 \lambda$

8. $2 - \sin^2 3\lambda = 1 + \cos^2 3\lambda$

9. $\dfrac{\sin \alpha}{\tan \alpha} + \dfrac{\cos \alpha}{\cot \alpha} = \cos \alpha + \sin \alpha$

10. $\dfrac{1}{1 + \cos 2\alpha} + \dfrac{1}{1 - \cos 2\alpha} = 2 \csc^2 2\alpha$

11. $\sec \beta + \cos \beta = \dfrac{2 - \sin^2 \beta}{\cos \beta}$

12. $2 \sin^2 3\beta - 1 = 1 - 2 \cos^2 3\beta$

13. $(\tan 5\beta - 1)(\tan 5\beta + 1) = \sec^2 5\beta - 2$

14. $\tan^2 7\alpha = (\sec 7\alpha - 1)(\sec 7\alpha + 1)$

15. $(\sin \gamma - \cos \gamma)^2 = 1 - 2 \sin \gamma \cos \gamma$

16. $(\sin \gamma + \cos \gamma)(\sin \gamma - \cos \gamma) = 2 \sin^2 \gamma - 1$

17. $\dfrac{\sin \gamma}{\cos \gamma} + \cot \gamma = \sec \gamma \csc \gamma$

18. $\tan 3\gamma + \cot 3\gamma = \sec 3\gamma \csc 3\gamma$

19. $(\sin A \cos A \cos B + \sin B \cos B \cos A) \sec A \sec B$
   $= \sin A + \sin B$

20. $(\cos A \cos B \tan A + \sin A \sin B \cot B) \csc A \sec B = 2$

21. $\sin \lambda = \dfrac{\sin^2 \lambda + \sin \lambda \cos \lambda + \sin \lambda}{\sin \lambda + \cos \lambda + 1}$

22. $\dfrac{1 + \cot 2\lambda \sec 2\lambda}{\tan 2\lambda + \sec 2\lambda} = \cot 2\lambda$

23. $\sin 2\alpha \cos 2\alpha (\tan 2\alpha + \cot 2\alpha) = 1$

24. $(\sin \beta - \cos \beta)^2 + (\sin \beta + \cos \beta)^2 = 2$

25. $\dfrac{1 - \cos \theta}{1 + \cos \theta} = \left(\dfrac{1 - \cos \theta}{\sin \theta} \right)^2$

26. $\dfrac{(\sec^2 \gamma + \tan^2 \gamma)^2}{\sec^4 \gamma - \tan^4 \gamma} = 1 + 2 \tan^2 \gamma$

27. $\dfrac{(\cos^2 \gamma - \sin^2 \gamma)^2}{\cos^4 \gamma - \sin^4 \gamma} = 2 \cos^2 \gamma - 1$

28. $(\sec 2\theta + \csc 2\theta)^2 = \dfrac{1 + 2 \sin 2\theta \cos 2\theta}{\cos^2 2\theta \sin^2 2\theta}$

29. $\dfrac{1}{\sec \theta + \tan \theta} = \sec \theta - \tan \theta$

30. $\dfrac{1 + \tan^3 \theta}{1 + \tan \theta} = \sec^2 \theta - \tan \theta$

31. $\dfrac{1 - \sec^3 \theta}{1 - \sec \theta} = \tan^2 \theta + \sec \theta + 2$

32. $\csc(-\theta) = -\csc \theta$

33. $\dfrac{\cot 3\theta - \sin 3\theta}{\sin 3\theta} = \dfrac{\cos^2 3\theta + \cos 3\theta - 1}{\sin^2 3\theta}$

34. $\cot \beta - \sin \beta = (\cos^2 \beta + \cos \beta - 1) \csc \beta$

35. $\dfrac{\tan(\alpha + \beta) - \tan \beta}{1 + \tan(\alpha + \beta) \tan \beta} = \tan \alpha$

36. $\dfrac{\tan \alpha - \tan(\alpha - \beta)}{1 + \tan \alpha \tan(\alpha - \beta)} = \tan \beta$

37. $\tan(\alpha + \beta + \gamma)$
   $= \dfrac{\tan \alpha + \tan \beta + \tan \gamma - \tan \alpha \tan \beta \tan \gamma}{1 - \tan \beta \tan \gamma - \tan \alpha \tan \gamma - \tan \alpha \tan \beta}$

38. $\cot(\alpha + \beta + \gamma)$
   $= \dfrac{\cot \alpha \cot \beta \cot \gamma - \cot \alpha - \cot \beta - \cot \gamma}{\cot \beta \cot \gamma + \cot \alpha \cot \gamma + \cot \alpha \cot \beta - 1}$

Find the exact value of each expression in Problems 39–48.

39. $\cos 345°$

40. $\sin(\theta - 60°)$ in terms of $\sin \theta$ and $\cos \theta$

41. $\sin 50° \cos 20° - \cos 50° \sin 20°$

42. $2 \cos^2 15° - 1$

43. $\dfrac{\tan 15° + \tan 30°}{1 - \tan 15° \tan 30°}$ **44.** $\dfrac{\tan 120° - \tan 60°}{1 + \tan 120° \tan 60°}$

45. $\dfrac{2 \tan 30°}{1 - \tan^2 30°}$ **46.** $\dfrac{1 - \cos 120°}{\sin 120°}$

47. $\sqrt{\dfrac{1 - \cos 270°}{2}}$ **48.** $-\sqrt{\dfrac{1 + \cos 270°}{2}}$

49. If $\sin \theta = \frac{4}{5}$ and θ is in Quadrant II, find the exact value of $\cos 2\theta$.

50. If $\cot 2\theta = \frac{4}{3}$, find the exact value of $\cos \theta$, $\theta > 90°$.

▌ Sunrise, Sunset, and Linear Interpolation

Courtesy of Sara Hunsaker

The most obvious example of a periodic phenomenon is the rotation of the earth, which produces our day/night cycles. Suppose we consult a table that gives the time of sunrise and sunset. These times depend on the latitude of the location, as illustrated by Table 3.1 on page 132, which gives the times of sunrise and sunset for selected cities. If the sunrise and sunset for latitude 35° N are plotted as shown in Figure 3.13, you can see that they are definitely periodic, although not quite sine or cosine curves. However, if you plot the length of daylight, you will find that the result is nearly a sine curve (see Problem 7 in this section).

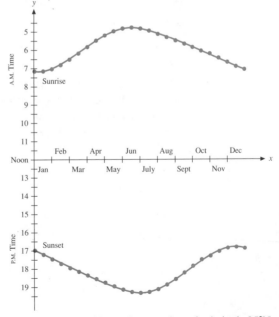

FIGURE 3.13 Sunrise and sunset times for latitude 35°N

TABLE 3.1 Times of Sunrise and Sunset for Various Latitudes

Date	Time of Sunrise						Time of Sunset					
	20°N LATITUDE (HAWAII)	30°N LATITUDE (NEW ORLEANS)	35°N LATITUDE (ALBUQUERQUE)	40°N LATITUDE (PHILADELPHIA)	45°N LATITUDE (MINNEAPOLIS)	60°N LATITUDE (ALASKA)	20°N LATITUDE (HAWAII)	30°N LATITUDE (NEW ORLEANS)	35°N LATITUDE (ALBUQUERQUE)	40°N LATITUDE (PHILADELPHIA)	45°N LATITUDE (MINNEAPOLIS)	60°N LATITUDE (ALASKA)
	h m	h m	h m	h m	h m	h m	h m	h m	h m	h m	h m	h m
Jan. 1	6 35	6 56	7 08	7 22	7 38	9 03	17 31	17 10	16 58	16 44	16 28	15 03
Jan. 15	6 38	6 57	7 08	7 20	7 35	8 48	17 41	17 22	17 12	16 59	16 44	15 31
Jan. 30	6 36	6 52	7 01	7 11	7 23	8 19	17 51	17 35	17 27	17 17	17 05	16 09
Feb. 14	6 30	6 41	6 48	6 55	7 03	7 42	17 59	17 48	17 42	17 34	17 26	16 48
Mar. 1	6 20	6 26	6 30	6 34	6 39	6 59	18 05	17 59	17 56	17 52	17 47	17 27
Mar. 16	6 08	6 09	6 10	6 11	6 11	6 15	18 10	18 09	18 08	18 08	18 07	18 04
Mar. 31	5 55	5 51	5 49	5 46	5 43	5 29	18 14	18 18	18 20	18 23	18 26	18 41
Apr. 15	5 42	5 34	5 28	5 23	5 16	4 44	18 18	18 27	18 32	18 38	18 45	19 18
Apr. 30	5 32	5 18	5 11	5 02	4 51	4 01	18 23	18 37	18 44	18 53	19 04	19 55
May 15	5 24	5 07	4 57	4 45	4 31	3 23	18 29	18 46	18 56	19 08	19 22	20 31
May 30	5 20	5 00	4 48	4 34	4 18	2 53	18 35	18 55	19 07	19 21	19 38	21 04
June 14	5 20	4 58	4 45	4 30	4 13	2 37	18 40	19 02	19 15	19 30	19 48	21 24
June 29	5 23	5 02	4 49	4 34	4 16	2 40	18 43	19 05	19 18	19 33	19 51	21 26
July 14	5 29	5 08	4 56	4 43	4 26	3 01	18 43	19 03	19 15	19 29	19 45	21 09
July 29	5 34	5 17	5 07	4 55	4 41	3 33	18 39	18 56	19 06	19 17	19 31	20 38
Aug. 13	5 39	5 26	5 18	5 09	4 59	4 09	18 30	18 43	18 51	19 00	19 10	19 59
Aug. 28	5 43	5 35	5 29	5 24	5 17	4 45	18 19	18 27	18 33	18 38	18 45	19 16
Sept. 12	5 47	5 43	5 40	5 38	5 35	5 20	18 06	18 10	18 12	18 14	18 17	18 31
Sept. 27	5 50	5 51	5 51	5 52	5 53	5 55	17 52	17 51	17 50	17 49	17 49	17 45
Oct. 12	5 54	6 00	6 03	6 07	6 11	6 31	17 39	17 33	17 29	17 26	17 21	17 01
Oct. 22	5 57	6 06	6 12	6 18	6 25	6 56	17 32	17 22	17 17	17 11	17 04	16 32
Nov. 6	6 04	6 18	6 26	6 35	6 45	7 35	17 23	17 10	17 02	16 52	16 42	15 52
Nov. 21	6 12	6 30	6 40	6 52	7 05	8 12	17 19	17 02	16 51	16 40	16 26	15 19
Dec. 6	6 22	6 42	6 54	7 07	7 23	8 44	17 20	17 00	16 48	16 35	16 19	14 58
Dec. 21	6 30	6 52	7 04	7 18	7 35	9 02	17 26	17 05	16 52	16 38	16 21	14 54

The data in this table are courtesy of the U.S. Naval Observatory. This table of times of sunrise and sunset may be used in any year of the 20th century with an error not exceeding two minutes and generally less than one minute. The times are fairly accurate anywhere in the vicinity of the stated latitude, with an error of less than on minute for each 9 miles from the given latitude.

If you wish to use Table 3.1 for your own town's latitude and it is not listed, you can use a procedure called **linear interpolation**. Your local Chamber of Commerce will probably be able to tell you your town's latitude. The latitude for Santa Rosa, California, is about 38°, which falls between 35° and 40° on the table (see Figure 3.14).

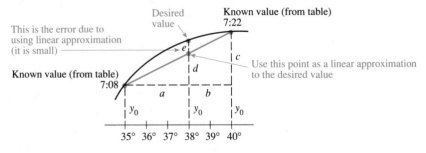

FIGURE 3.14 Linear interpolation

ANGLES	TIMES	
35°	7:08	Known, from Table 3.1
38°	x	Unknown; this is the latitude of the town for which the times of sunrise and sunset are not available.
40°	7:22	Known, from Table 3.1

Use the known information and a proportion to find x:

$$5 \left[3 \begin{bmatrix} 35° & 7:08 \\ 38° & x \\ \\ 40° & 7:22 \end{bmatrix} \begin{matrix} ? \\ \\ 14 \end{matrix} \right]$$

Since 38° is $\frac{3}{5}$ of the way between 35° and 40°, the sunrise time will be $\frac{3}{5}$ of the way between 7:08 and 7:22.

$$\frac{3}{5} = \frac{x}{14}$$

$$x = \frac{3}{5}(14)$$

$$= 8.4 \qquad \text{To the nearest minute, there is :08.}$$

The sunrise time is about 7:08 + :08 = 7:16.

PROBLEMS FOR FURTHER STUDY: SUNRISE, SUNSET, AND LINEAR INTERPOLATION

Use linear interpolation and Table 3.1 to approximate the times requested in each of Problems 1–6.

1. Sunset for Santa Rosa (lat. 38°N) on January 1.
2. Sunrise for Tampa, Florida (lat. 28°N), on June 29.
3. Sunset for Tampa, Florida (lat. 28°N), on June 29.
4. Sunset for Winnipeg, Canada (lat. 50°N), on October 12.
5. Sunrise for Winnipeg, Canada (lat. 50°N), on October 12.
6. Sunrise for Juneau, Alaska (lat. 58°N), on March 1.

7. Plot the length of daylight in Chattanooga, Tennessee (lat. 35°N).
8. Plot the time of sunrise and sunset for Seward, Alaska (lat. 60°N).
9. Plot the length of daylight for Seward, Alaska (lat. 60°N).
10. **Astronomy** Plot the sunrise and sunset for your town's latitude by drawing a graph similar to that in Figure 3.13.
11. **Astronomy** Plot the length of daylight for your town from the data in Problem 10.

▮ Leonhard Euler
(1707–1783)

▮ Leonhard Euler was one of the most prolific mathematical writers in history. He was widely known for his clarity, detail, and completeness. Even though he was blind for most of the last 17 years of his life, he made mathematical contributions all his life. The following quotation summarizes Euler's ability:

Euler calculated without any apparent effort, just as men breathe and as eagles sustain themselves in the air.

F. Argo in Howard Eves,
In Mathematical Circles, p. 47.

The aim of research is the discovery of the equations which subsist between the elements of phenomena.

ERNST MACH
Popular Scientific Lectures
(Chicago, 1910), p. 205.

Trigonometry was forever changed with the invention of the electronic calculator. As we have seen previously in the text, we no longer need to carry around and use tables of trigonometric values. However, calculators are important for much more than evaluation of trigonometric functions. For example, in the first section of this chapter, we are looking for the solution of the equation $\cos x = \frac{1}{2}$. On the Casio fx-7700G or the TI-81, or other graphing calculators, you can graph the functions $y = \cos x$ and $y = \frac{1}{2}$. This is shown below.

The solution of the equation $\cos x = \frac{1}{2}$ is found using the trace functions. Watch the cursor move along the line $y = .5$; the display when the cursor is closest to the points of intersection shows:

X = -306.94737°, X = -56.8421050°
X = 56.8421050°, X = 306.94737°

These values are solutions where x is measured in degrees. On the other hand, if you trace $y = \cos x$ and look for places where the cursor crosses $y = \frac{1}{2}$, you find the following ordered pairs:

$(-299.3684°, .4904235)$;
$(-56.84211°, .54694816)$;
$(56.842105°, .54694816)$;
$(299.36842°, .4904235)$

Since $y = .5$, you can see from these components that there is a great deal of approximation. Do not be misled about a calculator's accuracy because it shows you 10 or 12 decimal places. You can zoom in on the graphs around the points of intersection to obtain any reasonable degree of accuracy.

4

TRIGONOMETRIC EQUATIONS AND INVERSE FUNCTIONS

Contents

4.1 Introduction to Trigonometric Equations

4.2 Trigonometric Equations with Multiple Angles

4.3 Inverse Trigonometric Functions

4.4 A Reduction Identity

4.5 Summary and Review

Application for Further Study
Waves, Harmonic Motion, and Resonance

Cumulative Review for Chapters 3 and 4

Preview

In Chapter 3 we introduced the concept of a trigonometric identity. Suppose you are given an open equation with one or more trigonometric functions. Either the equation is true for all permissible replacements of the variable or it is not. That is, it is or is not an identity. If it is not an identity, then it is simply called a **trigonometric equation**, and in this chapter you will learn how to solve trigonometric equations. Remember, **to solve an equation** means to find all replacements for the variable that make the equation true.

Perspective

Calculus students must solve many types of equations. Linear and quadratic equations are generally routine, and the occasional polynomial equation can usually be solved without too much difficulty. However, the equation that frequently causes a great deal of difficulty is the trigonometric equation. In the example shown here, a calculus process called Newton's method is being discussed in the context of solving the trigonometric equation $\sin x = x^2$. We will solve this equation in this chapter. You should master the techniques for solving trigonometric equations given in this chapter.

EXAMPLE 3   Write and run a BASIC program based on Newton's method for estimating a solution of the equation $\sin x = x^2$.

Solution   The calculations to be made are these:

1. Start with $x_0 = 1$.
2. We want to make the function $f(x) = \sin x - x^2$ small, zero if possible, and its derivative is $f'(x) = \cos x - 2x$.
3. Given any x_n, the next value of x is

$$x_{n+1} = x_n - \frac{f(x_n)}{f'(x_n)}.$$

The computer does not recognize x^2 unless we write it as $x \char`^ 2$, as shown in line 30 of the program that follows. Likewise, $10 \char`^ (-6)$ in line 80 is how we write 10^{-6} in the program. The asterisk (*) in 2*x indicates multiplication—the machine does not recognize 2x as 2 times x.

Now, let's look again at line 80. We don't know how many iterations may be needed to find a "sufficiently accurate" estimate of the root of $f(x) = 0$, so we arbitrarily tell the computer to stop when it reaches a value of x for which the absolute value of f is less than 10^{-6}. The entire command in line 80 is known as an "IF...THEN" conditional. If the condition is satisfied, the machine goes to line 100; but if the condition is not satisfied, the

3.57   The curves $y = x^2$ and $y = \sin x$ cross at $x = 0$ and again near $x = 1$. Example 3 uses Newton's method to find this second solution of the equation $\sin x = x^2$.

From *Calculus* by Finney/Thomas (Reading: Addison-Wesley Publishing Company), p. 206.

4.1 Introduction to Trigonometric Equations

Consider the trigonometric equation

$$\cos \theta = \tfrac{1}{2}$$

It is a conditional equation, since some values of θ make it true and some make it false. For example,

$\theta = 40°$ is not a solution since $\cos 40° \neq \tfrac{1}{2}$.

$\theta = 420°$ is a solution since $\cos 420° = \tfrac{1}{2}$.

If you are given

$$\cos \theta = x$$

and wish to solve for θ, you must find the set of all values, or replacements, of θ that make the equation true. In other words, you want to find *the angle* or real number θ whose cosine is x. A notation has been invented to denote this idea:

$$\cdot \quad \theta = \cos^{-1} x \quad \text{or} \quad \theta = \arccos x$$

INVERSE
TRIGONOMETRIC
RELATIONS

> **$\theta = \arccos x$ and $\theta = \cos^{-1} x$** are equivalent to $\cos \theta = x$.
> θ is the angle whose cosine is equal to x.
>
> **$\theta = \arcsin x$ and $\theta = \sin^{-1} x$** are equivalent to $\sin \theta = x$
> θ is the angle whose sine is equal to x.
>
> **$\theta = \arctan x$ and $\theta = \tan^{-1} x$** are equivalent to $\tan \theta = x$.
> θ is the angle whose tangent is equal to x.
>
> **$\theta = \text{arcsec } x$ and $\theta = \sec^{-1} x$** are equivalent to $\sec \theta = x$.
> θ is the angle whose secant is equal to x.
>
> **$\theta = \text{arccsc } x$ and $\theta = \csc^{-1} x$** are equivalent to $\csc \theta = x$.
> θ is the angle whose cosecant is equal to x.
>
> **$\theta = \text{arccot } x$ and $\theta = \cot^{-1} x$** are equivalent to $\cot \theta = x$.
> θ is the angle whose cotangent is equal to x.

⊗ $(\cos x)^{-1} = \dfrac{1}{\cos x}$

but

$$\cos^{-1} x \neq \dfrac{1}{\cos x}$$

Remember, $\cos^{-1} x$ means inverse cosine, and does NOT represent the reciprocal of cosine. ⊗

The expression $\cos^{-1} x$ is pronounced "inverse cosine," and the other functions are similarly pronounced. Note the warning in the margin; the -1 notation is NOT an exponent, but is used as part of the notation for the inverse.

We use the inverse notation to solve trigonometric equations, but first we need to practice using this notation. We talk about **evaluating** an inverse trigonometric relation. It means the same thing as it does when referring to evaluating trigonometric functions—namely, to find the value. To do this we can use a table of exact values or a calculator.

Solving Equations with Exact Values

EXAMPLE 1 **Solving a trigonometric equation**

Solve $\sin \theta = \frac{1}{2}$.

SOLUTION Use the definition to write $\sin \theta = \frac{1}{2}$ as $\theta = \sin^{-1} \frac{1}{2}$. We have solved for θ, but have not yet evaluated (i.e., found the numerical value). To evaluate, notice that this means that θ is the angle whose sine is $\frac{1}{2}$. Since this is an exact value, we know $\theta = 30°$ or $\frac{\pi}{6}$, but we also need to remember that sine is positive (notice that $\frac{1}{2}$ is a positive number and $\sin \theta$ is equal to this positive number) in Quadrants I and II. Thus, we use our table of exact values to find the angle in Quadrant I and then the reduction principle to put the angle in Quadrant II: *What is the angle in Quadrant* II *whose reference angle is* 30°? The answer is 150°. You can check your work by finding $\sin 30° = \sin 150° = \frac{1}{2}$. Do these values form the complete solution set for θ? The answer is no; since we know that sine is periodic (with period 360°), we can add multiples of 360° to each of these values:

$$\theta = 30° + 360°k \quad \text{and} \quad \theta = 150° + 360°k$$

where k is any integer.

The positive angles less than one revolution are called the **primary angles**.

EXAMPLE 2 **Solving a trigonometric equation by first solving for the function**

Solve $2 \cos \theta = -1$.

SOLUTION Algebraically, solve first for the function $\cos \theta$ by dividing both sides by 2:

$$\cos \theta = -\frac{1}{2}$$

Next, use the definition of inverse cosine to solve for θ: $\theta = \cos^{-1}\left(-\frac{1}{2}\right)$. Compare this with Example 1 and you see it is very similar except that here the function is negative (notice $-\frac{1}{2}$ is a negative number and $\cos \theta$ is equal to this negative number). To evaluate we use our knowledge of exact values. In the first quadrant, $\cos 60° = \frac{1}{2}$ (from the table of exact values) and cosine is negative in Quadrants II and III, so we ask the following question: *What are the angles in Quadrants* II *and* III *whose reference angle is* 60°? We find these values (120° and 240°), and then add multiples of 360° to each:

$$\theta = 120° + 360°k \quad \text{and} \quad \theta = 240° + 360°k$$

where k is any integer.*

---

*In these problems it will be assumed that k is any integer. You can use this in your answers without saying "k is any integer" on each problem.

⊗ Note agreement about radian answers. ⊗

Even though we will sometimes solve trigonometric equations using degree measure (as in Examples 1 and 2), we will more commonly solve them using real numbers (radian measure). In fact, we agree from now on to give all answers using real numbers unless degree answers are specifically requested.

EXAMPLE 3 Solving a trigonometric equation of a reciprocal function

Solve $\sec \theta = -2$.

SOLUTION First, solve for θ: $\theta = \sec^{-1}(-2)$. We recognize this as an exact value, and in the first quadrant $\sec \frac{\pi}{3} = 2$. In this problem we see the secant is negative and therefore must be in Quadrants II and III. *What angles in Quadrants II and III have a reference angle of $\frac{\pi}{3}$?*

$$\theta = \tfrac{2\pi}{3} + 2\pi k \quad \text{and} \quad \theta = \tfrac{4\pi}{3} + 2\pi k$$

Notice that 2π has been added to the angles in the second and third quadrants.

CALCULATOR COMMENT

Y1=secX Y2=-2

Solving Equations by Calculator

In all the previous examples, the angle could be found by a knowledge of exact values. Not all equations, however, will involve exact values. For those that do not, we will use a calculator.

EXAMPLE 4 Solving a trigonometric equation using a calculator

Solve $\sin \theta = -.4695$.

SOLUTION First solve: $\theta = \sin^{-1}(-.4695)$.

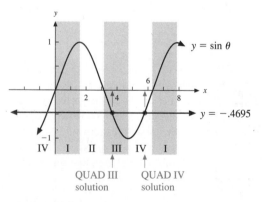

QUAD III solution   QUAD IV solution

⊗ Remember: Work is in radians unless degrees are specified. ⊗

By calculator, we will first find the reference angle by using .4695:

PRESS: .4695 INV sin *   DISPLAY: 0.4887244

---

*Many calculators will use the second-function key instead of INV. Other calculators require sin⁻¹ .4695. Because of these variations we will usually not show the exact keystrokes.

Quadrant III:

Use this reference angle ($\theta' \approx .4887244$) to place it in Quadrants III and IV (it is a good idea to store the displayed number and then use it for each quadrant):

$$\theta = 3.6303 + 2\pi k \quad \text{and} \quad \theta = 5.7945 + 2\pi k$$

You might ask why we use reference angles when finding the angle using a calculator. Why not simply input the negative value as follows:

$$\boxed{.4695} \;\boxed{+/-}\; \boxed{\text{INV}}\; \boxed{\sin} \quad \text{DISPLAY:} \quad -0.4887244$$

This answer (to four places) is $-.48887$, is negative, and is not equal to one of the primary angles 3.6303 or 5.7945; it is, however, in the list of solutions for Example 4:

$$5.7945 + 2\pi(-1) \approx -.4887$$

Quadrant IV:

This negative angle, however, does not make it easy to find the value in Quadrant III. We will discuss this more completely in Section 4.3, but you should note that the calculator does not give solutions to equations, but values for what are called the **inverse trigonometric functions**. These functions will be defined in Section 4.3, and your calculators have been preprogrammed to give the values of these functions. In the first quadrant, however, the values of the trigonometric functions give us the reference angles that we need to solve trigonometric equations.

EXAMPLE 5   **Solving a trigonometric equation of a multiple angle**

Solve $\cos 2\theta = 1$

SOLUTION   Notice the difference between this equation and the one given in Example 2.
First, solve for θ:

$$2\theta = \cos^{-1} 1$$

$$\theta = \tfrac{1}{2}\cos^{-1} 1$$

Next, evaluate this expression: since $\cos 0 = 1$ we see that $\cos^{-1} 1 = 0 + 2\pi k$; thus,

$$\theta = \tfrac{1}{2}(0 + 2\pi k) = \pi k$$

CALCULATOR COMMENT

Y1=cosX   Y2=1

Algebraic Solution for the Angle

In Examples 1–5, we solved some simple trigonometric equations. The next step is to practice the algebraic steps necessary to put them into a form that can be easily evaluated. You can see from Example 5 that this process can look complicated. You need to distinguish the function from the angle. Both of these must be distinguished from the unknown. Consider

The unknown is x.

$$\underbrace{\cos(\underbrace{2x + 1}_{}) = 0}$$

The angle is $2x + 1$.
The function is cosine.

The steps in solving a trigonometric equation are now given.

PROCEDURE FOR SOLVING TRIGONOMETRIC EQUATIONS

1. Solve for a single trigonometric function. You may use identities, factoring, or the quadratic formula.
2. Solve for the angle. You will use the definition of the inverse trigonometric functions for this step.
3. Solve for the unknown.

EXAMPLE 6 **Solving a trigonometric equation for the unknown**

Solve for the trigonometric function first; next, name the angle; finally, solve for the unknown. **Do not evaluate**.

a. $5 \sin(\theta + 1) = \frac{1}{2}$ **b.** $\sin 5(\theta + 1) = \frac{1}{2}$

c. $(3 \cos \theta + 2)(5 \cos \theta - 4) = 0$ **d.** $\tan^2 \theta - 5 \tan \theta - 4 = 0$

SOLUTION **a.** $5 \sin(\theta + 1) = \frac{1}{2}$

$\underbrace{\sin(\theta + 1)}_{\uparrow} = .1$ Divide both sides by 5.

The angle is $(\theta + 1)$.

$\theta + 1 = \sin^{-1} .1$

$\theta = \sin^{-1} .1 - 1$ This is solved for the unknown; notice that this is not the same as $\sin^{-1}(.1 - 1)$; to avoid this possible confusion, we usually write $\theta = -1 + \sin^{-1} .1$.

b. $\sin 5(\theta + 1) = \frac{1}{2}$ This is solved for the trigonometric function as given.

The angle is $5(\theta + 1)$.

$5(\theta + 1) = \sin^{-1} \frac{1}{2}$

$5\theta + 5 = \sin^{-1} \frac{1}{2}$

$5\theta = -5 + \sin^{-1} \frac{1}{2}$

$\theta = -1 + \frac{1}{5} \sin^{-1} \frac{1}{2}$ This is solved for the unknown.

c. $(3 \cos \theta + 2)(5 \cos \theta - 4) = 0$

This is solved by setting each factor equal to 0:

$3 \cos \theta + 2 = 0$ or $5 \cos \theta - 4 = 0$

$3 \cos \theta = -2$ $5 \cos \theta = 4$

$\cos \theta = -\frac{2}{3}$ $\cos \theta = \frac{4}{5}$ This is solved for the function.

The angle in both of these equations is the same as the unknown, namely, θ. Finally, solve for θ:

$$\theta = \cos^{-1}\left(-\tfrac{2}{3}\right) \quad \text{or} \quad \cos^{-1}\tfrac{4}{5}$$

d. $\tan^2\theta - 5\tan\theta - 4 = 0$ cannot be solved by factoring, so use the quadratic formula:*

$$\tan\theta = \frac{5 \pm \sqrt{25 - 4(1)(-4)}}{2}$$

$$= \frac{5 \pm \sqrt{41}}{2} \qquad \text{This is solved for the function.}$$

The angle is θ and is the same as the unknown:

$$\theta = \tan^{-1}\left(\frac{5 \pm \sqrt{41}}{2}\right)$$

Solving General Trigonometric Equations

We conclude this section by solving a variety of trigonometric equations. Notice that the procedure is the same as in Example 6 except for one added step—namely, the evaluation step. Since we know that multiple values are found by adding 2π to the angles found between 0 and 2π, we often ask for only those positive solutions less than one revolution.

EXAMPLE 7   **Solving a trigonometric equation by factoring**

Solve $2\cos\theta\sin\theta = \sin\theta$ for $0 \le \theta < 2\pi$ (exact values).

SOLUTION   This problem is solved by factoring.

Do not divide both sides by $\sin\theta$. Notice that if $\theta = 0$ or π, then $\sin\theta = 0$ and you cannot divide by 0. ⊗

$$2\cos\theta\sin\theta - \sin\theta = 0 \qquad \text{Obtain a 0 on one side.}$$
$$\sin\theta(2\cos\theta - 1) = 0 \qquad \text{Factor}$$
$$\sin\theta = 0 \quad \text{and} \quad 2\cos\theta - 1 = 0 \qquad \text{Set each factor equal to 0.}$$
$$\cos\theta = \frac{1}{2}$$
$$\theta = 0, \pi \qquad\qquad \theta = \frac{\pi}{3}, \frac{5\pi}{3} \qquad \text{Exact values}$$

$$\left\{0, \pi, \frac{\pi}{3}, \frac{5\pi}{3}\right\}$$

---

*Remember the quadratic formula: If $ax^2 + bx + c = 0$, and $a \ne 0$, then $x = \dfrac{-b \pm \sqrt{b^2 - 4ac}}{2a}$.

CALCULATOR COMMENT

You can use a graphing calculator to numerically solve (to a specified degree of accuracy) or to check an answer. For Example 7, we can graph both the left and right sides of the equation and then look for the points of intersection, as shown in Figure 4.1.

FIGURE 4.1 Graphs of Y1 $= 2 \cos x \sin x$ and Y2 $= \sin x$

On the interval $0 \le x < 2\pi$, we can check the results of Example 7; the points of intersection are at $x = 0, \dfrac{\pi}{3}, \pi, \dfrac{5\pi}{3}$. On a calculator such as the T1-81, you can approximate the solutions using the $\boxed{\text{TRACE}}$ as shown below:

```
X=0              Y=0
X=1.0263158      Y=.88614595
X=3.1473684      Y=.01155128        Xmin=0      Ymin=-1.5
X=5.2684211      Y=.8966146         Xmax=6.5    Ymax=1.5
X=6.2947368      Y=.02310101  ← (not less than 2π so it is not included)
```

Another alternative would be to solve the equation to obtain a zero on one side, and then estimate where the nonzero function crosses the x-axis. For Example 7, write

$$2 \cos \theta \sin \theta - \sin \theta = 0$$

The graph and estimates are shown below:

```
X=0              Y=0
X=1.015263       Y=.05322394
X=3.1157895      Y=-.0773838
X=5.2210526      Y=.0226899
X=6.3157895      Y=.03256374  (not less than 2π)
```

Compare these graphing solutions with the answer.

```
Xmin = 0   Ymin = -3
Xmax = 8   Ymax = 3
```

EXAMPLE 8   Solving a trigonometric equation using the quadratic formula

Solve $2 \sin^2 \theta = 1 + 2 \sin \theta$ for $0 \le \theta < 2\pi$ (real-number solution correct to four decimal places).

SOLUTION This problem is solved by the quadratic formula, since you cannot factor the left-hand side:

CALCULATOR COMMENT

$$2 \sin^2\theta - 2 \sin \theta - 1 = 0 \qquad \text{First obtain a 0 on one side.}$$

$$\sin \theta = \frac{2 \pm \sqrt{4 - 4(2)(-1)}}{2(2)} \qquad \text{Remember the quadratic formula.}$$

$$= \frac{1 \pm \sqrt{3}}{2}$$

$$\approx -.3660254038$$

$$1.366025404 \longleftarrow \qquad \text{Reject, since } -1 \le \sin \theta \le 1.$$

Solve $\theta' = \sin^{-1}(.3660254038)$ by using a calculator to find a reference angle of .3747. Since the sine is negative in Quadrants III and IV, the solutions are:

Quadrant III: Quadrant IV:

$$\theta = \pi + .3747 \qquad\qquad\qquad \theta = 2\pi - .3747$$
$$\approx 3.5163 \qquad\qquad\qquad\qquad \approx 5.9085$$

$\{3.5163, 5.9085\}$

In the process of solving for the function, you may need to use one or more identities, as illustrated by Example 9.

EXAMPLE 9   **Solving a trigonometric equation using identities and factoring**

Solve $\cos^2 x - \sin^2 x = \sin x$ for $0 \le x < 2\pi$ (exact values).

SOLUTION

$$\cos^2 x - \sin^2 x = \sin x$$
$$(1 - \sin^2 x) - \sin^2 x = \sin x$$
$$1 - 2 \sin^2 x = \sin x$$
$$2 \sin^2 x + \sin x - 1 = 0$$
$$(2 \sin x - 1)(\sin x + 1) = 0$$

CALCULATOR COMMENT

$$\sin x = \tfrac{1}{2} \qquad \text{and} \qquad \sin x = -1$$

$$x = \sin^{-1}\left(\tfrac{1}{2}\right) \qquad\qquad x = \sin^{-1}(-1)$$

$$= \tfrac{\pi}{6}, \tfrac{5\pi}{6} \qquad\qquad\qquad = \tfrac{3\pi}{2} \qquad \text{Remember } 0 \le x < 2\pi.$$

$$\left\{ \frac{\pi}{6}, \frac{5\pi}{6}, \frac{3\pi}{2} \right\}$$

As illustrated by Example 9, if different functions appear in the equation, use the fundamental identities to express the equation in terms of a single function. If it is necessary to multiply both sides by a trigonometric function or to square both sides, be sure to check for extraneous roots.

EXAMPLE 10 **Solving a trigonometric equation with an extraneous root**

Solve $\tan x + \sqrt{3} = \sec x$ for $0 \le x < 2\pi$ (exact values).

SOLUTION A better procedure is to square both sides rather than to introduce radicals:

$$\tan^2 x + 2\sqrt{3}\tan x + 3 = \sec^2 x$$

$$\tan^2 x + 2\sqrt{3}\tan x + 3 = 1 + \tan^2 x \qquad \text{Write the equations in terms of a single function, if possible.}$$

$$2\sqrt{3}\tan x = -2$$

$$\tan x = \frac{-1}{\sqrt{3}} = -\frac{1}{3}\sqrt{3}$$

Since the tangent is negative in Quadrants II and IV, use exact values to find $x' = \tan^{-1}\left(\frac{1}{3}\sqrt{3}\right) = \frac{\pi}{6}$, which means the reference angle is $\frac{\pi}{6}$. Place this reference angle in the appropriate quadrants:

$$x = \frac{5\pi}{6} \qquad \text{Quadrant II}\left(\pi - x' \text{ or } \pi - \frac{\pi}{6}\right)$$

$$x = \frac{11\pi}{6} \qquad \text{Quadrant IV}\left(2\pi - x' \text{ or } 2\pi - \frac{\pi}{6}\right)$$

Check for extraneous roots in the original equation by substituting possible solutions into the original equation:

$$x = \frac{5\pi}{6}: \quad \tan\frac{5\pi}{6} + \sqrt{3} = -\frac{\sqrt{3}}{3} + \sqrt{3} \quad \text{and} \quad \sec\frac{5\pi}{6} = \frac{-2\sqrt{3}}{3}$$

$$= \frac{2\sqrt{3}}{3}$$

These are not the same, so $\frac{5\pi}{6}$ is extraneous.

$$x = \frac{11\pi}{6}: \quad \tan\frac{11\pi}{6} + \sqrt{3} = -\frac{\sqrt{3}}{3} + \sqrt{3} \quad \text{and} \quad \sec\frac{11\pi}{6} = \frac{2\sqrt{3}}{3}$$

$$= \frac{2\sqrt{3}}{3}$$

These are the same, so $\frac{11\pi}{6}$ checks.

CALCULATOR COMMENT

$$\left\{\frac{11\pi}{6}\right\}$$

The third step in the procedure for solving trigonometric equations is used when the angle is not the same as the unknown. This process will be discussed in the next section.

4.1 Problem Set

A *Evaluate each of the expressions in Problems 1–4 to the nearest degree (use exact values when possible).*

1. a. $\sin^{-1}\frac{1}{2}$ **b.** $\arccos\frac{1}{2}$

2. a. $\cos^{-1}\left(\frac{-\sqrt{3}}{2}\right)$ **b.** $\arcsin\left(\frac{-\sqrt{2}}{2}\right)$

3. a. $\arcsin .3907$ **b.** $\arccos .2924$

4. a. $\arctan\sqrt{3}$ **b.** $\operatorname{arcsec}\sqrt{2}$

Evaluate each of the expressions in Problems 5–8 to the nearest hundredth of a radian (use exact values when possible).

5. a. $\arcsin\left(-\frac{1}{2}\right)$ **b.** $\cos^{-1}\left(-\frac{1}{2}\right)$

6. a. $\arccos\frac{\sqrt{3}}{2}$ **b.** $\sin^{-1}\frac{\sqrt{2}}{2}$

7. a. $\arcsin\left(-\frac{\sqrt{3}}{2}\right)$ **b.** $\cos^{-1}\left(-\frac{\sqrt{2}}{2}\right)$

8. a. $\operatorname{arccot}(-1)$ **b.** $\operatorname{arccsc}(-\sqrt{2})$

Solve each of the equations in Problems 9–17. Use exact values.

9. a. $\sin x = \frac{1}{2}$ $\left(0 \le x < \frac{\pi}{2}\right)$ **b.** $\sin x = \frac{1}{2}$ $(0° \le x < 90°)$

10. a. $\sin x = \frac{1}{2}$ (radians)   **b.** $\sin x = -\frac{1}{2}$ (radians)

11. a. $\sin x = \frac{1}{2}$ $(0 \le x < \pi)$ **b.** $\sin x = -\frac{1}{2}$ $(0 \le x < \pi)$

12. $(\sin x)(\cos x) = 0$ $(0° \le x < 360°)$

13. $(\sin x)(\tan x) = 0$ $(0° \le x < 360°)$

14. $(2\cos x + \sqrt{2})(2\cos x - 1) = 0$ $(0 \le x < 2\pi)$

15. $(3\tan x + \sqrt{3})(3\tan x - \sqrt{3}) = 0$ $(0 \le x < 2\pi)$

16. $(\sec x - 2)(2\sin x - 1) = 0$ $(0 \le x < 2\pi)$

17. $(\csc x - 2)(2\cos x - 1) = 0$ $(0 \le x < 2\pi)$

Solve each of the equations in Problems 18–28 for $0 \le x < 2\pi$. Use exact values where possible, but state approximate answers to four decimal places. The graphs are shown as a check on your work.

18. $\tan^2 x = \tan x$

19. $\tan^2 x = \sqrt{3}\tan x$

20. $2\cos^2 x = \frac{1}{2}$

21. $\sin^2 x = \frac{1}{2}$

22. $\tan^2 x = \frac{1}{2}$

23. $2\cos x \sin x = \sin x$

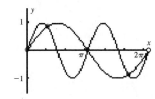

24. $3\sin x \cos x = \sin x$

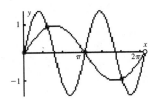

25. $\cos^2 x - 1 - \cos x = 0$

26. $\sin^2 x - \sin x - 2 = 0$

27. $\tan^2 x - 3\tan x + 1 = 0$

28. $4\tan^2 x - 8\tan x + 3 = 0$

REFRACTION   (*Problems* 29–30). If a light ray passes from one medium to another of greater or lesser density, the ray will bend. A measure of this tendency is called the *index of refraction, R,* and is defined by

$$R = \frac{\sin \alpha}{\sin \beta}$$

For example, if the angle of incidence, α, of a light ray entering a diamond is 30.0° and the angle of refraction, β, is 11.9°, then the diamond's index of refraction is:

$$R = \frac{\sin 30.0°}{\sin 11.9°} \approx 2.424781044$$

Since the index of refraction for any particular substance is a constant (see Table 4.1), we can compare our calculated number to the known index number to determine whether this diamond is an imitation. The index of refraction of an imitation diamond is rarely greater than 2. If a stone is a diamond and the angle of incidence is 45.0°, what should be its refracted angle?

$$R = \frac{\sin \alpha}{\sin \beta} \quad \text{where } R \approx 2.42, \ \alpha = 45°$$

TABLE 4.1

SUBSTANCE	INDEX OF REFRACTION
Diamond	2.42
Topaz	1.63
Garnet	1.78
Glass	1.50
Water	1.33
Ice	1.31

Solve for β and substitute known values to find the refracted angle β:

$$\sin \beta = \frac{\sin \alpha}{R}$$

$$\beta = \sin^{-1}\left(\frac{\sin \alpha}{R}\right) \approx 16.95°$$

(16.99° if you work with the rounded value for R)

To the nearest tenth, the refracted angle is 17.0°.

29. If a stone is a topaz and the angle of incidence is 20.0°, what should be its refracted angle (correct to the nearest tenth of a degree)?

30. If a stone is a garnet and the angle of incidence is 35.0°, what should be its refracted angle (correct to the nearest tenth of a degree)?

C *Solve each of the equations in Problems* 31–39 *for* $0 \le x < 2\pi$. *Use exact values where possible, but state approximate answers to four decimal places.*

31. $\sin^2 x - \sin x - 1 = 0$

32. $\cos^2 x - \cos x - 1 = 0$

33. $\sin x + 1 = \sqrt{3}$

34. $\cot x + \sqrt{3} = \csc x$

35. $\cos^2 x - 3 \sin x + 3 = 0$

36. $\sec^2 x - 1 = \sqrt{3} \tan x$

37. $2 \sin^2 x - 2 \cos^2 x = 1$

38. $\sin^2 x + \cos x = 0$

39. $2 \cos^2 x + \sin^2 x = 1$

*40. For $0 \le \theta < \frac{\pi}{2}$, solve

$$\left(\frac{16}{81}\right)^{\sin^2\theta} + \left(\frac{16}{18}\right)^{1 - \sin^2\theta} = \frac{26}{27}$$

41. How many solutions are there for the equation

$$\frac{x}{100} = \sin x$$

---

*Problem 40 is from The Mathematics Student Journal *of the National Council of Teachers of Mathematics, April 1973, 20(4). Reprinted by permission.*

4.2 Trigonometric Equations with Multiple Angles

There are two main procedures for solving trigonometric equations with multiple angles. The first is by substitution, and the second is by using identities. Neither method will be adequate for all equations, so you need to study both procedures.

If the unknown and the angle are not the same, you can let θ equal the angle and then solve the resulting equation as shown in the last section. In doing this, you will need to pay special attention to the domain, as shown in Example 1.

EXAMPLE 1   **Solving a trigonometric equation with multiple angles**

Solve for $0 \le x < 2\pi$, giving both exact values and calculator approximations correct to two decimal places.

a. $\sin x = \frac{1}{2}$ **b.** $\sin 2x = \frac{1}{2}$ **c.** $\sin 3x = -\frac{1}{2}$ **d.** $\sin(2x + 1) = \frac{1}{2}$

SOLUTION   **a.** $\sin x = \frac{1}{2}$ The reference angle (from the table of exact values) is $\frac{\pi}{6}$.
This reference angle should be placed in Quadrants I and II since sine is positive.

$$x = \frac{\pi}{6}, \frac{5\pi}{6} \longleftarrow \text{Quadrant II}$$
$$\uparrow\underline{\qquad\qquad}\text{Quadrant I}$$

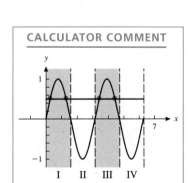

CALCULATOR COMMENT

These are the only values of x such that $0 \le x < 2\pi$. Giving the calculator approximations, the solution set is

$$\{.52, 2.62\}$$ All of these answers should be between 0 and 6.28 (approximate values correct to two decimal places).

b. The equation in part **a** was the type of equation solved in the last section. This example, $\sin 2x = \frac{1}{2}$, has an angle ($2x$) different from the unknown (which is x). **Consider the domain**: if

$$0 \le x < 2\pi$$

then $0 \le 2x < 4\pi$ Multiply all three parts of the inequality by 2.

You need to find values of $2x$ over a larger domain than that of x:

CALCULATOR COMMENT

$\ulcorner 2\pi$ added to the second-quadrant solution

$$2x = \frac{\pi}{6}, \frac{5\pi}{6}, \frac{13\pi}{6}, \frac{17\pi}{6}$$ Additional multiples of 2π are greater than 4π.

2π added to the first-quadrant solution

Now, solve for x:

$$2x = \frac{\pi}{6} \qquad 2x = \frac{5\pi}{6} \qquad 2x = \frac{13\pi}{6} \qquad 2x = \frac{17\pi}{6}$$

$$x = \frac{\pi}{12} \qquad x = \frac{5\pi}{12} \qquad x = \frac{13\pi}{12} \qquad x = \frac{17\pi}{12}$$

$$\{.26, 1.31, 3.40, 4.45\}$$ Notice that all of these roots are between 0 and 6.28.

Parts **a** and **b** suggest a pattern for the solution of the equation $\sin nx = \frac{1}{2}$. Since the period of $\sin nx$ is $2\pi/n$, we note that as the period decreases, the number of times the graph will intersect the line $y = \frac{1}{2}$ increases, as shown in Figure 4.2.

$n = 3$
6 roots

$n = 4$
8 roots

$n = 5$
10 roots

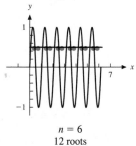

$n = 6$
12 roots

FIGURE 4.2 Graph of $y = \sin nx$ and $y = \frac{1}{2}$ for selected values of n

c. $\sin 3x = -\frac{1}{2}$; let $\theta = 3x$. Then $x = \frac{\theta}{3}$ and

$$0 \le x < 2\pi$$

$$0 \le \frac{\theta}{3} < 2\pi$$

$$0 \le \theta < 6\pi \qquad \text{Multiply by 3.}$$

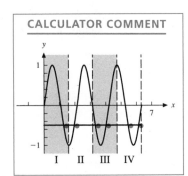
From part **a**, the reference angle is $\frac{\pi}{6}$, but this time it needs to be placed in Quadrants III and IV (because of the negative value):

Quadrant III, reference angle $\frac{\pi}{6}$
Quadrant IV, reference angle $\frac{\pi}{6}$

$$\theta = \frac{7\pi}{6}, \frac{11\pi}{6}, \frac{19\pi}{6}, \frac{23\pi}{6}, \frac{31\pi}{6}, \frac{35\pi}{6} \qquad \text{Additional multiples of } 2\pi \text{ are greater than } 6\pi.$$

$+2\pi \qquad +2\pi$

$+2\pi \qquad +2\pi$

Since $\theta = 3x$, replace θ by $3x$ and solve for x in each of the six equations to obtain $\left\{\frac{7\pi}{18}, \frac{11\pi}{18}, \frac{19\pi}{18}, \frac{23\pi}{18}, \frac{31\pi}{18}, \frac{35\pi}{18}\right\}$. The solution set using approximate values can now be found by using a calculator: $\{1.22, 1.92, 3.32, 4.01, 5.41, 6.11\}$.

d. $\sin(2x + 1) = \frac{1}{2}$; let $\theta = 2x + 1$. Then $\dfrac{\theta - 1}{2} = x$ and $0 \le x < 2\pi$.

$$0 \le \frac{\theta - 1}{2} < 2\pi$$

$$0 \le \theta - 1 < 4\pi \qquad \text{Multiply all parts by 2.}$$

$$1 \le \theta < 4\pi + 1 \qquad \text{Add 1 to all parts.}$$

This says that you will include only values of θ that are between 1 and about 13.57.

From part **a** ⌐ Multiples of 2π added to answers part **a**

$$\theta = \overline{\frac{\pi}{6}, \frac{5\pi}{6}, \frac{13\pi}{6}, \frac{17\pi}{6}, \frac{25\pi}{6}, \frac{29\pi}{6}}, \ldots$$

Include.

Reject (less than 1).

Reject $\frac{29\pi}{6} \approx 15.18$, which is greater than 13.57. Also reject additional larger multiples of 2π.

Now each of the included values of θ is used in an equation to find x:

$$\theta = \frac{5\pi}{6} \qquad\qquad \theta = \frac{13\pi}{6} \qquad\qquad \theta = \frac{17\pi}{6} \qquad\qquad \theta = \frac{25\pi}{6}$$

$$2x + 1 = \frac{5\pi}{6} \qquad 2x + 1 = \frac{13\pi}{6} \qquad 2x + 1 = \frac{17\pi}{6} \qquad 2x + 1 = \frac{25\pi}{6}$$

$$2x = \frac{5\pi}{6} - 1 \qquad 2x = \frac{13\pi - 6}{6} \qquad x = \frac{17\pi - 6}{12} \qquad x = \frac{25\pi - 6}{12}$$

$$x = \frac{5\pi - 6}{12} \qquad x = \frac{13\pi - 6}{12}$$

CALCULATOR COMMENT

The solution set (using approximate values) is

$$\{.81, 2.90, 3.95, 6.04\}$$ Notice that all of the roots are between 0 and 2π.   ∎

The second method for handling multiple angles is to use one or more trigonometric identities, as illustrated by Example 2.

EXAMPLE 2   **Solving a trigonometric equation by using a trigonometric identity**

Solve $\sin 2x = \cos x$ for $0 \le x < 2\pi$, giving both exact values and calculator approximations correct to two decimal places.

SOLUTION

CALCULATOR COMMENT

$$\sin 2x = \cos x \qquad \text{Given.}$$

$$2 \sin x \cos x = \cos x \qquad \text{Use the double-angle identity } \sin 2x = 2 \sin x \cos x.$$

$$2 \sin x \cos x - \cos x = 0$$

$$\cos x (2 \sin x - 1) = 0 \qquad \text{Complete the solution as shown in Section 4.1.}$$

$$\cos x = 0 \qquad \sin x = \frac{1}{2}$$

$$x = \frac{\pi}{2}, \frac{3\pi}{2} \qquad x = \frac{\pi}{6}, \frac{5\pi}{6}$$

$$\left\{\frac{\pi}{2}, \frac{3\pi}{2}, \frac{\pi}{6}, \frac{5\pi}{6}\right\} \quad \text{or} \quad \{.52, 1.57, 2.62, 4.71\}$$ ∎

4.2   Problem Set

A *Solve each of the equations in Problems 1–9 for exact values such that $0 \le x < 2\pi$.*

1. $\cos 2x = \frac{1}{2}$

2. $\cos 3x = \frac{1}{2}$

3. $\cos 2x = -\frac{1}{2}$

4. $\sin 2x = \frac{\sqrt{2}}{2}$

5. $\sin 2x = -\frac{\sqrt{3}}{2}$

6. $\sin 3x = \frac{\sqrt{2}}{2}$

7. $\tan 3x = 1$

8. $\tan 3x = -1$

9. $\sec 2x = -\frac{2\sqrt{3}}{3}$

14. $\tan 2x = \cot 2x$

15. $\sin \frac{x}{2} = \cos \frac{x}{2}$

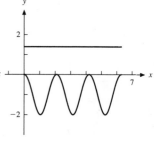

B *Solve each of the equations in Problems 10–30 for $0 \le x < 2\pi$ correct to two decimal places.*

10. $2 \cos 2x \sin 2x = \sin 2x$

11. $\sin 2x + 2 \cos x \sin 2x = 0$

16. $\sin 2x + 1 = \sqrt{3}$

17. $\cos 3x - 1 = \sqrt{2}$

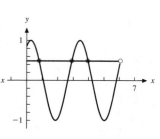

12. $\cos 3x + 2 \sin 2x \cos 3x = 0$

13. $\sin 2x = \cos 2x$

18. $\tan 2x + 1 = \sqrt{3}$

19. $\sin(2x + 1) = \frac{1}{2}$

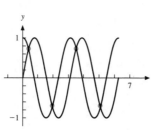

20. $\cos(3x - 1) = \frac{1}{2}$

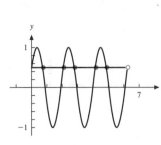

21. $\tan(2x + 1) = \sqrt{3}$

28. $\sin^2 3x + \sin 3x = \cos^2 3x - 1$

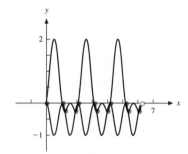

22. $\cos 2x = \cos x$

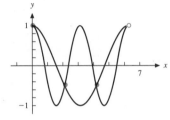

23. $\cos 2x = \sin x$

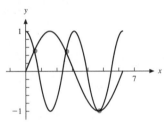

29. $3 \sin 4x - \sin 3x + \sqrt{3} = 2 \sin 2x - \sin 3x + 3 \sin 4x$

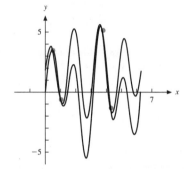

24. $\cos 2x + \cos x = 0$

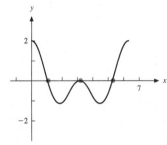

25. $1 - \sin x = \cos 2x$

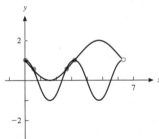

30. $3 \cos x - \cos 3x + 2 \cos 5x = 1 + 3 \cos x - \cos 3x$

26. $\cos x = \sin 2x$

27. $\sin^2 3x + \sin 3x + 1 = 1 - \sin^2 3x$

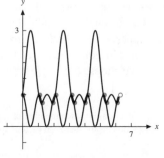

C

31. SOUND WAVES   A tuning fork vibrating at 264 Hz
($f = 264$) with an amplitude of .0050 cm produces C on the
musical scale and can be described by an equation of the
form

$$y = .0050 \sin 528\pi x$$

Find the smallest positive value of x (correct to four decimal
places) for which $y = .0020$.

32. ELECTRICAL   In a certain electric circuit, the electromotive
force V (in volts) and the time t (in seconds) are related by an
equation of the form

$$V = \cos 2\pi t$$

Find the smallest positive value for t (correct to three decimal places) for which $V = .400$.

33. SPACE SCIENCE   The orbit of a certain satellite alternates
above and below the equator according to the equation

$$y = 4{,}000 \sin\!\left(\tfrac{\pi}{45}t + \tfrac{5\pi}{18}\right)$$

where t is the time (in minutes) and y is the distance (in
kilometers) from the equator. Find the times the satellite
crosses the equator during the first hour and a half (that is,
for $0 \le t \le 90$).

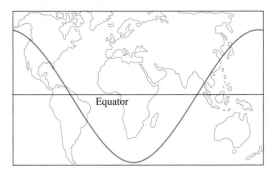

Equator

4.3   Inverse Trigonometric Functions

Principal Values

One of the troublesome aspects about solving trigonometric equations is the multiple
solutions. For example, in Section 4.1, we found that the solution of $\cos \theta = .5$ is $\theta =$
$\cos^{-1} .5 = \frac{\pi}{3}$ (Quadrant I), $\frac{\pi}{3}$ (Quadrant IV), $\frac{7\pi}{3}$ (Quadrant I, first value $+ 2\pi$), $\frac{11\pi}{3}$
(Quadrant IV, second value $+ 2\pi$), ... One of the simplifying agreements we used
was to limit our answers to positive angles less than one revolution. However, even with
that restriction, the inverse relation has more than one solution: $\theta = \cos^{-1} .5 = \frac{\pi}{3}, \frac{5\pi}{3}$.
Since there is more than one value, the inverse cosine is not a function.*

Our goal now is to limit the domain of the inverse cosine so that it *is* a function. We
begin by defining a **principal value** for each of the inverse relations. **The designation
that we want principal values rather than all values is denoted by using a capital
letter, as shown in Table 4.2.** When the capital letter is used, it tells us to give a single
value, and as such can properly be called an **inverse trigonometric function** to
distinguish it from an inverse trigonometric relation. With this definition we see
$\cos^{-1} .5 = \frac{\pi}{3}$ (.5 is positive, so choose the single value in Quadrant I).

---

*Remember the definition of a function. For each member of the domain there can be only one value in
the range. In everyday language, this means there can be only one answer if it is a function.

TABLE 4.2 Definition of Principal Values

INVERSE RELATION	INVERSE FUNCTION	QUADRANT OF PRINCIPAL VALUE, y		
		VALUE OF x CAN BE		
		POSITIVE	NEGATIVE	0
$y = \cos^{-1} x$ or $y = \arccos x$	$y = \text{Cos}^{-1} x$ or $y = \text{Arccos } x$	I	II	$\frac{\pi}{2}$
$y = \sin^{-1} x$ or $y = \arcsin x$	$y = \text{Sin}^{-1} x$ or $y = \text{Arcsin } x$	I	IV	0
$y = \tan^{-1} x$ or $y = \arctan x$	$y = \text{Tan}^{-1} x$ or $y = \text{Arctan } x$	I	IV	0
$y = \sec^{-1} x$ or $y = \text{arcsec } x$	$y = \text{Sec}^{-1} x$ or $y = \text{Arcsec } x$	I	II	undefined
$y = \csc^{-1} x$ or $y = \text{arccsc } x$	$y = \text{Csc}^{-1} x$ or $y = \text{Arccsc } x$	I	IV	undefined
$y = \cot^{-1} x$ or $y = \text{arccot } x$	$y = \text{Cot}^{-1} x$ or $y = \text{Arccot } x$	I	II	undefined

Answers in Quadrant I should be between 0 and $\frac{\pi}{2}$ (that is, between 0 and about 1.57 or, in degrees, between 0° and 90°); answers in Quadrant II should be between $\frac{\pi}{2}$ and π (that is, between about 1.57 and 3.14 or, in degrees, between 90° and 180°); answers in Quadrant IV should be negative numbers between $-\frac{\pi}{2}$ and 0 (that is, between about -1.57 and 0 or, in degrees, between $-90°$ and 0°).

The angle is y; it is in Quad I whenever x is positive.
y is in Quad II or IV whenever x is negative.

$y = \text{Cos}^{-1}x \Leftrightarrow \text{Cos } y = x$ $y = \text{Sin}^{-1}x \Leftrightarrow \text{Sin } y = x$ $y = \text{Tan}^{-1}x \Leftrightarrow \text{Tan } y = x$

Quad II	Quad I
x is negative	x is positive
$-1 < x < 0$	$0 < x < 1$
$\frac{\pi}{2} < y < \pi$	$0 < y < \frac{\pi}{2}$
	Quad IV

Quad II	Quad I
	x is positive
	$0 < x < 1$
	$0 < y < \frac{\pi}{2}$
	Quad IV
	x is negative
	$-1 < x < 1$
	$-\frac{\pi}{2} < y < 0$

Quad II	Quad I
	x is positive
	$x > 0$
	$0 < y < \frac{\pi}{2}$
	Quad IV
	x is negative
	$x < 0$
	$-\frac{\pi}{2} < y < 0$

$y = \text{Sec}^{-1}x \Leftrightarrow \text{Sec } y = x$ $y = \text{Csc}^{-1}x \Leftrightarrow \text{Csc } y = x$ $y = \text{Cot}^{-1}x \Leftrightarrow \text{Cot } y = x$

Quad II	Quad I
x is negative	x is positive
$x < -1$	$x > 1$
$\frac{\pi}{2} < y < \pi$	$0 < y < \frac{\pi}{2}$
	Quad IV

Quad II	Quad I
	x is positive
	$x > 1$
	$0 < y < \frac{\pi}{2}$
	Quad IV
	x is negative
	$-1 < x < 0$
	$-\frac{\pi}{2} < y < 0$

Quad II	Quad I
x is negative	x is positive
$x < 0$	$x > 0$
$\frac{\pi}{2} < y < \pi$	$0 < y < \frac{\pi}{2}$
	Quad IV

The definition of principal value means that instead of giving multiple values for the inverse relations, you will give only one value chosen from the appropriate quadrant as shown in Table 4.2.

EXAMPLE 1 **Evaluating inverse cosine function**

Evaluate the given inverse trigonometric functions.

a. $\text{Arccos} \frac{\sqrt{2}}{2}$ **b.** $\text{Arccos}\left(-\frac{\sqrt{2}}{2}\right)$ **c.** $\text{Cos}^{-1}(-.4695)$ **d.** $\text{Cos}^{-1} 2$

SOLUTION **a.** $\text{Arccos} \frac{\sqrt{2}}{2}$ Notice that x is positive so $\text{Arccos} \frac{\sqrt{2}}{2}$ is the first-quadrant angle whose cosine is $\frac{\sqrt{2}}{2}$; that is $\frac{\pi}{4} \approx .7854$.

b. $\text{Arccos}\left(-\frac{\sqrt{2}}{2}\right)$ Notice that x is negative so $\text{Arccos}\left(-\frac{\sqrt{2}}{2}\right)$ is the second-quadrant angle whose reference angle is $\frac{\sqrt{2}}{2}$; this is $\frac{3\pi}{4} \approx 2.356$.

c. $\text{Cos}^{-1}(-.4695)$ Notice that x is negative so $\text{Cos}^{-1}(-.4695)$ is the second-quadrant angle whose value is $-.4695$. Your calculator is pre-programmed to give you the correct principal values, so you can input negative values when using your calculator:

DISPLAY: `2.0595207`

d. $\text{Cos}^{-1} 2$ Notice that x is positive and also that this represents the angle whose cosine is 2; there is no such angle (remember, cosine is between -1 and 1). If you evaluate $\text{Cos}^{-1} 2$ by calculator it will indicate that an error has been made.

You can check your work by making sure you have appropriate answers for the inverse cosine function. All answers for the inverse cosine function should be between 0 and 3.14; more formally, if $y = \text{Arccos}\, x$, then $0 \leq y \leq \pi$ and $\cos y = x$. ▮

From Table 4.2 you can see that if the x value is positive, the inverse trigonometric functions are in the first quadrant. For negative values we need to specify either the second or the fourth quadrant, depending on the function. In Example 1 we saw that for negative x values the inverse cosine is a second-quadrant angle. In Example 2 we can see that for negative x values the inverse sine and inverse tangent must be in Quadrant IV rather than Quadrant II. If we tried, for example, to put the inverse sine in Quadrants I and II, we see that it would no longer be a function.

⊘ Remember, with the inverse *functions* we are looking for *one* value; that value is either positive or negative. ⊘

EXAMPLE 2 **Evaluating inverse sine and inverse tangent functions**

Evaluate the given inverse trigonometric functions.

a. $\text{Sin}^{-1}\left(\frac{\sqrt{3}}{2}\right)$ **b.** $\text{Sin}^{-1}\left(-\frac{\sqrt{3}}{2}\right)$ **c.** $\text{Arcsin}(-.4695)$ **d.** $\text{Arctan}\, 3.5$ **e.** $\text{Tan}^{-1}(-2.3)$

SOLUTION **a.** $\text{Sin}^{-1}\left(\frac{\sqrt{3}}{2}\right)$ Notice that x is positive so $\text{Sin}^{-1}\left(\frac{\sqrt{3}}{2}\right)$ is the first-quadrant angle whose sine is $\frac{\sqrt{3}}{2}$; this is $\frac{\pi}{3} \approx 1.0472$.

b. $\text{Sin}^{-1}\left(-\frac{\sqrt{3}}{2}\right)$ Notice that x is negative so $\text{Sin}^{-1}\left(-\frac{\sqrt{3}}{2}\right)$ is the fourth-quadrant angle whose reference angle is $\frac{\pi}{3}$; this is $-\frac{\pi}{3}$. Fourth-quadrant angles for the inverse sine function are negative numbers between $-\frac{\pi}{2}$ and 0.

c. Arcsin($-.4695$)   Notice that x is negative so Arcsin($-.4695$) is the fourth-quadrant angle whose value is $-.4695$. Your calculator is pre-programmed to give you the correct principal values, so you can input negative values when using your calculator:

DISPLAY: -0.488724398

d. Arctan 3.5 DISPLAY: 1.292496668

e. Tan$^{-1}(-2.3)$ DISPLAY: -1.160668986 ▮

To evaluate arcsecant, arccosecant, or arccotangent relations or functions when their exact values are not known, you need to use the reciprocal identities. First find the reciprocal, then find the inverse of the reciprocal. That is, find the inverse of the reciprocal function. This procedure is summarized by the followinig *inverse identities*.

INVERSE TRIGONOMETRIC IDENTITIES

Arccot $x = \pi +$ Arctan $\frac{1}{x}$ if x is negative is the most difficult of these identities to understand. Look at the picture of the angles:

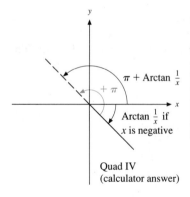

Quad IV
(calculator answer)

35. $\text{Arccot } x = \begin{cases} \text{Arctan } \dfrac{1}{x} & \text{if } x \text{ is positive} \\[2mm] \pi + \text{Arctan } \dfrac{1}{x} & \text{if } x \text{ is negative} \\[2mm] \dfrac{\pi}{2} & \text{if } x = 0 \end{cases}$

36. $\text{Arcsec } x = \text{Arccos } \dfrac{1}{x} \quad$ if $x \geq 1$ or $x \leq -1$

37. $\text{Arccsc } x = \text{Arcsin } \dfrac{1}{x} \quad$ if $x \geq 1$ or $x \leq -1$

PROOF OF IDENTITY 35   Let $\theta = \text{Arccot } x$, where x is positive. Then

$$\cot \theta = x, \qquad 0 < \theta < \tfrac{\pi}{2} \qquad \text{Definition of Arccot } x$$

$$\frac{1}{\tan \theta} = x, \qquad 0 < \theta < \tfrac{\pi}{2} \qquad \text{Fundamental identity}$$

$$\tan \theta = \frac{1}{x}, \qquad 0 < \theta < \tfrac{\pi}{2} \qquad \text{Solve for } \tan \theta.$$

We must reflect the angle into Quad II to satisfy the definition of Arccotangent.

$$\theta = \text{Tan}^{-1}\frac{1}{x}, \quad 0 < \theta < \tfrac{\pi}{2} \qquad \text{Definition of Arctan } x$$

Thus, $\text{Arccot } x = \text{Arctan } \frac{1}{x}$ where x is positive. The proof of the second part of Identity 35 is left as an exercise. It is needed because the range of the inverse tangent function does not coincide with the range of the inverse cotangent function. The third part of the identity is obvious since $\cot \frac{\pi}{2} = 0$.

Proofs of Identities 36 and 37 are left as exercises.

EXAMPLE 3   Evaluating inverse trigonometric functions that are reciprocals

Evaluate the given inverse functions using the inverse identities and a calculator.

   a. $\text{Arcsec}(-3)$ **b.** $\text{Arccsc } 7.5$

   c. $\text{Arccot } 2.4747$ **d.** $\text{Arccot}(-4.852)$

SOLUTION   **a.** $\text{Arcsec}(-3)$ uses Identity 36.

DISPLAY: `1.910633236`

   b. $\text{Arccsc } 7.5$ uses Identity 37.

DISPLAY: `0.1337315894`

   c. $\text{Arccot } 2.4747$ uses the first part of Identity 35 since 2.4747 is positive.

DISPLAY: `0.3840267299`

   d. $\text{Arccot}(-4.852)$ uses the second part of Identity 35 since -4.852 is negative.

DISPLAY: `2.938338095`

We can now state the calculator steps for the inverse trigonometric functions. The sequence of keystrokes may vary depending on the model.

INVERSE FUNCTION	CALCULATOR		
	ALGEBRAIC	OR	GRAPHING
$y = \text{Arccos } x$	$\boxed{x}\ \boxed{\text{INV}}\ \boxed{\text{COS}}$		$\boxed{\text{COS}^{-1}}\ \boxed{x}\ \boxed{\text{ENTER}}$
$y = \text{Arcsin } x$	$\boxed{x}\ \boxed{\text{INV}}\ \boxed{\text{SIN}}$		$\boxed{\text{SIN}^{-1}}\ \boxed{x}\ \boxed{\text{ENTER}}$
$y = \text{Arctan } x$	$\boxed{x}\ \boxed{\text{INV}}\ \boxed{\text{TAN}}$		$\boxed{\text{TAN}^{-1}}\ \boxed{x}\ \boxed{\text{ENTER}}$
$y = \text{Arcsec } x$	$\boxed{x}\ \boxed{1/x}\ \boxed{\text{INV}}\ \boxed{\text{COS}}$		$\boxed{\text{COS}^{-1}}\ \boxed{x}\ \boxed{x^{-1}}\ \boxed{\text{ENTER}}$
$y = \text{Arccsc } x$	$\boxed{x}\ \boxed{1/x}\ \boxed{\text{INV}}\ \boxed{\text{SIN}}$		$\boxed{\text{SIN}^{-1}}\ \boxed{x}\ \boxed{x^{-1}}\ \boxed{\text{ENTER}}$
$y = \text{Arccot } x$	If $x > 0$:		
	$\boxed{x}\ \boxed{1/x}\ \boxed{\text{INV}}\ \boxed{\text{TAN}}$		$\boxed{\text{TAN}^{-1}}\ \boxed{x}\ \boxed{x^{-1}}\ \boxed{\text{ENTER}}$
	If $x < 0$:*		
	$\boxed{x}\ \boxed{1/x}\ \boxed{\text{INV}}\ \boxed{\text{TAN}}\ \boxed{+}\ \boxed{\pi}\ \boxed{=}$		$\boxed{\text{TAN}^{-1}}\ \boxed{x}\ \boxed{x^{-1}}\ \boxed{\text{ENTER}}\ \boxed{+}\ \boxed{\pi}\ \boxed{\text{ENTER}}$

*Use 180 instead of π if working in degrees.

Graphs of the Inverse Trigonometric Functions

The graphs of $y = \cos x$ and $y = \cos^{-1} x$ are shown in Figure 4.3.

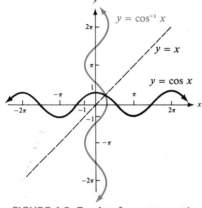

FIGURE 4.3 Graphs of $y = \cos x$ and $y = \cos^{-1} x$ (shown in color)

You can see from Figure 4.3 that $y = \cos x$ is a function but $y = \cos^{-1} x$ is not. However, if a suitable restriction is placed on the domain of $y = \cos x$, we will have a function whose *inverse is also a function*. This function with the restricted domain is denoted by $y = \text{Cos } x$ (note the capital letter) and is defined by $y = \cos x$ with the domain restricted to $0 \le x \le \pi$. The graph of $y = \text{Cos } x$ is shown in Figure 4.4 in color. Ordered pairs that satisfy $y = \text{Cos } x$ are called the *principal values* of the cosine function.

FUNCTION	DOMAIN	RANGE
$y = \cos x$	$-\infty < x < \infty$	$-1 \le y \le 1$
$y = \text{Cos } x$	$0 \le x \le \pi$ ←	$-1 \le y \le 1$

Restricted domain ┘

FIGURE 4.4 Graphs of $y = \cos x$ and $y = \text{Cos } x$ (shown in color)

The inverse of $y = \text{Cos } x$ is $y = \text{Cos}^{-1} x$ with domain $-1 \le x \le 1$ and range $0 \le y \le \pi$. The graphs of $y = \text{Cos } x$ and $y = \text{Cos}^{-1} x$ are shown in Figure 4.5.

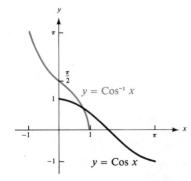

FUNCTION	DOMAIN	RANGE
$y = \text{Cos } x$	$0 \le x \le \pi$	$-1 \le y \le 1$
$y = \text{Cos}^{-1} x$	$-1 \le x \le 1$	$0 \le y \le \pi$

FIGURE 4.5 Graphs of $y = \text{Cos } x$ and $y = \text{Cos}^{-1} x$ (shown in color)

The inverse of the sine function, like the inverse cosine, is not a function, so a function $y = \text{Sin } x$ (note the capital letter) is defined as $y = \sin x$ with the domain restricted so that $-\frac{\pi}{2} \leq x \leq \frac{\pi}{2}$. The graph of $y = \sin x$ and the graph of $y = \text{Sin } x$ are compared in Figure 4.6. Ordered pairs that satisfy $y = \text{Sin } x$ are called the *principal values* of the sine function.

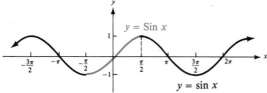

FUNCTION	DOMAIN	RANGE
$y = \sin x$	$-\infty \leq x \leq \infty$	$-1 \leq y \leq 1$
$y = \text{Sin } x$	$-\frac{\pi}{2} \leq x \leq \frac{\pi}{2}$	$-1 \leq y \leq 1$

FIGURE 4.6 Graphs of $y = \sin x$ and $y = \text{Sin } x$ (shown in color)

The inverse of $y = \text{Sin } x$ is $y = \text{Sin}^{-1} x$ with domain $-1 \leq x \leq 1$ and range $-\frac{\pi}{2} \leq y \leq \frac{\pi}{2}$. These graphs are shown in Figure 4.7.

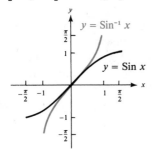

FUNCTION	DOMAIN	RANGE
$y = \text{Sin } x$	$-\frac{\pi}{2} \leq x \leq \frac{\pi}{2}$	$-1 \leq y \leq 1$
$y = \text{Sin}^{-1} x$	$-1 \leq x \leq 1$	$\frac{-\pi}{2} \leq y \leq \frac{\pi}{2}$

FIGURE 4.7 Graphs of $y = \text{Sin } x$ and $y = \text{Sin}^{-1} x$

The graphs of the other inverse relations and inverse functions are compared in Figure 4.8. Take special note of the range.*

CALCULATOR COMMENT

If you graph these on your calculator, only the part shown in color will be graphed.

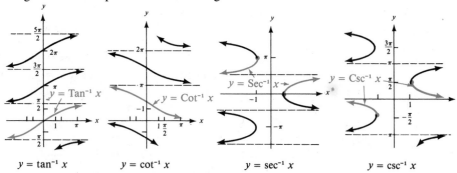

$y = \tan^{-1} x$	$y = \cot^{-1} x$	$y = \sec^{-1} x$	$y = \csc^{-1} x$
Domain: all reals	Domain: all reals	Domain: $x \geq 1$ or $x \leq -1$	Domain: $x \geq 1$ or $x \leq -1$
Range: $-\frac{\pi}{2} < y < \frac{\pi}{2}$	Range: $0 < y < \pi$	Range: $0 \leq y \leq \pi$, $y \neq \frac{\pi}{2}$	Range: $-\frac{\pi}{2} \leq y \leq \frac{\pi}{2}$, $y \neq 0$

FIGURE 4.8 Graphs of inverse trigonometric relations and inverse trigonometric functions

*There is no universal agreement about the range of the inverse secant and inverse cosecant functions. In this book, we used those that correspond to the preprogrammed values of your calculator. However, be sure to check these ranges carefully when you take calculus.

Evaluating Functions of Inverse Trigonometric Functions

Remember that the inverse trigonometric functions designate single angles, and as such can be used in any context that calls for angles.

EXAMPLE 4 **Evaluating functions of inverse trigonometric functions**

a. Arccos(cos 2)   **b.** cos(Arccos .5)

c. Arccos(cos 4)   **d.** sin(Arcsin 2.463)

SOLUTION When finding these values you must pay attention to the range of the inverse functions.

a. Since Arccosine and cosine are inverse functions and since the angle, 2, is between 0 and π, Arccos(cos 2) = 2.

b. Let θ = Arccos .5. By definition of Arccosine, cos θ = .5. Then, by substitution, cos(Arccos .5) = cos θ = .5.

c. The angle 4 radians is in the third quadrant, and, therefore, cos 4 is negative. The Arccosine of a negative angle will be a second-quadrant angle. This is the second-quadrant angle having the same reference angle as 4, as shown in Figure 4.9. The reference angle is

$$4 - \pi \approx .8584073464$$

In the second quadrant: π − reference angle ≈ 2.283185307

Therefore, Arccos(cos 4) ≈ 2.283185307. By calculator: 2.283185307.

d. sin(Arcsin 2.463) is not defined since 2.463 is not between −1 and +1.

This angle is Arccos(cos 4).

Reference angle

Reference angle

FIGURE 4.9

EXAMPLE 5 **Evaluation by using fundamental identities**

Evaluate $\cot[\text{Csc}^{-1}(-3)]$ without tables or a calculator.

SOLUTION Let θ = $\text{Csc}^{-1}(-3)$. θ is in Quadrant IV.

Then csc θ = −3.

Find $\cot[\text{Csc}^{-1}(-3)]$ = cot θ Substitute θ = $\text{Csc}^{-1}(-3)$.

$$= \pm\sqrt{\csc^2\theta - 1}$$ Fundamental identity

$$= -\sqrt{(-3)^2 - 1}$$ Substitute csc θ = −3; choose the negative value for the radical because θ is in Quadrant IV and cotangent is negative in Quadrant IV.

$$= -\sqrt{8}$$

$$= -2\sqrt{2}$$

EXAMPLE 6 **Evaluation by using the addition identities**

Evaluate $\cos\left(\text{Sin}^{-1}\frac{4}{5} + \text{Cos}^{-1}\frac{3}{5}\right)$ without tables or a calculator.

SOLUTION Let $x = \text{Sin}^{-1}\frac{4}{5}$ and $y\ \text{Cos}^{-1}\frac{3}{5}$; then

$$\sin x = \frac{4}{5} \quad \text{and} \quad \cos y = \frac{3}{5}$$

where x and y are in Quadrant I. Find $\cos(x + y) = \cos x \cos y - \sin x \sin y$.

We know $\sin x$, so we need to find $\cos x$:

$$\cos x = \sqrt{1 - \sin^2 x} \qquad \cos x \text{ is positive in Quadrant I.}$$
$$= \sqrt{1 - \frac{16}{25}}$$
$$= \frac{3}{5}$$

We also know $\cos y$, so we need to find $\sin y$:

$$\sin y = \sqrt{1 - \cos^2 y} \qquad \sin y \text{ is positive in Quadrant I.}$$
$$= \sqrt{1 - \frac{9}{25}}$$
$$= \frac{4}{5}$$

Therefore,
$$\cos(x + y) = \left(\frac{3}{5}\right)\left(\frac{3}{5}\right) - \left(\frac{4}{5}\right)\left(\frac{4}{5}\right)$$
$$= \frac{9}{25} - \frac{16}{25}$$
$$= -\frac{7}{25}$$

EXAMPLE 7 **Proving an identity involving an inverse function**

Let x be positive. Prove $\sin(\text{Arccos } x) = \sqrt{1 - x^2}$.

SOLUTION Let $\theta\ \text{Arccos } x$; then $\cos \theta = x$ and θ is in Quadrant I. Find

$$\sin(\text{Arccos } x) = \sin \theta \qquad \text{Substitution}$$
$$= \pm\sqrt{1 - \cos^2\theta} \qquad \text{Fundamental identity}$$
$$= \sqrt{1 - x^2} \qquad \text{Substitute csc } \theta = x \text{ and choose the positive value for the radical because } \theta \text{ is in Quadrant I.}$$

4.3 Problem Set

A *In each of Problems 1–6 obtain the given angle (in radians) from memory. Use exact values.*

1. a. $\text{Arcsin } 0$ **b.** $\text{Cos}^{-1}\frac{\sqrt{3}}{2}$

   c. $\text{Sin}^{-1} 1$ **d.** $\text{Arccos } 1$

2. a. $\text{Cos}^{-1}\frac{\sqrt{2}}{2}$ **b.** $\text{Arcsin }\frac{1}{2}$

   c. $\text{Arcsin}\left(\frac{1}{2}\sqrt{2}\right)$ **d.** $\text{Sin}^{-1}\frac{\sqrt{3}}{2}$

3. a. $\text{Arcsin}(-1)$ **b.** $\text{Arccos}\left(-\frac{\sqrt{2}}{2}\right)$

   c. $\text{Sin}^{-1}\left(-\frac{1}{2}\sqrt{2}\right)$ **d.** $\text{Cos}^{-1}\left(-\frac{1}{2}\right)$

4. a. $\text{Tan}^{-1}\frac{\sqrt{3}}{3}$ **b.** $\text{Arccot }\sqrt{3}$

   c. $\text{Arctan }\sqrt{3}$ **d.** $\text{Tan}^{-1} 1$

' 5. a. $\text{Arccot }\frac{\sqrt{3}}{3}$ **b.** $\text{Arccot } 1$

   c. $\text{Arccsc }\sqrt{2}$ **d.** $\text{Cot}^{-1}(-1)$

6. a. $\text{Arcsec } 2$ **b.** $\text{Arccsc}(-2)$

   c. $\text{Arcsec}\left(-\sqrt{2}\right)$ **d.** $\text{Arccot }\frac{\sqrt{3}}{3}$

Use a calculator to find the values in Problems 7–12 to the nearest degree.

7. a. $\text{Sin}^{-1}\,.2588$ **b.** $\text{Arccos}\,.3907$
   c. $\text{Cos}^{-1}\,.9397$

8. a. $\text{Sin}^{-1}(-.4695)$ **b.** $\text{Cos}^{-1}\,.3584$
   c. $\text{Arccos}(-.2672)$

9. a. $\text{Arccos}(-.9598)$ **b.** $\text{Arcsin}(-.6858)$
   c. $\text{Cos}^{-1}(-.3584)$

10. a. $\text{Tan}^{-1}\,.123$ **b.** $\text{Cot}^{-1}\,.341$
   c. $\text{Sec}^{-1}\,3.415$

11. a. $\text{Csc}^{-1}\,2.816$ **b.** $\text{Arccsc}\,3.945$
   c. $\text{Arccot}(-1)$

12. a. $\text{Arccot}(-2)$ **b.** $\text{Arctan}(-3)$
   c. $\text{Arcsec}(-6)$

Use a calculator to find the values (in radians) given in each of Problems 13–18. Round your answers to the nearest hundredth.

13. a. $\text{Arcsin}\,.20846$ **b.** $\text{Cos}^{-1}\,.83646$
   c. $\text{Arccos}\,.94604$

14. a. $\text{Sin}^{-1}(-.32404)$ **b.** $\text{Arcsin}(-.35227)$
   c. $\text{Cos}^{-1}(-.7074)$

15. a. $\text{Cos}^{-1}(-.36236)$ **b.** $\text{Arccos}(-.80210)$
   c. $\text{Arcsin}(-.38942)$

16. a. $\text{Tan}^{-1}\,1.489$ **b.** $\text{Cot}^{-1}\,3.451$
   c. $\text{Sec}^{-1}\,4.315$

17. a. $\text{Csc}^{-1}\,5.791$ **b.** $\text{Arccsc}\,2.985$
   c. $\text{Arccot}(-3)$

18. a. $\text{Arccot}(-4)$ **b.** $\text{Arctan}(-2)$
   c. $\text{Arcsec}(-5)$

B *Evaluate the expressions in Problems 19–30 without using tables or a calculator.*

19. a. $\text{Arccos}\left(\cos\frac{\pi}{6}\right)$ **b.** $\sin\left(\text{Arcsin}\frac{1}{3}\right)$
   c. $\text{Arcsin}\left(\sin\frac{2\pi}{15}\right)$

20. a. $\cos\left(\text{Arccos}\frac{2}{3}\right)$ **b.** $\sin(\text{Arcsin}\,.5)$
   c. $\cos(\text{Arccos}\,.4)$

21. $\tan[\text{Sec}^{-1}(-3)]$ **22.** $\tan[\text{Sec}^{-1}(-4)]$

23. $\cos\left[\text{Tan}^{-1}\left(-\frac{3}{4}\right)\right]$ **24.** $\sin\left[\text{Tan}^{-1}\left(-\frac{3}{4}\right)\right]$

25. $\sin\left[\text{Cot}^{-1}\left(-\frac{3}{4}\right)\right]$ **26.** $\cos\left[\text{Cot}^{-1}\left(-\frac{3}{4}\right)\right]$

27. $\sin\left(\text{Sin}^{-1}\frac{4}{5} - \text{Cos}^{-1}\frac{3}{5}\right)$ **28.** $\cos\left(\text{Sin}^{-1}\frac{3}{5} + \text{Cos}^{-1}\frac{4}{5}\right)$

29. $\cos\left(\text{Cos}^{-1}\frac{1}{4} - \text{Sin}^{-1}\frac{1}{3}\right)$ **30.** $\sin\left(\text{Cos}^{-1}\frac{2}{3} + \text{Sin}^{-1}\frac{1}{2}\right)$

C

31. Prove Identity 36.   **32.** Prove Identity 37.

33. Prove $\sec(\text{Arctan}\,x) = \sqrt{x^2 + 1}$ for $x \geq 0$.

34. Prove $\tan(\text{Arcsec}\,x) = \sqrt{x^2 - 1}$ for $x \geq 0$.

35. Prove $\cos(\text{Arctan}\,x) = 1/\sqrt{1 + x^2}$ for $x \geq 0$.

36. Prove $\text{Arccot}\,x = \pi + \text{Arctan}\frac{1}{x}$ if x is negative.

37. Graph $y = \text{Cot}^{-1}\,x$ on your graphing calculator. It is not the same as the graph shown in Figure 4.8. Reconcile those results.

4.4   A Reduction Identity

One application of inverse functions is useful in graphing, in calculus, and in certain engineering applications. It involves an identity for changing the form of a trigonometric expression.

REDUCTION IDENTITY

38. $a \sin\theta + b \cos\theta = \sqrt{a^2 + b^2}\,\sin(\theta + \alpha)$, where α is chosen so that

$$\cos\alpha = \frac{a}{\sqrt{a^2 + b^2}} \quad\text{and}\quad \sin\alpha = \frac{b}{\sqrt{a^2 + b^2}}$$

The proof of this identity is straightforward after you multiply by 1, written in the form

$$\frac{\sqrt{a^2 + b^2}}{\sqrt{a^2 + b^2}}$$

$$a \sin \theta + b \cos \theta = \frac{\sqrt{a^2 + b^2}}{\sqrt{a^2 + b^2}} \, (a \sin \theta + b \cos \theta)$$

$$= \sqrt{a^2 + b^2} \left(\frac{a}{\sqrt{a^2 + b^2}} \sin \theta + \frac{b}{\sqrt{a^2 + b^2}} \cos \theta \right)$$

$$= \sqrt{a^2 + b^2} \left(\sin \theta \, \frac{a}{\sqrt{a^2 + b^2}} + \cos \theta \, \frac{b}{\sqrt{a^2 + b^2}} \right)$$

Now, let $\alpha = \cos^{-1} \left(\dfrac{a}{\sqrt{a^2 + b^2}} \right)$ so that $\cos \alpha = \dfrac{a}{\sqrt{a^2 + b^2}}$. Then,

$$\sin^2 \alpha = 1 - \cos^2 \alpha$$

$$= 1 - \left(\frac{a}{\sqrt{a^2 + b^2}} \right)^2$$

$$= 1 - \frac{a^2}{a^2 + b^2}$$

$$= \frac{a^2 + b^2 - a^2}{a^2 + b^2}$$

$$= \frac{b^2}{a^2 + b^2}$$

and

$$\sin \alpha = \frac{b}{\sqrt{a^2 + b^2}}$$ If α is in Quadrant I or II, b is positive; if α is in Quadrant III or IV, b is negative; so $\pm$ signs are not necessary.

Therefore, by substitution,

$$a \sin \theta + b \cos \theta = \sqrt{a^2 + b^2} \, (\sin \theta \cos \alpha + \cos \theta \sin \alpha)$$

$$= \sqrt{a^2 + b^2} \, \sin(\theta + \alpha)$$

EXAMPLE 1   Using the reduction identity

Change $\dfrac{\sqrt{3}}{2} \sin \theta + \dfrac{1}{2} \cos \theta$ using the reduction identity.

SOLUTION   By inspection $a = \frac{\sqrt{3}}{2}$ and $b = \frac{1}{2}$, so $\sqrt{a^2 + b^2} = \sqrt{\frac{3}{4} + \frac{1}{4}} = 1$. Find α so that $\cos \alpha = \frac{\sqrt{3}}{2}$ and $\sin \alpha = \frac{1}{2}$. The reference angle α' is $\frac{\pi}{6}$, and since cosine and sine are both positive, α is in Quadrant I: $\alpha = \frac{\pi}{6}$. Therefore,

$$\frac{\sqrt{3}}{2} \sin \theta + \frac{1}{2} \cos \theta = \sin\left(\theta + \frac{\pi}{6} \right)$$

EXAMPLE 2   Using the reduction identity

Change $- \sin \dfrac{\theta}{2} - \cos \dfrac{\theta}{2}$ by using the reduction identity.

SOLUTION $a = -1$ and $b = -1$ so that $\sqrt{a^2 + b^2} = \sqrt{2}$. Find α so that $\cos \alpha = \dfrac{-1}{\sqrt{2}}$ and $\sin \alpha = \dfrac{-1}{\sqrt{2}}$. The reference angle α' is $\dfrac{\pi}{4}$, and since cosine and sine are both negative, α is in Quadrant III: $\alpha = \dfrac{5\pi}{4}$. Therefore,

$$- \sin \dfrac{\theta}{2} - \cos \dfrac{\theta}{2} = \sqrt{2} \sin\left(\dfrac{\theta}{2} + \dfrac{5\pi}{4}\right)$$

■

Graphing Sums of Sines and Cosines

You can use the reduction identity when graphing sums of sines and cosines, as shown in Example 3.

EXAMPLE 3   Graphing the reduction identity

Graph $y = \sqrt{3} \cos \theta - \sin \theta$ by first using the reduction identity.

SOLUTION   Compare $y = \sqrt{3} \cos \theta - \sin \theta$ to $a \sin \theta + b \cos \theta$ to see that $a = -1$ and $b = \sqrt{3}$. Thus, $\sqrt{a^2 + b^2} = \sqrt{1 + 3} = 2$. Find α so that

$$\cos \alpha = \dfrac{-1}{2} \quad \text{and} \quad \sin \alpha = \dfrac{\sqrt{3}}{2}$$

The reference angle for α is $\dfrac{\pi}{3}$, and sine is positive while cosine is negative, so α must be in Quadrant II: $\alpha = \dfrac{2\pi}{3}$. Thus,

$$y = \sqrt{3} \cos \theta - \sin \theta = 2 \sin\left(\theta + \dfrac{2\pi}{3}\right)$$

The graph is now easy to frame and complete, as shown in Figure 4.10.

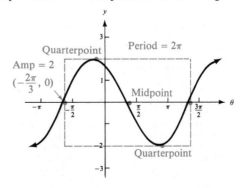

FIGURE 4.10 Graph of $y = \sqrt{3} \cos \theta - \sin \theta$

■

4.4 Problem Set

A *Change each expression in Problems 1–12 by using the reduction identity to write it as a sine function.*

1. $\sin \theta + \cos \theta$

2. $\sin \theta - \cos \theta$

3. $-\sin \frac{\theta}{2} + \cos \frac{\theta}{2}$

4. $\sin \frac{\theta}{2} - \cos \frac{\theta}{2}$

5. $\frac{1}{2}\sin \theta + \frac{\sqrt{3}}{2}\cos \theta$

6. $\frac{1}{2}\sin \theta - \frac{\sqrt{3}}{2}\cos \theta$

7. $-\frac{1}{2}\sin \theta + \frac{\sqrt{3}}{2}\cos \theta$

8. $\cos \theta - \sin \theta$

9. $-\cos \theta - \sin \theta$

10. $\cos 2\theta - \sqrt{3}\sin 2\theta$

11. $\sqrt{3}\cos \pi \theta - \sin \pi \theta$

12. $\frac{\sqrt{3}}{2}\cos \pi \theta - \frac{1}{2}\sin \pi \theta$

B *Change each expression in Problems 13–21 by using the reduction identity. Write α as a real number rounded to the nearest tenth.*

13. $2\sin \theta + 3\cos \theta$

14. $3\sin \theta - 2\cos \theta$

15. $-3\sin \theta + 2\cos \theta$

16. $3\cos \theta - 4\sin \theta$

17. $4\cos \theta - 3\sin \theta$

18. $8\sin \theta + 15\cos \theta$

19. $-8\sin \theta - 15\cos \theta$

20. $15\cos \theta - 8\sin \theta$

21. $5\cos \theta - 2\sin \theta$

Use the reduction identity when graphing each equation in Problems 22–33.

22. $y = \sin \theta + \cos \theta$

23. $y = \sin \theta - \cos \theta$

24. $y = \frac{1}{2}\sin \theta + \frac{\sqrt{3}}{2}\cos \theta$

25. $y = \frac{\sqrt{3}}{2}\sin \theta - \frac{1}{2}\cos \theta$

26. $y = 2\sin \theta - 3\cos \theta$

27. $y = 3\sin \theta + 2\cos \theta$

28. $y = \cos \theta - \sin \theta$

29. $y = -\sin \theta - \cos \theta$

30. $y = 8\sin \theta + 15\cos \theta$

31. $y = -8\sin \theta - 15\cos \theta$

32. $y = 2\sin \theta + 3\cos \theta$

33. $y = 3\sin \theta - 2\cos \theta$

4.5 Summary and Review

Objectives	*Pages / Examples / Problems*
4.1 Introduction to Trigonometric Equations	
1. Solve trigonometric equations in linear form.	pp. 136–141; Examples 1–6; Problems 1–11, 33.
2. Solve trigonometric equations by factoring.	pp. 140–143; Examples 6, 7, and 9; Problems 12–19, 23–24.
3. Solve trigonometric equations by using the quadratic formula.	pp. 140–143; Examples 6 and 8; Problems 20–22, 25–28, 31–32.
4. Solve trigonometric equations by using trigonometric identities.	pp. 143–144; Examples 9 and 10; Problems 29–30, 34–39.
4.2 Trigonometric Equations with Multiple Angles	
5. Solve trigonometric equations with multiple angles.	pp. 147–149; Example 1; Problems 1–27.
6. Solve trigonometric equations with multiple angles by using trigonometric identities.	p. 149; Example 2; Problems 28–33.
4.3 Inverse Trigonometric Functions	
7. Evaluate inverse trigonometric functions by knowing the quadrants for principal values.	pp. 152–156; Examples 1–3; Problems 1–18.
8. Define the inverse trigonometric relations and the inverse trigonometric functions, including the domain and the range. Draw a quick sketch of each.	p. 136 and pp. 153–158.
9. Evaluate functions of inverse trigonometric functions.	pp. 159–160; Examples 4–6; Problems 19–30.

4.4 **A Reduction Identity**

10. Use the reduction identity to simplify certain trigonometric equations.

pp. 161–163; Examples 1 and 2; Problems 1–21.

11. Use the reduction identity to graph certain trigonometric equations.

p. 163; Example 3; Problems 22–33.

TERMS

arccosecant [4.1]	arccotangent [4.1]	arcsine [4.1]	Inverse trigonometric functions [4.3]
Arccosecant [4.3]	Arccotangent [4.3]	Arcsine [4.3]	primary angles [4.1]
arccosine [4.1]	arcsecant [4.1]	arctangent [4.1]	Reduction identity [4.4]
Arccosine [4.3]	Arcsecant [4.3]	Arctangent [4.3]	

CHAPTER 4 REVIEW OF OBJECTIVES

The problem numbers correspond to the objectives listed in Section 4.5.

1. Find all real values of θ so that $0 \le \theta < 2\pi$ and $\tan \theta = 1.557407725$.

2. Solve $4 \cos^2\theta = 1$ for $0 \le \theta < 2\pi$.

3. Solve $\tan^2\theta - 2 \tan \theta - 3 = 0$ for $0 \le \theta < 2\pi$ using radians correct to two decimal places.

4. Solve $\cos^2\theta = 3 \sin \theta$ for $0 \le \theta < 2\pi$ using real numbers correct to two decimal places.

5. Solve $3 \cos^2 2\theta = 1$ for $0 \le \theta < 6.28$, correct to two decimal places.

6. Solve $2 \sin^2 2\theta - 2 \cos^2 2\theta = 1$ for $0 \le \theta < 2\pi$ (exact values).

7. Simplify the expressions; in parts **a–d** use exact values in degrees. In parts **e–h** use exact values in radians. In parts **i** and **j** use real numbers correct to two decimal places.

a. $\text{Arccos} \dfrac{\sqrt{3}}{2}$ **b.** $\text{Arctan } 1$

c. $\text{Arccot} \dfrac{\sqrt{3}}{3}$ **d.** $\text{Cos}^{-1} \dfrac{\sqrt{2}}{2}$

e. $\text{Cos}^{-1}\left(-\dfrac{\sqrt{2}}{2}\right)$ **f.** $\text{Cot}^{-1}\left(-\sqrt{3}\right)$

g. $\text{Arcsin} \dfrac{1}{2}$ **h.** $\text{Arctan} \dfrac{\sqrt{3}}{3}$

i. $\text{Tan}^{-1} 2.310$ **j.** $\text{Arcsec } 3.485$

8. In the table below, fill in the blanks indicated by the lower-case letters.

FUNCTION	SIGN OF x	QUADRANT OF θ	APPROXIMATE VALUE OF θ
$\theta = \text{Arccos } x$	Positive	I	θ between 0 and 1.57
$\theta = \text{Arccos } x$	Negative	II	θ between 1.57 and 3.14
$\theta = \text{Arcsin } x$	Positive	I	θ between 0 and 1.57
$\theta = \text{Arcsin } x$	Negative	IV	θ between 1.57 and 0 (a negative number or angle)
$\theta = \text{Arctan } x$	Positive	**a.** _____	θ between 0 and 1.57
$\theta = \text{Arctan } x$	Negative	IV	**b.** _____
$\theta = \text{Arcsec } x$	Positive	**c.** _____	**d.** _____
$\theta = \text{Arcsec } x$	Negative	**e.** _____	θ between 1.57 and 3.14
$\theta = \text{Arccsc } x$	Positive	I	**f.** _____
$\theta = \text{Arccsc } x$	Negative	**g.** _____	**h.** _____
$\theta = \text{Arccot } x$	Positive	I	θ between 0 and 1.57
$\theta = \text{Arccot } x$	Negative	**i.** _____	**j.** _____

9. Evaluate $\tan(\operatorname{Sec}^{-1} 3)$ without using tables or a calculator.

10. In the reduction identity

$$a \sin \theta + b \cos \theta = \sqrt{a^2 + b^2} \sin(\theta + \alpha),$$

what is the proper quadrant for α, given the conditions in parts **a–d**?

a. a and b are both positive.   **b.** a and b are both negative.

c. a is positive, b is negative.

d. a is negative, b is positive.

e. Use the reduction identity to write $12 \sin \theta - 5 \cos \theta$ as a function of sine; approximate α to the nearest tenth.

11. Use the results of Problem 10 to graph
$y = 12 \sin \theta - 5 \cos \theta$

DISCUSSION PROBLEMS

1. Compare and contrast trigonometric equations and identities.

2. Outline your "method of attack" for solving a trigonometric equation.

3. In your own words explain inverse trigonometric relations and inverse trigonometric functions.

4. Find a mathematics book on trigonometry written in a foreign language that you do not read or understand. Locate a page in that book that contains some trigonometry that you do understand, and explain in your own words what is going on.

5. Write a 500-word paper based on the historical note found on page 171.

MISCELLANEOUS PROBLEMS

Fill in the domain and range for each of the given inverse trigonometric functions in Problems 1–6.

	DOMAIN	RANGE
1. $y = \operatorname{Arccos} x$	_____	_____
2. $y = \operatorname{Arcsin} x$	_____	_____
3. $y = \operatorname{Arctan} x$	_____	_____
4. $y = \operatorname{Arcsec} x$	_____	_____
5. $y = \operatorname{Arccsc} x$	_____	_____
6. $y = \operatorname{Arccot} x$	_____	_____

Evaluate the expressions in Problems 7–30. Use exact values (radians) where possible; otherwise give your answer as a real number correct to two decimal places.

7. $\sin^{-1}\left(-\frac{\sqrt{3}}{2}\right)$ **8.** $\arccos \frac{1}{2}$

9. $\arcsin .813$ **10.** $\arccos(-4.521)$

11. $\cos^{-1}(-.3592)$ **12.** $\arctan(-1)$

13. $\operatorname{Arccot}(-1)$ **14.** $\operatorname{Cot}^{-1} \pi$

15. $\operatorname{Cot}^{-1}(-\pi)$ **16.** $\operatorname{Tan}^{-1}(-\pi)$

17. $\operatorname{Tan}^{-1} \pi$ **18.** $\operatorname{Arcsec}(-\sqrt{2})$

19. $\operatorname{Arccsc}(-\sqrt{2})$ **20.** $\operatorname{Arccsc}(-3.4589)$

21. $\operatorname{Sec}^{-1}(-4.3298)$ **22.** $\operatorname{Sec}^{-1}\frac{1}{2}$

23. $\operatorname{Arccot}\left(-\frac{3}{5}\right)$ **24.** $\operatorname{Arctan}\left(-\frac{3}{5}\right)$

25. $\operatorname{Arcsin}(\sin 4)$ **26.** $\operatorname{Arccos}(\cos 5)$

27. $\operatorname{Arcsin}(\sin 6)$ **28.** $\sin(\operatorname{Arcsin} .7568)$

29. $\cos(\operatorname{Arccos} .2836)$ **30.** $\cos(\operatorname{Arccos} .4567)$

Solve the equations in Problems 31–50 for $0 \le \theta < 2\pi$. Use exact values (radians) where possible. Otherwise give your answer as a real number correct to two decimal places.

31. $\sin \theta = \frac{1}{2}$ **32.** $\frac{1}{2} \sin \theta = 0$

33. $\sin\left(\frac{1}{2}\theta\right) = 0$ **34.** $\cos \theta = \frac{1}{5}$

35. $4 \cos \theta = 1$ **36.** $\cos 4\theta = 1$

37. $(3 \cos \theta - 1)(2 \cos \theta + 1) = 0$

38. $(4 \sin \theta + 1)(3 \sin \theta - 1) = 0$

39. $\sin^2\theta - 1 = 0$

40. $2 \cos^2\theta = \cos \theta$

41. $6 \cos^2\theta - 7 \cos \theta + 2 = 0$

42. $12 \sin^2\theta - \sin \theta - 6 = 0$

43. $6 \sec^2\theta - 7 \sec \theta + 2 = 0$

44. $12 \tan^2\theta - \tan \theta - 6 = 0$

45. $3 \sin^2\theta - 8 \sin \theta + 4 = 0$

46. $5 \cos^2\theta + 14 \cos \theta - 4 = 0$

47. $3 \cot^2\theta - 8 \cot \theta + 4 = 0$

48. $5 \csc^2\theta + 14 \csc \theta - 4 = 0$

49. a. $2 \cos \theta - 1 = 0$ **b.** $2 \cos(\theta - 1) = 0$

50. a. $5 \sin \theta + 2 = 0$ **b.** $5 \sin(\theta + 2) = 0$

▌ Waves, Harmonic Motion, and Resonance

Courtesy of Photo Researchers, Inc.

Water Waves

Wave motion is one of the most common examples of periodic motion. There are water waves, sound waves, and radio waves, as well as many other electromagnetic waves. Sometimes waves are classified by considering how the motions of the particles of matter are related to the direction of the waves themselves. This classification system yields waves called **transverse** waves and others called **longitudinal** waves. In a transverse wave, the particles vibrate at right angles to the direction in which the wave itself is propagated. In a longitudinal wave, the particles vibrate in the same direction as that in which the wave is propagated. Both types of waves are shown in Figure 4.11.

Transverse waves   Longitudinal waves

FIGURE 4.11 Transverse and longitudinal waves

CALCULATOR COMMENT

If we assume that x is held constant (say, $x = 50$ft), then the following graph represents the water level in Example 3.

If we hold t constant (photograph waves when $t = 5$ sec), we obtain a picture at the waves:

If we look at water waves, we find the wave action to be a combination of both transverse and longitudinal waves. The water molecules move back and forth and, at the same time, up and down. The path of the molecules makes a circle or an ellipse. (See Figure 4.12.) In their simplest form, water waves can be shown to be sine curves represented by an equation of the form

$$y = \alpha \sin \frac{2\pi}{\lambda}(x - \lambda nt)$$

where

a = amplitude of the wave;

n = frequency (defined as the number of waves passing any given point per second; it is customary to express frequency in vibrations per second or in cycles per second);

λ = wavelength (representing the distance between two adjacent points in the wave that have the same phase);

t = time;

x = distance from source.

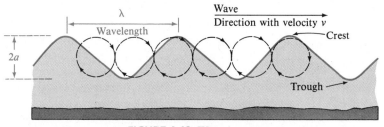

FIGURE 4.12 Water waves

If T is the period of the wave, then

$$T = \frac{1}{n}$$

and it can be shown that the phase velocity (the speed of the wave) is

$$v = \sqrt{\frac{g\lambda}{2\pi}} \quad \text{or} \quad v = \frac{\lambda}{T}$$

in feet per second, where $g = 32$ ft/sec^2. Thus,

$$v \approx \sqrt{5.09\lambda} \quad \text{or} \quad v = \lambda n$$

EXAMPLE 1   Solving a wave equation

The wave equation can also be written as

$$y = a \sin \frac{2\pi}{\lambda}(x - vt)$$

Solve for v.

SOLUTION

$$\frac{y}{a} = \sin\frac{2\pi}{\lambda}(x - vt)$$

$$\sin^{-1}\left(\frac{y}{a}\right) = \frac{2\pi}{\lambda}(x - vt)$$

$$\lambda \sin^{-1}\frac{y}{a} = 2\pi(x - vt)$$

$$\lambda \sin^{-1}\frac{y}{a} = 2\pi x - 2\pi vt$$

$$2\pi vt = 2\pi x - \lambda \sin^{-1}\left(\frac{y}{a}\right)$$

$$v = \frac{2\pi x - \lambda \sin^{-1}\left(\frac{y}{a}\right)}{2\pi t}$$

EXAMPLE 2   Solving a trigonometric equation

Find λ in terms of T.

SOLUTION $v \approx \sqrt{5.09\lambda}$ and $v = \frac{\lambda}{T}$, so

$$\frac{\lambda}{T} \approx \sqrt{5.09\lambda}$$

$$\frac{\lambda^2}{T^2} \approx 5.09\lambda$$

$$\lambda^2 - 5.09\lambda T^2 \approx 0$$

$$\lambda(\lambda - 5.09T^2) \approx 0$$

$$\lambda \approx 0, \lambda \approx 5.09T^2$$

$\underset{\text{Trivial solution}}{\uparrow\!\!\!\!\!\_\_\_}$

EXAMPLE 3   Finding an equation for ocean waves

Suppose you are watching waves pass a pier piling, and you count 20 waves per minute, each with an amplitude of 2 ft. Find the wavelength and phase velocity (in mph), and write the equation for one of these waves.

SOLUTION   20 waves per minute is $\frac{1}{3}$ wave per second, so $n = \frac{1}{3}$ and $T = 3$. Then, from Example 2,

$$\lambda = 5.09(3^2)$$

$$\approx 45.81 \qquad \text{This is the wavelength (in ft).}$$

and

$$v = \frac{\lambda}{T}$$

$$= \frac{45.81}{3}$$

$$= 15.27 \text{ ft/sec} \qquad \text{This is the phase velocity.}$$

To change to miles per hour, write

$$\frac{15.27 \text{ ft}}{\text{sec}} = \frac{15.27 \text{ ft}}{\text{sec}} \cdot \frac{60 \text{ sec}}{1 \text{ min}} \cdot \frac{60 \text{ min}}{1 \text{ hr}} \cdot \frac{1 \text{ mi}}{5{,}280 \text{ ft}}$$

These fractions are all one, so multiplication does not change the original value.

$$= \frac{15.27 \cdot 60 \cdot 60}{5{,}280} \cdot \frac{\text{sec}}{\text{sec}} \cdot \frac{\text{min}}{\text{min}} \cdot \frac{\text{ft}}{\text{ft}} \cdot \frac{\text{mi}}{\text{hr}}$$

Use a calculator.

These are all one. They are from a rearrangement of the factors from the line above.

This is abbreviated mph.

$$\approx 10.41 \text{ mph} \qquad \text{This is the phase velocity (in mph).}$$

The equation is

$$y = 2 \sin \frac{2\pi}{45.81} (x - 15.27t)$$

$$y = 2 \sin .14(x - 15.27t)$$

Harmonic Waves

In Section 2.4, we spoke of modulation and beats. Harmonic wave motion and resonance are related to these ideas. The displacement of a particle that can be expressed in terms of an equation of the form

$$y = a \cos(\omega t + h)$$

or

$$y = a \sin(\omega t + h)$$

where a, ω, and h are constants, is said to be in **simple harmonic motion**. However, many moving bodies do not move back and forth between precisely fixed limits because friction slows down their motion. Such motions are called **damped harmonic motions**. The period T of a harmonic motion is the time required to complete one round trip of the motion, and the frequency n is the number of oscillations or cycles per unit of time so that

$$n = \frac{1}{T} = \frac{\omega}{2\pi}$$

HISTORICAL NOTE*

It might be interesting to look at a comparison of the sun and the moon as tide-generating bodies. Even though the gravitational attraction between the earth and the sun is much greater than between the earth and moon (over 177 times greater), the moon dominates the tides, as shown by the calculation below, which uses the formula

$$\text{tide-generating force} \approx \frac{m_1 m_2}{r^3}$$

where m_1 and m_2 are the masses of the bodies and r is the distance between the bodies.

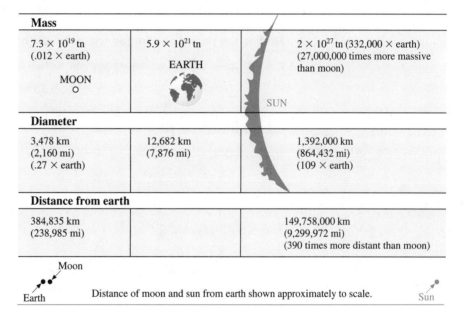

Mass		
7.3×10^{19} tn (.012 × earth) **MOON** ○	5.9×10^{21} tn **EARTH**	2×10^{27} tn (332,000 × earth) (27,000,000 times more massive than moon) SUN
Diameter		
3,478 km (2,160 mi) (.27 × earth)	12,682 km (7,876 mi)	1,392,000 km (864,432 mi) (109 × earth)
Distance from earth		
384,835 km (238,985 mi)		149,758,000 km (9,299,972 mi) (390 times more distant than moon)

Moon
Earth   Distance of moon and sun from earth shown approximately to scale.   Sun

DETERMINATION OF TIDE-GENERATING FORCE OF SUN RELATIVE TO MOON

$$\frac{\text{tide-generating force of sun}}{\text{tide-generating force of moon}} = \frac{\text{mass of sun/mass of moon}}{(\text{sun's distance/moon's distance})^3}$$

$$\approx \frac{27 \text{ million}}{(390)^3}$$

$$\approx .46 \text{ or } 46\%$$

The sun has 46% the tide-generating force of the moon.

*Historical Note is from *Introductory Oceanography,* 5th ed., by H. V. Thurman, p. 248. Reprinted with permission of Merrill, an imprint of Macmillan Publishing Company. Copyright ©1988 Merrill Publishing Company, Columbus, Ohio.

The effect of damping can be seen by looking at the graphs shown in Figure 4.13.

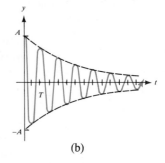

(a)   (b)

FIGURE 4.13 (a) Harmonic motion (amplitude a constant) and
(b) damped harmonic motion (amplitude a decreasing variable)

EXAMPLE 4   **Finding the period and frequency for simple harmonic motion**

If a particle is moving in simple harmonic motion according to the equation

$$y = 2 \cos 18\pi t$$

what are the period, frequency, and maximum distance from the origin when $t = 0$?

SOLUTION   By inspection, $\omega = 18\pi$, so

$$n = \frac{\omega}{2\pi} = \frac{18\pi}{2\pi} = 9$$

and

$$T = \frac{1}{n} = \frac{1}{9}$$

Therefore the period is $\frac{1}{9}$, and the frequency is 9 cycles per unit of time. If $t = 0$, then $y = 2$, so the maximum distance from the origin when $t = 0$ is 2 units.

> **CALCULATOR COMMENT**
>
> If you graph the equation of Example 4 you will need to pay attention to the scale.
>
> ```
> Xmin=0      Ymin=-8
> Xmax=.2     Ymax=8
> ```
>
>
>
> Period $= \frac{1}{9}$

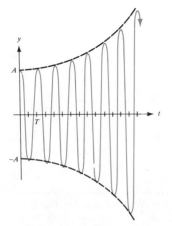

FIGURE 4.14 Resonance:
Amplitude an increasing variable

Resonance

Notice in Figure 4.13 that, for damped harmonic motion, the amplitude decreases toward zero. The reverse situation, when the amplitude increases to a maximum, is called **resonance** and has a graph as shown in Figure 4.14.

In physics it is shown that the amplitude of the vibrations is large compared to the amount of static displacement. For example, it would be possible for soldiers marching in step across a suspension bridge to make it vibrate with a very large destructive amplitude, provided that the frequency of their steps happened to be the same as the natural frequency of the bridge. This is why marching soldiers break step when crossing a bridge.

A striking example of the destructive force of resonance is the collapse of the Tacoma Narrows Bridge, which was opened on July 1, 1940. Early on the morning of November 7, 1940, the wind velocity was 40 to 45 mi/hr—perhaps a greater velocity than any previously encountered by the bridge. Traffic was shut down. By 9:30 A.M. the span was vibrating in eight or nine segments, with frequency 36 vib/min and amplitude about 3 ft. While measurements were under way, at about 10:00 A.M. the main span abruptly began to vibrate torsionally in two segments with frequency 14 vib/min. The amplitude of torsional vibrations quickly built up to 35° each direction from horizontal. The main span broke up shortly after 11:00 A.M. During most of the catastrophic torsional vibration there was a transverse nodal line at midspan and a longitudinal nodal line down the center of the roadway.* (See Figure 4.15.)

Courtesy of Wide World Photos, Inc.

FIGURE 4.15 The Tacoma Narrows Bridge at Puget Sound, Washington

PROBLEMS FOR FURTHER STUDY: WAVES

1. **Physics** If a water wave has an amplitude of 3 ft and you count 10 waves per minute, what are the wavelength, phase velocity (in mph), and equation of the wave?

2. **Physics** If a water wave has an amplitude of 35 ft with a period of 10 seconds, what are its wavelength, its phase velocity (in mph), and its equation?

3. **Physics** If a water wave has the equation

$$y = 12 \sin 1.32(x - 4.75t)$$

how high is the wave from trough to crest? What are the wavelength and phase velocity (in mph) of the wave?

4. **Physics** If a water wave has the equation

$$y = 18 \sin .62(x - 4.50t)$$

how high is the wave from trough to crest? What are the wavelength and phase velocity (in mph)?

5. Solve the equation given in Problem 3 for t.

6. Solve the equation given in Problem 4 for t.

7. **Physics** If a particle is moving in simple harmonic motion according to the equation

$$y = \sqrt{3} \sin\left(10\pi t - \frac{\pi}{2}\right)$$

what are the period and frequency, and what is the maximum distance from the origin when $t = 0$?

8. **Physics** If a particle is moving in simple harmonic motion with period 4π and constant $h = 0$, write an equation describing the distance y of the particle from the origin if the maximum distance from the origin is 5 units.

*A filmstrip *The Tacoma Narrows Bridge Collapse* (#80-218) is available from The Ealing Corporation, Cambridge, Mass. 02140, which provided the information about the bridge.

9. Physics Graph the given equations for $t \geq 0$, and classify each motion as harmonic motion, damped harmonic motion, or resonance.

a. $y = 3 \cos 4\pi t$ **b.** $y = 3t \cos 4\pi t$

c. $y = \dfrac{3}{t} \cos 4\pi t$

10. In calculus it is shown that if a particle is moving in simple harmonic motion according to the equation

$$y = a \cos(\omega t + h)$$

then the velocity of the particle at any time t is given by

$$v = a\omega \sin(\omega t + h)$$

Graph this velocity curve, where the frequency is $\frac{1}{4}$ cycle per unit of time, $h = \frac{\pi}{6}$, and $a = \frac{120}{\pi}$.

11. Use Problem 10 to find a maximum value and a minimum value for the velocity.

Graph the curves given in Problems 12–17, and state whether there is damping, resonance, neither, or both.

12. $y = \cos x + x \sin x$

13. $y = \sin x - x \cos x$

14. $y = \dfrac{9}{4} \sin \dfrac{9}{8} x + \sin \dfrac{1}{8} x$

15. $y = \dfrac{9}{2} \cos \dfrac{9}{4} x + \cos \dfrac{1}{4} x$

16. $y = \dfrac{13}{5} \cos x - \sin x - \dfrac{8}{5} \cos \dfrac{3}{2} x$

17. $y = 2 \sin\left(5x - \dfrac{\pi}{2}\right) + \sin x$

▮ Chapters 3–4

Sample Test I: Objective Form

1. *State* the eight fundamental identities.

2. *Prove* $\cot^2\theta + 1 = \csc^2\theta$.

3. *Given* a unit circle, as shown in the figure below,
    a. What are the coordinates of the points A and P_θ?
    b. Use the distance formula to find the length of chord AP_θ.

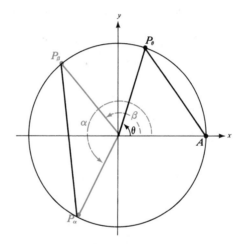

4. *Prove* $(\cos\beta - \cos\alpha)^2 + (\sin\beta - \sin\alpha)^2 = 2 - 2(\cos\alpha\cos\beta + \sin\alpha\sin\beta)$.

5. *Prove*
$$\frac{\sin 3\theta + \cos 3\theta + 1}{\cos 3\theta} = \tan 3\theta + \sec 3\theta + 1.$$

6. *Prove* $\sec A - \tan^2 A = \sec^2 A(\cos^2 A + \cos A - 1)$.

7. *Prove* or *disprove* $\tan 2\theta \cos 2\theta = 2\sin\theta\cos\theta$.

8. *Solve* $\cos^2\theta = 3\sin\theta$ for $0 \le \theta < 2\pi$ (in radians correct to three places).

9. *Graph* $y - 3 = 2\sin\left(\frac{\pi}{3} - x\right)$.

10. *Find* the other circular functions so that $\cos\theta = \frac{4}{5}$ when $\sin\theta < 0$.

Sample Test II: Fill In and Multiple Choice

Fill in Problems 1–5 with the word or words to make the statements complete and correct.

1. The three reciprocal identities are _____ , _____ , and _____ .

2. The two ratio identities are _____ and _____ .

3. The three Pythagorean identities are _____ , _____ , and _____ .

4. The first step in the proof of $\cos^2\theta + \sin^2\theta = 1$ is _____ .

5. To multiply $\dfrac{\cos\theta}{1 - \sin\theta}$ by $\dfrac{1 + \sin\theta}{1 + \sin\theta}$ is called _____ .

*Choose the **best** answer from the choices given.*

6. If $\sin\theta = -\sqrt{1 - \cos^2\theta}$ then θ could be in Quadrants

    A. I and II   B. III and IV   C. I and IV   D. II and III   E. none of these

7. $\cot^2 173° - \csc^2 173°$ is

    A. 16.35   B. 0   C. 1   D. 2   E. none of these

8. If $\csc\theta = -\sqrt{\dfrac{10}{3}}$ and $\cos\theta > 0$ then $\tan\theta$ is

    A. $-\dfrac{1}{3}$ B. $\dfrac{3}{10}\sqrt{10}$ C. $-\dfrac{\sqrt{21}}{3}$ D. $-\dfrac{\sqrt{21}}{7}$ E. none of these

9. If $\dfrac{\cos^2\theta(\tan^2\theta + 1)}{\cot^2\theta}$ is changed to sines and cosines and then completely simplified, the result is

    A. $\sin^4\theta + \sin^2\theta$ B. 1   C. $\dfrac{\cos^4\theta}{\sin^4\theta}$ D. $\dfrac{\sin^2\theta}{\cos^2\theta}$ E. none of these

10. $\dfrac{1}{1 + \sin 2\theta} + \dfrac{1}{1 - \sin 2\theta}$ is identical to

    A. $2\sec^2 2\theta$ B. $2\csc^2 2\theta$

    C. 1   D. $\dfrac{2}{\cos^4\theta}$

    E. none of these

11. $\sec^2 A(\cos^2 A + \cos A - 1)$ is identical to

    A. $\sec A - \tan^2 A$ B. $\sec^2 A - \sec A - 1$

    C. $1 + \sec A - 2\sec^2 A$ D. $\sec A - \tan^2 A - 1$

    E. none of these

12. $\dfrac{\cot\theta + \tan\theta}{\sec\theta\csc\theta}$ is identical to

A. 1   B. $\cos 2\theta$ C. $\sin\theta - \cos\theta$ D. $\sin\theta\cos\theta$ E. none of these

13. If $\cos^2\theta + \sin\theta = 0$, then a solution is

A. $\dfrac{1 \pm \sqrt{5}}{2}$ B. 3.808   C. .666   D. 38.2   E. none of these

14. Which of the following statements are false?

A. $\cos^2\theta + \sin^2\theta = 1$ B. $\tan^2\theta = \sec^2\theta - 1$

C. $\cos\theta\sec\theta = 1$ D. $\cot^2\theta - 1 = \csc^2\theta$

E. none of these

15. If $\cos(\alpha + \beta) = \cos\alpha\cos\beta - \sin\alpha\sin\beta$, then an exact value for $\cos 165°$ is

A. $\dfrac{\sqrt{2} - \sqrt{6}}{4}$ B. $\dfrac{\sqrt{6} - \sqrt{2}}{4}$

C. $\dfrac{\sqrt{2} + \sqrt{6}}{4}$ D. $-.9659$

E. none of these

16. If $\alpha = \frac{\pi}{2}$ and $\beta = -\theta$ and these values are substituted into the identity given in Problem 15, then the result is

A. $\cos\left(\frac{\pi}{2} - \theta\right) = \sin(-\theta)$ B. $\cos\left(\frac{\pi}{2} - \theta\right) = -\cos\theta$

C. $\cos\left(\frac{\pi}{2} - \theta\right) = -\sin(-\theta)$ D. $\cos\left(\frac{\pi}{2} - \theta\right) = -\sin\theta$

E. none of these

17. If you use the identity given in Problem 15 to simplify $\cos 3\theta$ the result is

A. $3\cos\theta$ B. $\cos 2\theta\cos\theta - \sin 2\theta\sin\theta$

C. $2\cos^3\theta - 2\sin^2\theta - 1$ D. $2\cos^3\theta - \sin 2\theta\sin\theta - 1$

E. none of these

18. The leftmost point (yet still within the frame) of the graph of $y = -\cos\theta$ is

A. $\left(0, \frac{\pi}{2}\right)$ B. $\left(0, -\frac{\pi}{2}\right)$ C. $(0, 1)$ D. $(0, -1)$ E. none of these

19. The (h, k) point for the graph of $y + 2 = \sin\left(\frac{\pi}{3} - \theta\right)$ is

A. $(0, -2)$ B. $\left(\frac{\pi}{3}, -2\right)$ C. $\left(-\frac{\pi}{3}, -2\right)$ D. $\left(\frac{\pi}{3}, 2\right)$ E. none of these

20. The amplitude for the graph of $y = \sin\theta + \cos\theta$ is

A. 1   B. $\sqrt{2}$ C. $\frac{\pi}{4}$ D. $-\frac{\pi}{4}$ E. none of these

❚ Benjamin Banneker
(1731–1806)

❚ I apprehend you will embrace every opportunity to eradicate that train of absurd and false ideas and opinions which so generally prevail with respect to us [black people] and that your sentiments are concurrent with mine, which are: that one universal Father hath given being to us all; and He not only made us all of one flesh, but that He hath also without partiality afforded us all with these same faculties and that, however diversified in situation or color, we are all the same family and stand in the same relation to Him.

BENJAMIN BANNEKER
in *Black Mathematicians and Their Works*

Analytic trigonometry has its roots in geometry and in the practical application of surveying. Surveying is the technique of locating points on or near the surface of the earth. The Egyptians are credited with being the first to do surveying. The annual flooding of the Nile and its constant destruction of property markers led the Egyptians to the principles of surveying, as noted by the Greek historian Herodotus:

"Sesostris . . . made a division of the soil of Egypt among the inhabitants. . . . If the river carried away any portion of a man's lot . . . the king sent persons to examine, and determine by measurement the exact extent of the loss. . . . From this practice, I think, geometry first came to be known in Egypt, whence it passed into Greece."

The first recorded measurements to determine the size of the earth were also made in Egypt when Eratosthenes measured a meridian arc in 230 B.C. (See Problem 33, Section 1.2.) A modern-day example of a surveyor's dream is the city of Washington, D.C. The center of the city is the Capitol, and the city is divided into four sections: Northwest, Northeast, Southwest, and Southeast. These sections are separated by North, South, and East Capitol streets, which all converge at the Capitol. The initial surveying of this city was done by a group of mathematicians and surveyors who included the first distinguished black mathematician in U.S. history, Benjamin Banneker.

"Benjamin Banneker, born in 1731 in Maryland, was the first American black to be recognized as a mathematician and astronomer. He exhibited unusual talent in mathematics, and with little formal education, produced an almanac and invented a clock. Banneker was born free in a troubled time of black slavery. The right to vote was his by birth. It is important to note that this right to vote was denied him as well as other free black Americans by 1803 in his native state of Maryland. Yet, history must acknowledge him as the first American black man of science."*

*From Virginia K. Newell, Joella H. Gipson, L. Waldo Righ, and Beauregard Stubblefield, *Black Mathematicians and Their Works* (Admore: Dorrance), p. xiv.

SOLVING TRIANGLES

Contents

5.1 Right Triangles

5.2 Law of Cosines

5.3 Law of Sines

5.4 Areas and Volumes

5.5 Vector Triangles

5.6 Vector Operations

5.7 Summary and Review

Application for Further Study
    Parking-Lot Design

Preview

In the first four chapters, we were concerned with one use of trigonometry—using identities to treat the circular functions as important mathematical relationships. In this chapter, we are concerned with another use of trigonometry—solving triangles to find unknown parts. This procedure of **solving a triangle** is useful in surveying, in measuring distances indirectly, in navigation and aviation, as well as in finding areas and volumes. We begin by solving right triangles using the Pythagorean theorem. We then turn to the solution of **oblique** triangles (triangles that are not right triangles). We will derive two trigonometric laws, the *Law of Cosines* and the *Law of Sines*, that allow us to solve certain triangles using a calculator. There is a third trigonometric law that lends itself to logarithmic calculation—the *Law of Tangents*, which is useful when no calculator is at hand. This law is discussed in Problems 41–43 of Problem Set 5.3. The triangle shown in Figure 5.1 illustrates the way we designate the parts of a triangle, with points A, B, and C; sides a, b, and c; and angles α, β and γ.

Perspective

From the time of the Egyptian *rope stretchers* in the 6th century B.C. to celestial navigation of spacecraft in the 20th century, we have used triangles and trigonometry to measure inaccessible distances. The diagram below shows the *gradient* of a road, which uses the tangent function. For example, a gradient of 8% means a difference in height of 8 m over a map distance of 100 m.

From *The VNR Concise Encyclopedia of Mathematics,* W. Gellert, H. Kustner, M. Hellwich, and H. Kastner, ed., p. 221. ©VEB Bibliographisches Institut Leipzig, 1975.

Cosine. On maps the *projections e on the horizontal plane* of the inclined segments s appear as map distances (Fig.), The ratio e/s is also a function of the angle α, called the cosine: $\cos\alpha = e/s$, $e = s\cos\alpha$. The values of the cosine function for an acute angle α decrease as the angle α increases. One obtains graphically $\cos 40° \approx 7.70\ cm/10\ cm = 0.770$ (Fig. 10.1-3).

10.1-4 Projection of an inclined line segment s onto the plane of the map (plan)

5.1 Right Triangles

Solving Triangles

One of the most important uses of trigonometry is for the solution of triangles. Recall from geometry that every triangle has three sides and three angles, which are called the six *parts* of the triangle. We say that a **triangle is solved** when all six parts are known and listed. Typically, three parts will be given, or known, and it will be our task to find the other three parts.

A triangle is usually labeled as shown in Figure 5.1. The vertices are labeled A, B, and C, with the sides opposite those vertices labeled a, b, and c, respectively. The angles are labeled α, β, and γ, respectively. In this section, we assume that γ denoted the right angle and c the hypotenuse.

FIGURE 5.1
Correctly labeled triangle

VERTICES	SIDES	ANGLES
A	a	α
B	b	β
C	c	γ

Right Triangle Definition

According to the definition of the trigonometric functions, the angle under consideration must be in standard position. This requirement is sometimes inconvenient, so we use that definition to give us the following special case, which applies to any acute angle θ of a right triangle. Notice from Figure 5.1 that θ might be α or β but would not be γ, since γ is not an acute angle. Also notice that the **hypotenuse** is one of the sides of both acute angles. The other side making up the angle is called the **adjacent side**. Thus side a is adjacent to β and side b is adjacent to α. The third side of the triangle (the one that is not part of the angle) is called the **opposite side**. Thus side a is opposite α and side b is opposite β.

RIGHT TRIANGLE DEFINITION OF THE TRIGONOMETRIC RATIOS

If θ is an acute angle in a right triangle,

$$\cos\theta = \frac{\text{adjacent side}}{\text{hypotenuse}} \qquad \sec\theta = \frac{\text{hypotenuse}}{\text{adjacent side}}$$

$$\sin\theta = \frac{\text{opposite side}}{\text{hypotenuse}} \qquad \csc\theta = \frac{\text{hypotenuse}}{\text{opposite side}}$$

$$\tan\theta = \frac{\text{opposite side}}{\text{adjacent side}} \qquad \cot\theta = \frac{\text{adjacent side}}{\text{opposite side}}$$

When solving triangles, you will not need to use the secant, cosecant, or cotangent ratios, so it is necessary to remember only the cosine, sine, and tangent ratios.

EXAMPLE 1   Using the right triangle definition of the trigonometric ratios

Let $\theta = \beta$ in Figure 5.1. Write the trigonometric ratios in terms of a, b, and c.

SOLUTION

$$\cos \beta = \frac{\text{adjacent side}}{\text{hypotenuse}} = \frac{a}{c} \qquad \sec \beta = \frac{\text{hypotenuse}}{\text{adjacent side}} = \frac{c}{a}$$

$$\sin \beta = \frac{\text{opposite side}}{\text{hypotenuse}} = \frac{b}{c} \qquad \csc \beta = \frac{\text{hypotenuse}}{\text{opposite side}} = \frac{c}{b}$$

$$\tan \beta = \frac{\text{opposite side}}{\text{adjacent side}} = \frac{b}{a} \qquad \cot \beta = \frac{\text{adjacent side}}{\text{opposite side}} = \frac{a}{b}$$

EXAMPLE 2   **Finding the trigonometric ratios for a given triangle**

Given the triangle with sides 3, 4, and 5, as shown in Figure 5.2, find the six trigonometric ratios of the angle β.

SOLUTION $\cos \beta = \dfrac{3}{5} = .6 \qquad \sec \beta = \dfrac{5}{3} = 1.\overline{6}$

$\sin \beta = \dfrac{4}{5} = .8 \qquad \csc \beta = \dfrac{5}{4} = 1.25$

$\tan \beta = \dfrac{4}{3} = 1.\overline{3} \qquad \cot \beta = \dfrac{3}{4} = .75$

FIGURE 5.2

When solving triangles, you are dealing with measurements that are never exact. In "real life," measurements are made with a certain number of digits of precision, and results are not claimed to have more digits of precision than the least accurate number in the input data. This limitation is sometimes confusing when comparing linear measure for the lengths of sides of a triangle with angle measurement in that same triangle. In this book, we assume the following relationship between **accuracy** of the measurements of sides and angles.

ACCURACY IN SIDES	EQUIVALENT ACCURACY IN ANGLES
Two significant digits	Nearest degree
Three significant digits	Nearest tenth of a degree
Four significant digits	Nearest hundredth of a degree

Ⓩ Note agreement about equivalent accuracies. Ⓩ

The definition of significant digits, as well as a set of rules for working with significant digits, is given in Appendix B. We do not want the focus of this chapter to be significant digits (which is why it is in the appendix and not here), but instead the focus is on the trigonometry. Just remember, your answers should not be more accurate than any of the given measurements.

EXAMPLE 3   **Significant digits**

Round each set of triangle descriptions to the least accurate of the measurements.

a. $a = 46.6$ $\qquad \alpha = 7.27°$ $\qquad$ **b.** $a = 803$ $\qquad \alpha = 110°$
$\quad\ \ b = 35.3$ $\qquad \beta = 46.3°$ $\qquad\qquad\ \ b = 455$ $\qquad \beta = 32.07896°$
$\quad\ \ c = 42.7$ $\qquad \gamma = 61°$ $\qquad\qquad\ \ c = 521.7952$ $\quad \gamma = 37.52103°$

c. $a = 68.123$ $\quad \alpha = 31.29033°$ $\quad$ **d.** $a = 34.6$ $\qquad \alpha = 8.5°$
$\quad\ \ b = 95.3$ $\qquad \beta = 46.70967°$ $\qquad\quad\ \ b = 27.43995$ $\quad \beta = 6.731806°$
$\quad\ \ c = 128.0556$ $\quad \gamma = 46.70967°$ $\qquad\ \ c = 61.5$ $\qquad \gamma = 164.76682°$

SOLUTION   **a.** The least accurate measurement is $\gamma = 61°$, which is the nearest degree, so all answers involving angles should be rounded to the nearest degree, and all answers involving sides should be rounded to two significant digits:

$$a = 47, \quad b = 35, \quad c = 43, \quad \alpha = 73°, \quad \beta = 46°, \quad \gamma = 61°$$

b. The least accurate measurement is $\alpha = 110°$, which is the nearest degree, so all answers involving angles should be rounded to the nearest degree, and all answers involving sides should be to two significant digits:

$$a = 800, \quad b = 460, \quad c = 520, \quad \alpha = 110°, \quad \beta = 32°, \quad \gamma = 38°*$$

c. The least accurate measurement is $b = 95.3$, which is three significant digits. This means that the angles should be rounded to the nearest tenth of a degree:

$$a = 68.1, \quad b = 95.3, \quad c = 128, \quad \alpha = 32.3°, \quad \beta = 46.7°, \quad \gamma = 46.7°$$

Note that after rounding, the sum of the angles of a triangle may not be exactly 180°.

d. The least accurate measurements are $a = 34.6$, $c = 61.5$, and $\alpha = 8.5°$, which all indicate that we should be rounding answers to three significant digits and to the nearest tenth of a degree:

$$a = 34.6, \quad b = 27.4, \quad c = 61.5, \quad \alpha = 8.5°, \quad \beta = 6.7°, \quad \gamma = 164.8°$$ ∎

⊗ Do not round your answers when doing your calculations; you should round only when stating your answers. ⊗

Another important rounding principle discussed in Appendix B is often overlooked: DO NOT ROUND AT THE BEGINNING OR IN THE MIDDLE OF THE PROBLEM. You should round only when you are stating your answers.

EXAMPLE 4   **Solving a right triangle given a side and an angle**

Solve the right triangle with $a = 50.0$ and $\alpha = 35°$.

SOLUTION   Begin by drawing a triangle (not necessarily to scale) to see the given relationships. When solving triangles, you should state all six parts, including those parts given in the problem. These parts should be stated to the correct number of significant digits. These parts are shown in color below.

FIGURE 5.3

$\alpha = 35°$ **Given.**

$\beta = 55°$ Since $\alpha + \beta = 90°$ for any right triangle with right angle at C

$\gamma = 90°$ **Given.**

$a = 50$ **Given.** Notice that we write the answer to the nearest unit because the given angle is measured to the nearest degree, and we use the *least accurate* of the given measurements to determine the accuracy of the answer (see Appendix B).

---

*As we point out in Appendix B, the answer $a = 800$ is ambiguous regarding the number of significant digits (is it one or two?), and to be clear we would need scientific notation. We do not wish to require scientific notation in this book, so we allow the ambiguity in answers such as $a = 800$. The rest of the parts of the triangle will let us know that we are rounding to two significant digits.

b: $\tan 35° = \dfrac{50}{b}$ See Figure 5.3; tangent is opp/adj. Do not forget
                                                                that c is the hypotenuse, which forces b to be the
                                                                adjacent side.

                      $b = \dfrac{50}{\tan 35°}$ By calculator: DISPLAY: 71.40740034

The correct answer should be stated to two significant digits:
$b = 71$.

c: $\sin 35° = \dfrac{50}{c}$ See Figure 5.3; sine is opp/hyp.

                      $c = \dfrac{50}{\sin 35°}$ The intermediate steps of finding c by calculator are
                                                                left for you.

                      $c \approx 87$ To two significant digits   ▌

EXAMPLE 5   Solving a right triangle given two sides

Solve the right triangle with $a = 30.0$ and $c = 70.0$.

SOLUTION   Begin by drawing a triangle (not necessarily to scale) to see the given relationships.
We begin by listing the answer (in color).

$a = 30.0$ **Given.**

$c = 70.0$ **Given.**

b: Since this is a right triangle,
$$a^2 + b^2 = c^2$$
$$(30.0)^2 + b^2 = (70.0)^2$$
$$b^2 = 4{,}000$$
$$b = 20\sqrt{10}$$ By calculator: DISPLAY: 63.2455532

The answer to the correct number of significant digits is $b = 63.2$.

$\gamma = 90.0°$ **Given;** a right triangle so $\gamma = 90.0°$

α: Use the definition of sine: $\sin \alpha = \dfrac{\text{OPP}}{\text{HYP}} = \dfrac{30}{70}$

Thus, $\alpha = \sin^{-1}\left(\dfrac{3}{7}\right)$ By calculator: DISPLAY: 25.37693353

The answer to the correct number of significant digits is $d = 25.4°$.

β: Since the sum of the angles of a triangle is 180°, we can find β:
$$\beta = 180° - \alpha - \gamma \approx 64.62306647$$

The answer to the correct number of significant digits is $\beta = 64.6°$.   ▌

EXAMPLE 6   Finding the distance across an inaccessible path

Find the distance *PA* across the river shown in Figure 5.4.

FIGURE 5.4 Right-triangle trigonometry for finding the distance across a river

SOLUTION   You can find the distance across a river by finding some reference point *P* on the other side, as shown in Figure 5.4. Next, fix a point *A* directly across from *P* and measure out a fixed distance (say, 50 m) to a point *D* so that $\angle PAD$ is a right angle. Now measure the angle θ (say, 38°) and use the definition of tangent to write

$$\tan \theta = \frac{PA}{DA}$$

$$\tan 38° = \frac{PA}{50}$$

$$PA = 50 \tan 38° \quad \text{By calculator:} \quad \text{DISPLAY:} \quad \text{39.06428133}$$

To two significant digits, the distance across the river is 39 m.

At a 25% incline, many vehicles will lose traction.

Angles of Depression and Elevation

A common application of right triangles involves an observer looking at an object. The **angle of depression** is the acute angle measured from a horizontal line down to the line of sight, as shown in Figure 5.5. On the other hand, if we measure from a horizontal line up to the line of sight, we call that angle the **angle of elevation,** also shown in Figure 5.5.

FIGURE 5.5 Angles of depression and elevation

EXAMPLE 7   Using angle of elevation to find inaccessible height

The angle of elevation of a tree from a point
on the ground 42 m from its base is 33°. Find
the height of the tree (see Figure 5.6).

SOLUTION   Let θ = angle of elevation; h = height of tree:

$$\tan \theta = \frac{h}{42}$$

$$h = 42 \tan 33°$$

$$\approx 27.28$$

FIGURE 5.6

The tree is 27 m tall.

Bearing

Another application of the solution of right triangles involves the **bearing** of a line,
which is defined as an acute angle made with a north–south line. When giving the
bearing of a line, first write N or S to determine whether to measure the angle from the
north or the south side of a point on the line. Then give the measure of the angle
followed by E or W, denoting which side of the north–south line you are measuring.
Some examples are shown in Figure 5.7.

FIGURE 5.7

HISTORICAL NOTE

You may have seen surveyors using an
instrument called a transit, or perhaps
you have wondered how one could
decide on the angle of elevation for a
problem like Example 7. A transit is a
telescope that allows for accurate line-
of-sight measurements. After fixing the
line of sight, you can read directions
from both horizontal and vertical circles. Look at the transit pictured at the
left and find these horizontal and vertical circles. These circles are graduated
every 10', 20', or 30' (depending on the
instrument). A transit also contains a
device called a *vernier*, which is used to
measure fractional parts of the scale
shown on the horizontal and vertical
circles. Finally, the transit contains a
level bubble to ensure that the transit is
level when readings are taken.

EXAMPLE 8   Using bearing to find an inaccessible distance

To find the width $|AB|$ of a canyon, a surveyor measures 100 m from A in the direction of N42.6°W to locate point C. The surveyor then determines that the bearring of CB is N73.5°E. Find the width of the canyon if the point B is situated so that $\angle BAC =$ 90.0°. (See Figure 5.8.)

SOLUTION   Let $\theta = \angle BCA$ in Figure 5.8:

$$\angle BCE' = 16.5°$$ Complementary angles
$$\angle ACS' = 42.6°$$ Alternate interior angles
$$\angle E'CA = 47.4°$$ Complementary angles
$$\theta = \angle BCA = \angle BCE' + \angle E'CA$$
$$= 16.5° + 47.4°$$
$$= 63.9°$$

Since $\tan \theta = \dfrac{AB}{AC}$,

$$AB = AC \tan \theta$$
$$= 100 \tan \theta$$
$$= 100 \tan 63.9°$$
$$\approx 204.125$$

The canyon is 204 m across.

FIGURE 5.8 Surveying problem

5.1   Problem Set

A *Solve each of the right triangles (γ is a right angle) in Problems 1–18.*

1. $a = 80; \beta = 60°$ **2.** $b = 37; \alpha = 69°$
3. $a = 29; \alpha = 76°$ **4.** $b = 90; \beta = 13°$
5. $a = 49; \alpha = 45°$ **6.** $b = 82; \alpha = 50°$
7. $c = 28.3; \alpha = 69.2°$ **8.** $\beta = 57.4°; a = 70.0$
9. $\alpha = 56.00°; b = 2,350$ **10.** $\beta = 16.4°; b = 2,580$
11. $\alpha = 42°; b = 350$ **12.** $c = 75.4; \alpha = 62.5°$
13. $c = 28.7; \alpha = 67.8°$ **14.** $c = 343.6; \beta = 32.17°$
15. $c = 418.7; \beta = 61.05°$ **16.** $b = .8024; \beta = 24.16°$
17. $b = .316; \alpha = 48.32°$ **18.** $b = .3596; \beta = 76.06°$

19. SURVEYING   Find the distance PA across the river shown in Figure 5.4 if $\theta = 42°$.

20. SURVEYING   Find the distance PA across the river shown in Figure 5.4 if $\theta = 16°$.

21. NAVIGATION   The most powerful lighthouse is on the coast of Brittany, France, and is 50 m tall. Suppose you are in a boat just off the coast, as shown in Figure 5.9. Determine your distance from the base of the lighthouse if $\theta = 12°$.

FIGURE 5.9 Distance from a boat to the *Créac'h d'Ouessant* on the coast of Brittany, France

22. NAVIGATION Use Problem 21 to determine your distance from the lighthouse if $\theta = 32°$.

23. SURVEYING The angle of evelation of a building from a point on the ground 30 m from its base is 38°. Find the height of the building.

24. SURVEYING The angle of elevation of the top of the Great Pyramid of Khufu (or Cheops) from a point on the ground 351 ft from a point directly below the top is 52.0°. Find the height of the pyramid.

Courtesy of Photo Researchers, Inc.

Problems 25–26 refer to the baseball stadium shown in Figure 5.10. Find the requested distances to the nearest foot.

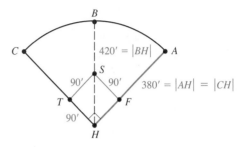

FIGURE 5.10 Baseball stadium

25. Right field (point A) to left field (point C)

26. Right field (point A) to center field (point B)

B

27. SURVEYING To find the northern boundary of a piece of land, a surveyor must divert his path from point C on the boundary by proceeding due south for 300 ft to a point A. Point B, which is due east of point C, is now found to be in the direction of N49°E from point A. What is the distance CB?

HISTORICAL NOTE

The angle of elevation for a pyramid is the angle between the edge of the base and the slant height, the line from the apex of the pyramid to the midpoint of any side of the base. It is the maximum possible angle of ascent for anyone trying to climb the pyramid to the top. In an article "Angles of Elevation of the Pyramids of Egypt," in *The Mathematics Teacher* (February 1982, pp. 124–127), author Arthur F. Smith notes that the angle of elevation of these pyramids is either 43.58° or 52°. Why did the Egyptians build pyramids using these angles of elevation? See Problem 34, Problem Set 5.4, page 209.

Smith states that (according to Kurt Mendelssohn, *The Riddle of the Pyramids*; New York: Praeger Publications, 1974) Egyptians might have measured long horizontal distances by means of a circular drum with some convenient diameter such as one cubit. The circumference would then have been pi cubits. In order to design a pyramid of convenient and attractive proportions, the Egyptians used a 1:4 ratio for the rise relative to one revolution of the drum. Smith then shows that the angle of elevation of the slant height is

$$\text{Arctan} \frac{4}{\pi} \approx 51.9°$$

If a small angle of elevation was desired (as in the case of the Red Pyramid), a 1 : 3 ratio might have been used. In that case, the angle of elevation of the slant height is

$$\text{Arctan} \frac{3}{\pi} \approx 43.7°$$

28. SURVEYING   To find the distance across a river that runs east–west, a surveyor locates points *P* and *Q* on a north–south line on opposite sides of the river. She then paces out 150 ft from *Q* due east to a point *R*. Next she determines that the bearing of *RP* is N58°W. How far is it across the river (distance *PQ*)?

29. SURVEYING   To find the boundary of a piece of land, a surveyor must divert his path from a point *A* on the boundary for 500 ft in the direction S50°E. He then determines that the bearing of a point *B* located directly south of *A* is S40°W. Find the distance *AB*.

30. SURVEYING   To find the distance across a river, a surveyor locates points *P* and *Q* on either side of the river. Next she measures 100 m from point *Q* in the direction of S35.0°E to point *R*. Then she determines that point *P* is now in the direction of N25.0°E from point *R* and that angle *PQR* is a right angle. Find the distance across the river.

31. SURVEYING   On the top of the Empire State Building is a TV tower. From a point 1,000 ft from a point on the ground directly below the top of the tower, the angle of elevation to the bottom of the tower is 51.34° and to the top of the tower is 55.81°. What is the length of the TV tower?

32. SURVEYING   The distance across a canyon can be determined from an airplane. Using the information given in Figure 5.11, what is the distance across the canyon?

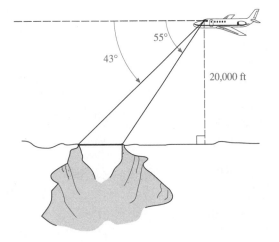

FIGURE 5.11 Determining the distance across a canyon from the air

33. In the movie *Close Encounters of the Third Kind,* Devil's Tower in Wyoming figured prominently. There was a scene in which the star, Richard Dreyfuss, was approaching the Tower. He could have determined his distance from Devil's Tower by stopping at a point *P* and estimating the angle *P*, as shown in Figure 5.12.

FIGURE 5.12 Procedure for determining the height of Devil's Tower

After moving 100 m toward Devil's Tower, he could estimate the angle *N*, as shown in Figure 5.12. How far away from Devil's Tower is point *N*?

34. Using Problem 33, determine the height of Devil's Tower.

FIGURE 5.13 Planetary positions

C

35. ASTRONOMY   If the distance from the earth to the sun is 92.9 million miles and the angle formed between Venus, the earth, and the sun (as shown in Figure 5.13) is 47.0°, find the distance from the sun to Venus.

36. ASTRONOMY   Use the information in Problem 35 to find the distance from the earth to Venus.

37. The largest ground area covered by any office building is that of the Pentagon in Arlington, Virginia. If the radius of the circumscribed circle is 783.5 ft, find the length of one side of the Pentagon.

38. Use the information in Problem 37 to find the radius of the circle inscribed in the Pentagon.

39. In algebra, the slope of a nonvertical line L passing through points $P(x_1, y_1)$ and $Q(x_2, y_2)$ is defined by

$$m = \frac{y_2 - y_1}{x_2 - x_1}$$

as shown in Figure 5.14. If we let θ be the angle that the line L makes with the x-axis such that $0 \le \theta < 180°$, then θ is called the **angle of inclination** of L. It follows that

$$m = \tan \theta$$

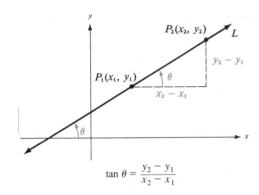

$$\tan \theta = \frac{y_2 - y_1}{x_2 - x_1}$$

FIGURE 5.14 Slope of a line

a. Find the slope of a line whose angle of inclination is 35°.

b. If the equation of a line passing through (h, k) is $y - k = m(x - h)$, find the equation of a line through $(2, -3)$ having an angle of inclination of 60°.

c. If the equation of a line passing through (h, k) is $y - k = m(x - h)$, find the equation of a line through $(4, -1)$ having an angle of inclination of 30°.

40. Show that, in every right triangle, the value of c lies between $\dfrac{a + b}{\sqrt{2}}$ and $a + b$.

41. Show that the area of a right triangle is given by
$$\tfrac{1}{2} bc \sin A.$$

42. A given triangle ABC is *not* a right triangle. Draw BD perpendicular to AC forming right triangles ABD and BDC (with right angle at D). Show that
$$\frac{\sin A}{a} = \frac{\sin C}{c}$$

5.2   LAW OF COSINES

Oblique Triangles

FIGURE 5.15
Correctly labeled triangle

In the last section we solved right triangles. We now extend that study to triangles with no right angles. Such triangles are called **oblique triangles.** Consider a triangle labeled as in Figure 5.15, and notice that γ is not now restricted to 90°.

In general, you will be given three parts of a triangle and be asked to find the remaining three parts. But can you do so given *any* three parts? Consider the possibilities.

1. SSS—By SSS we mean that you are given three sides and want to find the three angles.

2. SAS—You are given two sides and an included angle.

3. AAA—You are given three angles.

4. SSA—You are given two sides and the angle opposite one of them.

5. ASA or AAS—You are given two angles and a side.

We will consider these possibilities one at a time.

SSS

To solve a triangle given SSS, it is necessary for the sum of the lengths of the two smaller sides to be greater than the length of the largest side. In this case, we use a generalization of the Pythagorean theorem called the **Law of Cosines.**

LAW OF COSINES

In triangle ABC,

$$c^2 = a^2 + b^2 - 2ab \cos \gamma$$

PROOF Let γ be an angle in standard position with A on the positive x-axis, as shown in Figure 5.16. The coordinates of the vertices are as follows.

FIGURE 5.16

$C(0, 0)$	Since C is in standard position
$A(b, 0)$	Since A is on the x-axis a distance of b units from the origin
$B(a \cos \gamma, a \sin \gamma)$	Let $B(x, y)$; then by definition of the trigonometric functions, $\cos \gamma = \frac{x}{a}$ and $\sin y = \frac{y}{a}$. Thus, $x = a \cos \gamma$ and $y = a \sin \gamma$.

Use the distance formula for the distance between $A(b, 0)$ and $B(a \cos \gamma, a \sin \gamma)$:

$$\begin{aligned}
c^2 &= (a \cos \gamma - b)^2 + (a \sin \gamma - 0)^2 \\
&= a^2 \cos^2\gamma - 2ab \cos \gamma + b^2 + a^2 \sin^2\gamma \\
&= a^2(\cos^2\gamma + \sin^2\gamma) + b^2 - 2ab \cos \gamma \\
&= a^2 + b^2 - 2ab \cos \gamma
\end{aligned}$$

Notice that for a right triangle, $\gamma = 90°$. This means that

$$c^2 = a^2 + b^2 - 2ab \cos 90° \quad \text{or} \quad c^2 = a^2 + b^2$$

since $\cos 90° = 0$. But this last equation is simply the Pythagorean theorem.

By letting A and B, respectively, be in standard position, it can also be shown that

$$a^2 = b^2 + c^2 - 2bc \cos \alpha \qquad \text{and} \qquad b^2 = a^2 + c^2 - 2ac \cos \beta$$

To find the angles when you are given three sides, solve for α, β, or γ.

**LAW OF COSINES
ALTERNATE FORMS**

$$a^2 = b^2 + c^2 - 2bc \cos \alpha \qquad \cos \alpha = \frac{b^2 + c^2 - a^2}{2bc}$$

$$b^2 = a^2 + c^2 - 2ac \cos \beta \qquad \cos \beta = \frac{a^2 + c^2 - b^2}{2ac}$$

$$c^2 = a^2 + b^2 - 2ab \cos \gamma \qquad \cos \gamma = \frac{a^2 + b^2 - c^2}{2ab}$$

EXAMPLE 1 **Finding a part of a triangle given three sides (SSS)**

What is the smallest angle of a triangle patio whose sides measure 25, 18, and 21 feet?

SOLUTION If γ represents the smallest angle, then c (the side opposite γ) must be the smallest side, so $c = 18$. Then

$$\cos \gamma = \frac{a^2 + b^2 - c^2}{2ab}$$

$$= \frac{25^2 + 21^2 - 18^2}{2(25)(21)}$$

$$\gamma = \cos^{-1}\left(\frac{25^2 + 21^2 - 18^2}{2(25)(21)}\right) \quad \text{By calculator: DISPLAY: } \text{45.03565072}$$

To two significant digits, the answer is 45°.

CALCULATOR COMMENT

There are many computer programs that will help you solve triangles. One such program is Converge from JEMware and is typical. You begin with the choice on the pull-down menu called Alg/Trig—Solving Triangles and then input the given values. Next, YOU must choose the correct method of solution:

1. SSS   2. SAS   3. SSA   4. AAS or ASA   5. AAA

Then, if you make the correct choice, the program forces you to select one of the following:

1. No solution.   2. Infinitely many solutions.   3. One solution using the Law of Cosines.

4. One solution using the Law of Sines.*   5. Two solutions using the Law of Sines.*

Finally, after making the correct choice here (or being tutored if you do not), you can request not only the solution, but also a diagram of the given triangle. For Example 1, the solution output by this program is

$\alpha \approx 79.32807°$ $\beta \approx 55.63628°$ $\gamma \approx 45.03565°$

FRANÇOIS VIÈTE
1540–1603

The transition from Renaissance mathematics to modern mathematics was aided, in large part, by the Frenchman François Viète (1540–1603). He was a lawyer who practiced mathematics as a hobby. In his book *Canon mathematicus*, he solved oblique triangles by breaking them into right triangles. He was probably the first person to develop and use the Law of Tangents. He also was the first to prove (in 1579) the Law of Cosines. When he proved this he did not use the form used in this book. Instead,

he used:

$$\frac{2ab}{a^2 + b^2 - c^2} = \frac{1}{\sin(90° - C)}$$

However, the Law of Cosines was first stated by Euclid in Book II of his *Elements*.

---

*Law of Sines is discussed in the next section.

SAS

The second possibility listed for solving oblique triangles is that of being given two sides and an included angle. It is necessary that the given angle be less than 180°. Again use the Law of Cosines for this possibility, as shown by Example 2.

EXAMPLE 2   Finding a part of a triangle given two sides and an included angle (SAS)

Find c where $a = 52.0$, $b = 28.3$, and $\gamma = 28.5°$.

SOLUTION   By the Law of Cosines:

$$c^2 = a^2 + b^2 - 2ab \cos \gamma$$
$$= (52.0)^2 + (28.3)^2 - 2(52.0)(28.3) \cos 28.5°$$

By calculator:

$$c^2 \approx 918.355474$$
$$c \approx 30.30438044$$

To three significant digits, $c = 30.3$.

FIGURE 5.17 Similar triangles

AAA

The third case supposes that three angles are given. However, from what you know of similar triangles (see Figure 5.17)—that they have the same shape but not necessarily the same size, and thus their corresponding angles have equal measure—you can conclude that the triangle cannot be solved without knowing the length of at least one side.

SSA

Using the Law of Cosines to solve triangles in which the given angle is not included requires that we use the quadratic formula.

EXAMPLE 3   Solving a triangle given two sides and an angle that is not included (SSA)

Solve the triangle where $a = 3.0$, $b = 2.0$, and $\alpha = 110°$.

SOLUTION   By the Law of Cosines:

$$a^2 = b^2 + c^2 - 2bc \cos \alpha$$
$$9 = 4 + c^2 - 4c \cos 110°$$
$$c^2 - 4 \cos 110° \, c - 5 = 0$$
$$c = \frac{4 \cos 110° \pm \sqrt{(4 \cos 110°)^2 - 4(1)(-5)}}{2}$$
$$\approx 1.654316212, \; -3.022396785$$

·We need to reject the negative value (since c represents a length). We can now use this value c to find the angles:

$$\cos \beta = \frac{a^2 + c^2 - b^2}{2ac} \approx \frac{9 + 1.654316212^2 - 4}{2(3)(1.654316212)} \approx .7794521661$$

Therefore, $\beta \approx 38.78955642°$.

Finally, $\gamma = 180° - \alpha - \beta \approx 31.2104436°$.

We state the solution (to the required degree of accuracy):

$$
\begin{aligned}
a &= 3.0 & \alpha &= 110° \\
b &= 2.0 & \beta &= 39° \\
c &= 1.7 & \gamma &= 31°
\end{aligned}
$$

It may have occurred to you when working Example 3 that, since we are using the quadratic formula, it is certainly possible to obtain two solutions. We consider this possibility with the next example.

EXAMPLE 4   Solving an SSA triangle for which there are two solutions

Solve the triangle where $a = 1.50$, $b = 2.00$, and $\alpha = 40.0°$.

SOLUTION   By the Law of Cosines:

$$a^2 = b^2 + c^2 - 2bc \cos \alpha$$
$$1.50^2 = 2.00^2 + c^2 - 2(2)c \cos 40°$$
$$c^2 - 4 \cos 40° \, c + 1.75 = 0$$
$$c = \frac{4 \cos 40° \pm \sqrt{(-4 \cos 40°)^2 - 4(1)(1.75)}}{2}$$
$$\approx 2.30493839, .7592393826$$

How can this be? How can we have two solutions? Look at Figure 5.18. There are two possible values of c (and consequently two triangles) that satisfy the conditions of the problem.

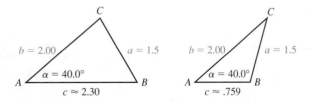

FIGURE 5.18 Given $a = 1.50$, $b = 2.00$, and $\alpha = 40.0°$, there are two possible triangles.

Since there are two triangles, we must now double our work:

SOLUTION I

$a = 1.50$ (given)

$b = 2.00$ (given)

$c_1 = 2.30$ (first solution)

SOLUTION II

$a = 1.50$ (given)

$b = 2.00$ (given)

$c_2 = .759$ (second solution)

⧐ Do not work with rounded values when finding the other parts. ⊗

$\alpha = 40.0°$ (given)

$$\beta = \cos^{-1}\left(\frac{a^2 + c^2 - b^2}{2ac}\right)$$

$$= \cos^{-1}\left(\frac{1.5^2 + 2.30493839^2 - 4}{2(1.5)(2.30493839)}\right)$$

$$\approx 58.98696953°$$

$\alpha = 40.0°$ (given)

$$\beta = \cos^{-1}\left(\frac{a^2 + c^2 - b^2}{2ac}\right)$$

$$= \cos^{-1}\left(\frac{1.5^2 + .7592393826^2 - 4}{2(1.5)(.7592393826)}\right)$$

$$\approx 121.0130305$$

Thus, (to the correct accuracy)

$\beta_1 = 59.0°$ (first solution)

$\gamma_1 = 81.0°$ $\gamma = 180° - \alpha - \beta$

$\beta_2 = 121.0°$ (second solution)

$\gamma_2 = 19.0°$ (second solution)   ❙

CALCULATOR COMMENT

It is very easy to write a program for solving a quadratic formula. Each calculator model is somewhat different, but it is instructive to show the process on the TI-81. Press the ⬚PRGM⬚ key. Then, input the formula for the two roots (from the quadratic formula). Finally, display the answer.

$$:(-B+\sqrt{(B^2 - 4AC)})/(2A)$$
$$:\text{Disp Ans}$$
$$:(-B-\sqrt{(B^2 - 4AC)})/(2A)$$
$$:\text{Disp Ans}$$

For Example 4, input the A, B, and C values as follows:

⬚1⬚ ⬚STO→⬚ ⬚A⬚   ⬚-4⬚ ⬚COS⬚ ⬚40⬚ ⬚ENTER⬚ ⬚STO→⬚   ⬚B⬚ ⬚1.75⬚ ⬚STO→⬚ ⬚C⬚

Then run the program for the DISPLAY: 2.30493839
                                        .7592393826

Since there is more than one solution in Example 4, this possibility (namely, two sides with an angle that is not included) is sometimes referred to as **the ambiguous case**) because under certain circumstances two triangles result. However, as long as we solve this problem by the Law of Cosines, we simply need to interpret the results of the Law of Cosines to decide whether there are one or two solutions.

EXAMPLE 5   Solving an SSA triangle with no solution

Solve the triangle given by $a = 3.0$, $b = 1.5$, and $\beta = 40°$.

SOLUTION We see that we are given two sides and an angle that is not included. There *may* be two solutions, but we will simply use the Law of Cosines:

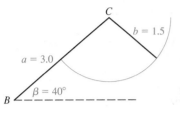

$$b^2 = a^2 + c^2 - 2ac \cos \beta$$
$$1.5^2 = 3.0^2 + c^2 - 2(3.0)c \cos 40°$$
$$c^2 - 6 \cos 40° \, c + 6.75 = 0$$
$$c = \frac{6 \cos 40° \pm \sqrt{(-6 \cos 40°)^2 - 4(1)(6.75)}}{2(1)}$$

FIGURE 5.19 No solution

Since the discriminant (the number under the radical) is negative, we see there is no solution. Look at Figure 5.19 to see why there is no solution. The side opposite the 40° angle is not as long as the height of the triangle with a 40° angle. ▮

EXAMPLE 6   Solving an SSA triangle with one solution

Solve the triangle given by $a = 2.0$, $b = 3.0$, and $\beta = 40°$.

SOLUTION We see that we are given two sides and an angle that is not included. There *may* be two solutions, but we will simply use the Law of Cosines:

$$b^2 = a^2 + c^2 - 2ac \cos \beta$$
$$3.0^2 = 2.0^2 + c^2 - 2(2.0)c \cos 40°$$
$$c^2 - 4 \cos 40° \, c - 5 = 0$$
$$c = \frac{4 \cos 40° \pm \sqrt{(-4 \cos 40°)^2 - 4(1)(-5)}}{2(1)}$$
$$\approx 4.242678556, \; -1.178500783$$

Finally, we need to find another angle:

$$\cos \alpha = \frac{b^2 + c^2 - a^2}{2bc}$$
$$= \frac{3.0^2 + c^2 - 2.0^2}{2(3.0)c}$$
$$\approx .9035298898$$

so that $\alpha \approx \cos^{-1}(.9035298898) \approx 25.37399393°$. We can now state the answer to the correct number of significant digits:

$$a = 2.0 \qquad \alpha = 25°$$
$$b = 3.0 \qquad \beta = 40°$$
$$c = 4.2 \qquad \gamma = 115° \quad (\gamma = 180° - \alpha - 40°)$$

▮

We will discuss ASA and AAS in the next section.

5.2 Problem Set

A *Solve triangle ABC for the parts requested in Problems 1–12.*

1. $a = 7.0; b = 8.0; c = 2.0$. Find α.

2. $a = 7.0; b = 5.0; c = 4.0$. Find β.

3. $a = 10; b = 4.0; c = 8.0$. Find γ.

4. $a = 4.0; b = 5.0; c = 6.0$. Find α.

5. $a = 11; b = 9.0; c = 8.0$. Find the largest angle.

6. $a = 12; b = 6.0; c = 15$. Find the smallest angle.

7. $a = 18; b = 25; \gamma = 30°$. Find c.

8. $a = 18; c = 11; \beta = 63°$. Find b.

9. $a = 15; b = 8.0; \gamma = 38°$. Find c.

10. $b = 21; c = 35; \alpha = 125°$. Find a.

11. $b = 14; c = 12; \alpha = 82°$. Find a.

12. $a = 31; b = 24; \gamma = 120°$. Find c.

Solve triangle ABC in Problems 13–28. If two triangles are possible, give both solutions. If the triangle does not have a solution, tell why.

13. $a = 14.2; b = 16.3; \beta = 115.0°$

14. $a = 14.2; c = 28.2; \gamma = 135.0°$

15. $a = 5.0; b = 4.0; \alpha = 125°$

16. $a = 5.0; b = 4.0; \alpha = 80°$

17. $b = 82.5; c = 52.2; \gamma = 32.1°$

18. $a = 151; b = 234; c = 416$

19. $a = 10.2; b = 11.8; \alpha = 47.0°$

20. $\alpha = 82.5°; \beta = 16.9°; \gamma = 80.6°$

21. $a = 123; b = 225; c = 351$

22. $a = 27.2; c = 35.7; \alpha = 43.7°$

23. $b = 5.2; c = 3.4; \alpha = 54.6°$

24. $a = 81; c = 53; \beta = 85.2°$

25. $a = 214; b = 320; \gamma = 14.8°$

26. $a = 18; b = 12; c = 23.3$

27. $a = 140; b = 85.0; c = 105$

28. $b = 428; c = 395; \gamma = 28.4°$

B

29. AVIATION   New York City is approximately 210 mi N9°E of Washington, D.C., and Buffalo, New York, is N49°W of Washington, D.C. How far is Buffalo from New York City if the distance from Buffalo to Washington, D.C., is approximately 285 mi?

30. AVIATION   New Orleans, Louisiana, is approximately 1,080 mi S56°E of Denver, Colorado, and Chicago, Illinois, is N76°E of Denver. How far is Chicago from New Orleans if it is approximately 950 mi from Denver to Chicago?

31. SURVEYING   A vertical tower is located on a hill whose inclination is 6°. From a point P 100 ft down the hill from the base of the tower, the angle of elevation to the top of the tower is 28°. What is the height of the tower? [*Hint:* Use right triangles.]

32. BALLISTICS   An artillery-gun observer must determine the distance to a target at point T. He knows that the target is 5.20 miles from point I on a nearby island. He also knows that he (at point H) is 4.30 miles from point I. If $\angle HIT$ is 68.4°, how far is he from the target? (See Figure 5.20.)

FIGURE 5.20 Determining the distance to a target

33. CONSUMER   A buyer is interested in purchasing a triangular lot with vertices *LOT*, but unfortunately, the marker at point L has been lost. The deed indicates that *TO* is 453 ft and *LO* is 112 ft, and that the angle at L is 82.6°. What is the distance from L to T?

34. GEOMETRY   A dime, a penny, and a quarter are placed on a table so that they just touch each other, as shown. Let D, P, and Q be the respective centers. Solve $\triangle DPQ$. [*Hint:* If you measure the coins, the diameters are 1.75 cm, 2.00 cm, and 2.50 cm.]

35. GEOMETRY An equilateral triangle is inscribed in a circle of radius 5.0 in. Find the perimeter of the triangle.

36. GEOMETRY A square is inscribed in a circle of radius 5.0 in. Find the area of the square.

C

37. Prove that $a^2 = b^2 + c^2 - 2bc \cos \alpha$.

38. Prove that $b^2 = a^2 + c^2 - 2ac \cos \beta$.

39. In $\triangle ABC$, use the Law of Cosines to prove
$a = b \cos \gamma + c \cos \beta$.

40. In $\triangle ABC$, use the Law of Cosines to prove
$b = a \cos \gamma + c \cos \alpha$.

41. In $\triangle ABC$, use the Law of Cosines to prove
$c = a \cos \beta + b \cos \alpha$.

42. From the equation given in Problem 38 show that

$$\cos \beta + 1 = \frac{(a + c - b)(a + b + c)}{2ac}$$

43. From the equation given in Problem 37 show that

$$\cos \alpha + 1 = \frac{(b + c - a)(a + b + c)}{2bc}$$

44. *Historical Question* In January 1978, Martin Cohen, Terry Goodman, and John Benard published an article, "SSA and the Law of Cosines," in *The Mathematics Teacher*. In this article, they used the quadratic formula to solve the Law of Cosines for *c*:

$$c = b \cos \alpha \pm \sqrt{a^2 - b^2 \sin^2 \alpha}$$

Show this derivation.

45. Suppose $\alpha < 90°$. If $d = a^2 - b^2 \sin^2 \alpha$ in Problem 44, show that if $d < 0$, there is no solution and this equation corre-

sponds to no triangle. If $d = 0$, there is one solution, which corresponds to one triangle; if $d > 0$, there are two solutions and two triangles are formed.

46. COMPUTER PROBLEM On page 191 in a Calculator Comment, we mentioned an example of computer software that would solve triangles. The logic used by such software is summarized in Table 5.1 of this section. Write a flowchart or use pseudocode to write such a program. Even if you do not know computer programming, you could write a set of directions telling exactly how to proceed with solving triangles in order to help you understand Table 5.1.

47. Pythagorean triplets are well known and are defined to be integers that satisfy the Pythagorean theorem—namely, *a*, *b*, and *c* integers that satisfy.

$$c^2 = a^2 + b^2$$

We define a **primitive quadruple** as integers *a*, *b*, and *c*, along with an angle θ (measured in degrees) that is also an integer that satisfy the Law of Cosines. For example, if $a = 57, b = 55, c = 97$, and $\theta = 120°$, then

$$c^2 = a^2 + b^2 - 2ab \cos \theta$$

is satisfied since

$$97^2 = 57^2 + 55^2 - 2(57)(55)\cos 120°$$

Find another primitive quadruple.

48. COMPUTER PROGRAM Use a computer to print out the primitive quadruples for positive integers *a*, *b*, *c* less than 100. (See Problem 47.)

49. *Historical Question* In *The Mathematics Teacher*, November 1958 (pp. 544–546), Howard Eves published an article, "Pappus's Extension of the Pythagorean Theorem." In this article, he points out that if you combine Propositions 12 and 13 of Book II and Proposition 47 of Book I of Euclid's *Elements,* you can derive the Law of Cosines from the Pythagorean theorem. Do some research and write a paper showing how this can be done.

5.3 Law of Sines

In the last section, we listed five possibilities for given information in solving oblique triangles: SSS, SAS, AAA, SSA, and ASA or AAS. We noted that there is no solution if we are given three angles (AAA), and found that if we have at least two sides given (SSS, SAS, and SSA), we use the Law of Cosines. However, if we know only one side, then there is insufficient information to use the Law of Cosines.

ASA or AAS

If we are given two angles, then we essentially are given all three angles of the triangle since the sum of the angles of a triangle is 180°. We also note that for a triangle to be formed, the sum of the two given angles must be less than 180°. We now state and prove a result, called the Law of Sines, that solves triangles when only one side is known.

LAW OF SINES

In any $\triangle ABC$,

$$\frac{\sin \alpha}{a} = \frac{\sin \beta}{b} = \frac{\sin \gamma}{c}$$

The equation in the Law of Sines means that you can use any of the following equations to solve a triangle:

$$\frac{\sin \alpha}{a} = \frac{\sin \beta}{b} \qquad \frac{\sin \alpha}{a} = \frac{\sin \gamma}{c} \qquad \frac{\sin \beta}{b} = \frac{\sin \gamma}{c}$$

PROOF   Consider any oblique triangle, as shown in Figure 5.21.

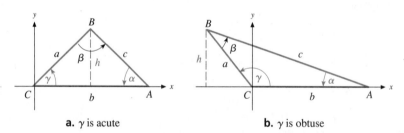

a. γ is acute   **b.** γ is obtuse

FIGURE 5.21   Oblique triangles for Law of Sines

Let h = height of the triangle with base CA. Then

$$\sin \alpha = \frac{h}{c} \quad \text{and} \quad \sin \gamma = \frac{h}{a}$$

Solving for h, $h = c \sin \alpha \quad \text{and} \quad h = a \sin \gamma$

Thus, $c \sin \alpha = a \sin \gamma$

Dividing by ac, $\dfrac{\sin \alpha}{a} = \dfrac{\sin \gamma}{c}$

Repeat these steps for the height of the same triangle with base AB:

$$\frac{\sin \alpha}{a} = \frac{\sin \beta}{b}$$

EXAMPLE 1   Solving a triangle given two angles

Solve the triangle in which $a = 20$, $\alpha = 38°$, and $\beta = 121°$.

SOLUTION $\alpha = 38°$ Given.

$\beta = 121°$ Given.

$\gamma = 21°$ Since $\alpha + \beta + \gamma = 180°$, $\gamma = 180° - 38° - 121° = 21°$.

$a = 20$ Given.

$b = 28$ Use the Law of Sines: $\dfrac{\sin 38°}{20} = \dfrac{\sin 121°}{b}$; then

$\qquad\qquad b = \dfrac{20 \sin 121°}{\sin 38°}$ Use a calculator.

$\qquad\qquad \approx \dfrac{20(.8572)}{.6157}$ Use a calculator.

$\qquad\qquad \approx 27.8454097$ Give the answer to two significant digits.

$c = 12$ Use the Law of Sines: $\dfrac{\sin 38°}{20} = \dfrac{\sin 21°}{c}$; then

$\qquad\qquad c = \dfrac{20 \sin 21°}{\sin 38°}$

$\qquad\qquad \approx 11.64172078$

Notice that the answers are given to two significant digits.

EXAMPLE 2   Finding a rate by using trigonometry

A boat traveling at a constant rate due west passes a buoy that is 1.0 kilometer from a lighthouse. The lighthouse is N30°W of the buoy. After the boat has traveled for one-half hour, its bearing to the lighthouse is N74°E. How fast is the boat traveling?

SOLUTION   The angle at the lighthouse (see Figure 5.22) is $180° - 60° - 16° = 104°$.

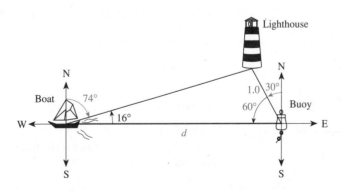

FIGURE 5.22 Finding the rate of travel

Therefore, by the Law of Sines,

$$\frac{\sin 104°}{d} = \frac{\sin 16°}{1.0}$$

$$d = \frac{\sin 104°}{\sin 16°} \approx 3.520189502 \qquad \text{Do not round answers in your work;}$$
$$\text{round only when stating the answer.}$$

After one hour the distance is $2d$, so the rate of the boat is about 7.0 kph.

Solving Triangles: Summary

The most important skill to be learned from this section is the ability to select the proper trigonometric law when given a particular problem. Refer to Table 5.1.

TABLE 5.1 Summary for Solving Triangles

GIVEN	CONDITIONS ON THE GIVEN INFORMATION		METHOD OF SOLUTION
1. AAA			No solution
2. SSS	**More than one side given:**		**Law of Cosines**
	a.	The sum of the lengths of the two smaller sides is less than or equal to the length of the larger side.	No solution
	b.	The sum of the lengths of the two smaller sides is greater than the length of the larger side.	Law of Cosines
3. SAS	a.	The given angle is greater than or equal to 180°.	No solution
	b.	The given angle is less than 180°.	Law of Cosines
3. SSA	a.	The given angle is greater than or equal to 180°.	No solution
	b.	There are no solutions, one solution, or two solutions as determined by the quadratic formula.	Law of Cosines
4. ASA or AAS	**More than one angle given:**		**Law of Sines**
	a.	The sum of the given angles is greater than or equal to 180°.	No Solution
	b.	The sum of the given angles is less than 180°.	**Law of Sines**

EXAMPLE 3   Determining the time to reach a destination

An airplane is 100 miles N40°E of a Loran station, traveling due west at 240 miles per hour. How long will it take (to the nearest minute) before the plane is 90 miles from the Loran station?

SOLUTION   We begin by drawing a picture, and label the parts as shown in Figure 5.23. Note that we are given two sides and an angle that is not included (SSA), so we expect to use the quadratic formula with the Law of Cosines.

We find $\beta = 50°$, so we have

$$b^2 = a^2 + c^2 - 2ac\cos\beta$$
$$90^2 = a^2 + 100^2 - 2a(100)\cos 50°$$
$$a^2 - 200\cos 50° \, a + 1{,}900 = 0$$

FIGURE 5.23

From the quadratic formula,

$$a = \frac{200\cos 50° \pm \sqrt{(-200\cos 50°)^2 - 4(1)(1{,}900)}}{2}$$

$$\approx 111.5202587, \ 17.0372632$$

We see there are two solutions. The one shown in Figure 5.23 is $a \approx 17.04$. To find the times, we use the distance formula, $d = rt$, for the two distances found in the quadratic formula and a given rate of 240 mph.

	SOLUTION I	SOLUTION II
$d = rt$	$a_1 = 240t_1$	$a_2 = 240t_2$
	$t_1 \approx \dfrac{17.0372632}{240}$	$t_2 \approx \dfrac{111.5202587}{240}$
	$\approx .0709885967$	$\approx .4646677447$

To convert these times (which are in hours) to minutes, multiply each by 60, and round to the nearest minute:

$$t_1 = 4 \text{ minutes} \qquad t_2 = 28 \text{ minutes}$$

The plane will be 90 miles from the Loran station in 4 minutes (location shown in Figure 5.23), and again in 28 minutes. Can you draw the location of the plane at this time?

5.3 Problem Set

A *Solve triangle ABC with the parts given in Problems 1–12. If the triangle cannot be solved, tell why.*

1. $a = 10$; $\alpha = 48°$; $\beta = 62°$
2. $a = 18$; $\alpha = 65°$; $\gamma = 115°$
3. $a = 30$; $\beta = 50°$; $\gamma = 100°$
4. $b = 23$; $\alpha = 25°$; $\beta = 110°$
5. $b = 40$; $\alpha = 50°$; $\gamma = 60°$
6. $b = 90$; $\beta = 85°$; $\gamma = 25°$
7. $c = 53$; $\alpha = 82°$; $\beta = 19°$
8. $c = 43$; $\alpha = 120°$; $\gamma = 7°$
9. $c = 115$; $\beta = 81.0°$; $\gamma = 64.0°$
10. $\alpha = 18.3°$; $\beta = 54.0°$; $a = 107$
11. $\beta = 85°$; $\gamma = 24°$; $b = 223$
12. $a = 85$; $\alpha = 48.5°$; $\gamma = 72.4°$

B *Solve △ ABC with the parts given in Problems 13–24. If the triangle cannot be solved, tell why.*

13. $a = 41.0$; $\alpha = 45.2°$; $\beta = 21.5°$
14. $b = 55.0$; $c = 92.0$; $\alpha = 98.0°$
15. $b = 58.3$; $\alpha = 120°$; $\gamma = 68.0°$
16. $c = 123$; $\alpha = 85.2°$; $\beta = 38.7°$
17. $a = 26$; $b = 71$; $c = 88$
18. $\alpha = 48°$; $\beta = 105°$; $\gamma = 27°$
19. $a = 25.0$; $\beta = 81.0°$; $\gamma = 25.0°$
20. $a = 25.0$; $b = 45.0$; $c = 102$
21. $a = 80.6$; $b = 23.2$; $\gamma = 89.2°$
22. $b = 1{,}234$; $\alpha = 85.26°$; $\beta = 24.45°$
23. $c = 28.36$; $\beta = 42.10°$; $\gamma = 102.30°$
24. $a = 481$; $\beta = 28.6°$; $\gamma = 103.0°$

25. SURVEYING In San Francisco, a certain hill makes an angle of 20° with the horizontal and has a tall building at the top. At a point 100 ft down the hill from the base of the building, the angle of elevation to the top of the building is 72°. What is the height of the building?

26. In the movie *Star Wars*, the hero, Luke, must hit a small target on the Death Star by flying a horizontal distance to reach the target. When the target is sighted, the onboard computer calculates the angle of depression to be 28.0°. If after 150 km the target has an angle of depression of 42.0°, how far is the target from Luke's spacecraft at that instant?

27. The Galactic Empire's computers on the Death Star are monitoring the positions of the invading forces (see Problem 26). At a particular instant, two observation points 2,500 m apart make a fix on Luke's spacecraft, which is between the observation points and in the same vertical plane. If the angle of elevation from the first observation point is 3.00° and the angle of elevation from the second is 1.90°, find the distance from Luke's spacecraft to each of the observation points.

28. Solve Problem 27 if both observation points are on the same side of the spacecraft and all the other information is unchanged.

29. Johnny Carson is standing on the deck of his Malibu home and is looking down a hillside with an inclination of 35° to a small boat on the beach (see Figure 5.24). If the angles of depression to the bow (point *B*) and stern (point *S*) are $\beta = 32°$ and $\theta = 25°$, how long is the boat if it is known that distance *JF* is 50 ft and *F*, *B*, and *S* are on level ground?

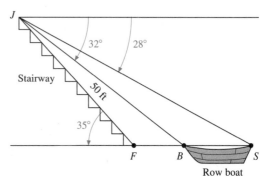

FIGURE 5.24

30. NAVIGATION A UFO is sighted by people in two cities 2,300 miles apart. The UFO is between and in the same vertical plane as the two cities. The angle of elevation of the UFO from the first city is 10.48° and from the second is 40.79°. At what altitude is the UFO flying? What is the actual distance of the UFO from each city?

31. ENGINEERING At 500 ft in the direction that the Tower of Pisa is leaning, the angle of elevation is 20.24°. If the tower leans at an angle of 5.45° from the vertical, what is the length of the tower?

32. ENGINEERING What is the angle of elevation of the leaning Tower of Pisa (described in Problem 31) if you measure from a point 500 ft in the direction exactly opposite from the way it is leaning?

33. SURVEYING The world's longest deepwater jetty is at Le Havre, France. Since access to the jetty is restricted, it was necessary for me to calculate its length by noting that it forms an angle of 85.0° with the shoreline. After pacing out 1,000 ft along the line making an 85.0° angle with the jetty, I calculated the angle to the end of the jetty to be 83.6°. What is the length of the jetty?

34. NAVIGATION From a blimp, the angle of depression to the top of the Eiffel Tower is 23.2° and to the bottom is 64.6°. After flying over the tower at the same height and at a distance of 1,000 ft from the first location, you determine that the angle of depression to the top of the tower is now 31.4°. What is the height of the Eiffel Tower given that these measurements are in the same vertical plane? (See Figure 5.25.)

FIGURE 5.25 Determining the height of the Eiffel Tower

35. Given $\triangle ABC$, show that $\dfrac{\sin \alpha}{a} = \dfrac{\sin \beta}{b}$.

36. Given $\triangle ABC$, show that $\dfrac{\sin \beta}{b} = \dfrac{\sin \gamma}{c}$.

37. Given $\triangle ABC$, show that $\dfrac{\sin \alpha}{\sin \beta} = \dfrac{a}{b}$.

C

38. HARPSICHORD PROBLEM A harpsichord maker, reproducing a particular historical instrument, needs to join two pieces of wood of different thicknesses to give a neat miter at a certain angle. The pieces of wood have thicknesses w and W, and the necessary angle at the joint is θ. Find the angles γ and ϕ,

as shown in Figure 5.26, at which the two pieces must be cut, in terms of w, W, and θ.   [*Hint:* $\theta = \gamma + \phi$]

FIGURE 5.26 Construction of a harpsichord

39. Using Problem 35, show that

$$\frac{\sin \alpha - \sin \beta}{\sin \alpha + \sin \beta} = \frac{a - b}{a + b}$$

40. Using Problem 39 and the formulas for the sum and difference of sines, show that

$$\frac{2 \cos \frac{1}{2}(\alpha + \beta) \sin \frac{1}{2}(\alpha - \beta)}{2 \sin \frac{1}{2}(\alpha + \beta) \cos \frac{1}{2}(\alpha - \beta)} = \frac{a - b}{a + b}$$

41. Using Problem 40, show that

$$\frac{\tan \frac{1}{2}(\alpha - \beta)}{\tan \frac{1}{2}(\alpha + \beta)} = \frac{a - b}{a + b}$$

This result is known as the **Law of Tangents**.

42. Another form of the Law of Tangents (see Problem 41) is

$$\frac{\tan \frac{1}{2}(\beta - \gamma)}{\tan \frac{1}{2}(\beta + \gamma)} = \frac{b - c}{b + c}$$

Derive this formula.

43. A third form of the Law of Tangents (see Problems 41 and 42) is

$$\frac{\tan\frac{1}{2}(\alpha - \gamma)}{\tan\frac{1}{2}(\alpha + \gamma)} = \frac{a - c}{a + c}$$

Derive this formula.

44. Newton's formula involves all six parts of a triangle. It is not useful in solving a triangle, but it is helpful in checking your results after you have done so. Show that

$$\frac{a + b}{c} = \frac{\cos\frac{1}{2}(\alpha - \beta)}{\sin\frac{1}{2}\gamma}$$

45. Mollweide's formula involves all six parts of a triangle. It is not useful in solving a triangle, but it is helpful in checking

your results after you have done so. Show that

$$\frac{a - b}{c} = \frac{\sin\frac{1}{2}(a - \beta)}{\cos\frac{1}{2}\gamma}$$

46. An article "Calculating the distance to the sun by observing the trail of a meteor," by Jearl Walker is discussed in the March 1987 issue of *Scientific American* (p. 122). In the article it was necessary to solve several triangles. Using this article as a basis, give one triangle that must be solved and then show the solution of that triangle.

47. The following letter to the editor appeared in the February 1977 issue of *Popular Science.** See if you can solve this puzzle. As a hint, you might guess from its placement in this book that it uses the Law of Cosines or the Law of Sines.

Thirty-Seven-Year-Old Puzzle*

For more than 37 years I have tried, on and off, to solve a problem that appeared in *Popular Science*.

All right, I give up. What's the answer? In the September, 1939 issue (page 14) the following letter appeared:

"Have enjoyed monkeying with the problems you print, for the last couple of years. Here's one that may make the boys scratch their heads a little:

A man always drives at the same speed (his car probably has a governor on it). He makes it from A direct to C in 30 minutes; from A through B to C in 35 minutes; and from A through D to C in 40 minutes. How fast does he drive?
                                    D.R.C., Sacramento, Calif."

It certainly looked easy, and I started to work on it. By 3 A.M. I had filled a lot of sheets of paper, both sides, with notations. In subsequent issues of *Popular Science*, all I could find on the subject was the following, in the November, 1939 issue:

"In regard to the problem submitted by D.R.C. of Sacramento, Calif. in the September issue. According to my calculations, the speed of the car must be 38.843 miles an hour.
                                    W.L.B., Chicago, Ill."

But the reader gave no hint of how he had arrived at that figure. It may or may not be correct. Over the years, I probably have shown the problem to more than 500 people. The usual reaction was to nod the head knowingly and start trying to find out what the hypotenuse of the right angle is. Bot those few who remembered the formula then found there is little you can do with it when you know the hypotenuse distance.

Now I'd like to challenge *Popular Science*. If your readers cannot solve this little problem (and prove the answer), how about your finding the answer, if there is one, editors?
                                    R.F. Davis, Sun City, Ariz.

Darrell Huff, master of the pocket calculator and frequent PS contributor [Feb. '76, June '75, Dec. '74] has come up with a couple of solutions. We'd like to see what your readers can do before we publish his answers.

5.4 Areas and Volumes

Areas

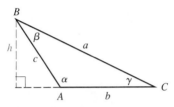

In elementary school you learned that the area, K, of a triangle is $K = \frac{1}{2}bh$.* Now you sometimes need to find the area of a triangle given the measurements for the sides and angles but not for the height, and, therefore, you need trigonometry to find the area.

Using the orientation shown in Figure 5.27,

$$K = \tfrac{1}{2}bh \quad \text{and} \quad \sin \alpha = \frac{h}{c} \quad \text{or} \quad h = c \sin \alpha$$

Thus,

$$K = \tfrac{1}{2}bc \sin \alpha$$

We could use any other pair of sides to derive the following area formulas.

FIGURE 5.27 Oblique triangles

AREA OF A TRIANGLE
(two sides and an included angle are known)

$$K = \tfrac{1}{2}bc \sin \alpha \qquad K = \tfrac{1}{2}ac \sin \beta \qquad K = \tfrac{1}{2}ab \sin \gamma$$

EXAMPLE 1   Finding the area of a SAS triangle

Find the area of $\triangle ABC$ where $\alpha = 18.4°$, $b = 154$ ft, and $c = 211$ ft.

SOLUTION
$$K = \tfrac{1}{2}(154)(211) \sin 18.4°$$
$$= 5{,}128.349903$$

To three significant figures, the area is 5,130 sq ft.   ▌

The area formulas given above are useful for finding the area of a triangle when two sides and an included angle are known. Suppose, however, that you know the angles but only one side—say, side a. Then you can use either formula involving a and replace the other variable by using the Law of Sines:

$$\frac{\sin \alpha}{a} = \frac{\sin \beta}{b} \quad \text{so} \quad b = \frac{a \sin \beta}{\sin \alpha}$$

Therefore, $K = \dfrac{1}{2}ab \sin \gamma = \dfrac{1}{2}a\,\dfrac{a \sin \beta}{\sin \alpha}\sin \gamma = \dfrac{a^2 \sin \beta \sin \gamma}{2 \sin \alpha}$

The other area formulas shown in the following box can be similarly derived.

*In elementary school, you no doubt used A for area. In trigonometry, we use K for area because we have already used A to represent a vertex of a triangle.

AREA OF A TRIANGLE
(two angles and an included
side are known)

$$K = \frac{a^2 \sin \beta \sin \gamma}{2 \sin \alpha} \qquad K = \frac{b^2 \sin \alpha \sin \gamma}{2 \sin \beta} \qquad K = \frac{c^2 \sin \alpha \sin \beta}{2 \sin \gamma}$$

EXAMPLE 2   Finding the area of an ASA triangle

Find the area of a triangle with angles 20°, 50°, and 110° if the side opposite the 50° angle is 24 in. long.

SOLUTION

$$K = \frac{24^2 \sin 20° \sin 110°}{2 \sin 50°}$$

$$\approx 120.8303469$$

To two significant figures, the area is 120 sq in.

If three sides (but none of the angles) are known, you will need another formula to find the area of a triangle. The formula is derived from the Law of Cosines; this derivation is left as an exercise but is summarized in the following box. The result is known as **Heron's formula.**

AREA OF A TRIANGLE
(three sides known)

$$K = \sqrt{s(s - a)(s - b)(s - c)} \qquad \text{where } s = \tfrac{1}{2}(a + b + c)$$

EXAMPLE 3   Finding the area of an SSS triangle

Find the area of a triangle having sides of 43 ft, 89 ft, and 120 ft.

SOLUTION Let $a = 43, b = 89,$ and $c = 120.$ Then

$$s = \frac{1}{2}(43 + 89 + 120) = 126$$

Thus,
$$K = \sqrt{126(126 - 43)(126 - 89)(126 - 120)}$$
$$\approx 1,523.704696$$

To two significant digits, the area is 1,500 sq ft.

Area of a Sector

A **sector of a circle** is the portion of the interior of a circle cut by a central angle, θ, as shown in Figure 5.28. Since the area of a circle is $A = \pi r^2$, the area of a sector is the fraction $\frac{\theta}{2\pi}$ of the entire circle.

FIGURE 5.28
Sector of a circle (shaded portion)

In general,

$$\text{area of sector} = (\text{fractional part of circle})(\text{area of circle})$$

$$= \frac{\theta}{2\pi} \cdot \pi r^2$$

AREA OF A SECTOR	Area of a sector = $\frac{1}{2}\theta r^2$ where θ is measured in radians

EXAMPLE 4   Finding the area of a sector given a central angle measured in radians

What is the area of the sector of a circle of radius 12 in. whose central angle is 2?

SOLUTION   area of sector $= \frac{1}{2} \cdot 2 \cdot (12 \text{ in.})^2 = 144 \text{ in.}^2$ ▌

EXAMPLE 5   Finding the area of a sector given a central angle measured in degrees

What is the area of the sector of a circle of radius 420 m whose central angle is 2°?

SOLUTION   First change 2° to radians:

$$2° = 2\left(\frac{\pi}{180}\right) = \frac{\pi}{90} \text{ radians}$$

$$\text{area of sector} = \frac{1}{2} \cdot \left(\frac{\pi}{90}\right) (420 \text{ m})^2 = 980\pi \text{ m}^2 \approx 3{,}079 \text{ m}^2$$ ▌

Volume of a Cone

A third application in this section is finding the volume of a cone when you do not know its height. From geometry,

$$V = \frac{\pi r^2 h}{3}$$

Now suppose you know the radius of the base but not the height. If you know the angle of elevation α (see Figure 5.29), then

$$\tan \alpha = \frac{h}{r} \quad \text{or} \quad h = r \tan \alpha$$

FIGURE 5.29

A cone with volume $V = \frac{1}{3}\pi r^2 h$

Thus, the volume is

$$V = \frac{\pi r^3 \tan \alpha}{3} = \frac{1}{3}\pi r^3 \tan \alpha$$

EXAMPLE 6   Finding the volume of a cone

If sand is dropped from the end of a conveyor belt, the sand will fall in a conical heap such that the angle of elevation is about 33° (see Figure 5.30). Find the volume of sand when the radius is exactly 10 ft.

SOLUTION $V = \frac{1}{3}\pi(10)^3 \tan 33°$ DISPLAY: ЬႦ0.0580413

≈ 680

FIGURE 5.30

5.4   Problem Set

A *Find the area of each triangle in Problems 1–12.*

1. $a = 15; b = 8.0; \gamma = 38°$

2. $a = 18; c = 11; \beta = 63°$

3. $b = 14; c = 12; \alpha = 82°$

4. $b = 21; c = 35; \alpha = 125°$

5. $a = 30; \beta = 50°; \gamma = 100°$

6. $b = 23; \alpha = 25°; \beta = 110°$

7. $b = 40; \alpha = 50°; \gamma = 60°$

8. $b = 90; \beta = 85°; \gamma = 25°$

9. $a = 7.0; b = 8.0; c = 2.0$

10. $a = 10; b = 4.0; c = 8.0$

11. $a = 11; b = 9.0; c = 8.0$

12. $a = 12; b = 6.0; c = 15$

B *Find the area of each triangle in Problems 13–24. If two triangles are formed, give the area of each, and if no triangle is formed, so state.*

13. $a = 12.0; b = 9.00; \alpha = 52.0°$

14. $a = 7.0; b = 9.0; \alpha = 52°$

15. $a = 10.2; b = 11.8; \alpha = 47.0°$

16. $a = 8.629973679; b = 11.8; \alpha = 47.0°$

17. $b = 82.5; c = 52.2; \gamma = 32.1°$

18. $a = 352; b = 230; c = 418$

19. $\beta = 15.0°; \gamma = 18.0°; b = 23.5$

20. $b = 45.7; \alpha = 82.3°; \beta = 61.5°$

21. $a = 68.2; \alpha = 145°; \beta = 52.4°$

22. $a = 151; b = 234; c = 416$

23. $a = 124; b = 325; c = 351$

24. $a = 27.2; c = 35.7; \alpha = 43.7°$

25. If the central angle subtended by the arc of a segment of a circle is 1.78 and the area is 54.4 sq cm, what is the radius of the circle?

26. If the area of a sector of a circle is 162.5 sq cm and the angle of the sector is .52, what is the radius of the circle?

27. A field is in the shape of a sector of a circle with a central angle of 20° and a radius of 320 m. What is the area of the field?

28. CONSUMER A level lot has the dimensions shown in Figure 5.31. What is the total cost of treating the area for poison oak if the fee is $45 per acre (1 acre = 43,560 sq ft)?

FIGURE 5.31 Area of a lot

29. If vulcanite is dropped from a conveyor belt, it will fall in a conical heap such that the angle of elevation is about 36°. Find the volume of vulcanite when the radius is 30 ft.

30. The volume of a slice cut from a cylinder (see Figure 5.32) is found from the formula

$$V = hr^2\left(\frac{\theta}{2} - \sin\frac{\theta}{2}\cos\frac{\theta}{2}\right)$$

for θ measured in radians such that $0 \le \theta \le \pi$. Find the volume of a slice cut from a log with a 6.0-in. radius that is 3.0 ft long when $\theta = \frac{\pi}{3}$.

FIGURE 5.32 Slice of a cylinder

C

31. CONSTRUCTION A 50-ft culvert carries water under a road. If there is 2 ft of water in a culvert with a 3-ft radius, use Problem 30 to find the volume of water in the culvert.

32. Derive the formula for finding the area when you know three sides of a triangle.

33. Derive the formula given in Problem 30.

34. *Historical Question* If we define the angle of elevation of a pyramid as the angle between the edge of the base and the slant height, we find that the major Egyptian pyramids are $43\frac{1}{2}°$ or $52°$. Write a paper explaining why the Egyptians built pyramids using angles of elevation of $43\frac{1}{2}°$ or $52°$.
References:
Billard, Jules B., ed. *Ancient Egypt.* Washington, D.C.: National Geographic Society, 1978.

Edwards, I. E. S. *The Pyramids of Egypt.* New York: Penguin Books, 1961.
Fakhry, Ahmed. *The Pyramids* (2nd ed.). Chicago: University of Chicago Press, 1974.
Mendelssohn, Kurt. *The Riddle of the Pyramids.* New York: Praeger Publications, 1974.
Smith, Arthur F. "Angles of Elevation of the Pyramids of Egypt," *The Mathematics Teacher,* February 1982, pp. 124–127.

35. *Historical Question* This problem derives the formula of Brahmagupta.* Given a quadrilateral with sides measuring a, b, c, and d, let θ and ϕ be the angles as shown in Figure 5.33. Let $\mu = \theta + \phi$.

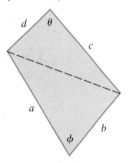

FIGURE 5.33

a. Show that the area of the quadrilateral is

$$K = \frac{ab}{2}\sin\phi + \frac{cd}{2}\sin\theta$$

b. Show $a^2 + b^2 - 2ab\cos\phi = c^2 + d^2 - 2cd\cos\theta$.

c. Using parts **a** and **b** derive Brahmagupta's formula where $p = \dfrac{a + b + c + d}{2}$ and $\mu = \theta + \phi$:

$$K^2 = (p - a)(p - b)(p - c)(p - d) - abcd\cos^2\left(\frac{\mu}{2}\right)$$

36. Use Brahmagupta's formula (Problem 35) to derive Heron's formula.

5.5 # Vector Triangles

Many applications of mathematics involve quantities that have *both* magnitude and direction, such as forces, velocities, accelerations, and displacements.

---

*This elegant problem was given to me over lunch with my friend Les Lang. I extend many thanks for this problem and the many other mathematical ideas he has given to me.

Definition of a Vector

EXAMPLE 1   Illustrating magnitude and direction

State the magnitude and direction, and draw a diagram for each.

SOLUTION   **a.** An airplane travels with an airspeed of 790 mph and a direction of 230°. [*Note:* In aviation, direction is measured clockwise from the north. This is called the **heading.** The *true course* is the direction of the path over the ground, which is also measured clockwise from the north.]

Magnitude: 790

Direction: 230° measured from the north

b. A car travels at 42 mph due west.

Magnitude: 42

Direction: west

c. The wind blows at 46.1 mph in the direction of S63°W.

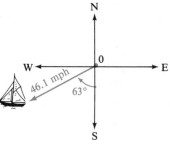

Magnitude: 46.1

Direction: S63°W

d. An object on an inclined plane has a mass of 850 kg.

Magnitude: 850

Direction: down (gravity)

e. A man is lifting a 130-lb weight by using a pulley and a rope being pulled through the pulley at a 42° angle from the vertical.

Magnitude: 130

Direction: 42° from the vertical

Vectors are used to describe quantities having both magnitude and direction. That is, a **vector** is a directed line segment specifying both a magnitude and a direction. The **magnitude** of the vector represents the *length* of the quantity being represented, and the **direction** of the vector represents the *direction* of the quantity. Two vectors are *equal* if they have the same magnitude and direction.

Suppose we choose a point O in the plane and call it the origin. A vector may be represented by a directed line segment from O to a point $P(x, y)$ in the plane. This vector is denoted by $\overrightarrow{OP}$ or **v**. In the text we use **v**; in your work you will write $\vec{v}$. The magnitude of $\overrightarrow{OP}$ is denoted by $|\mathbf{OP}|$ or $|\mathbf{v}|$. The vector from O to O is called the zero vector, **0**.

If **v** and **w** represent any two vectors having different (but not opposite) directions, they can be drawn so that they have the same base point. If they are represented in this fashion, they determine a geometrical figure called a *parallelogram* by drawing two other sides parallel to the given vectors forming a quadrilateral with the opposite sides parallel and equal in length, as shown in Figure 5.34.

FIGURE 5.34 A parallelogram formed by vectors **v** and **w**

Segments $\overline{OB}$ and $\overline{AC}$ are called the *diagonals* of the parallelogram, and angles having a common side are called *adjacent angles*. It is easy to show that the sum of two adjacent angles in the parallelogram is 180°.

Sum of Vectors (Resultant)

The **sum** or **resultant** of two vectors **v** and **w** is the vector that is the diagonal of the parallelogram having **v** and **w** as the adjacent sides and O as its base, as shown in Figure 5.35. The vectors **v** and **w** are called **components.**

There are basically two types of vector problems dealing with addition of vectors. The first is to find the resultant vector. To do this you use a right triangle, Law of Sines, or Law of Cosines. The second problem is to **resolve** a given vector into two component vectors. If the two vectors form a right angle, they are called **rectangular components.** You will usually resolve a vector into rectangular components.

FIGURE 5.35
Resultant of vectors **v** and **w**

EXAMPLE 2   Finding a resultant vector

Consider two forces, one with magnitude 3.0 in a N20°W direction and the other with magnitude 7.0 in a S50°W direction. Find the resultant vector.

SOLUTION　Sketch the given vectors and draw the parallelogram formed by these vectors. The diagonal is the resultant vector, as shown in Figure 5.36.

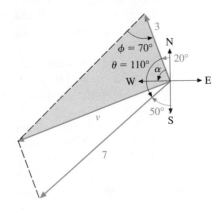

FIGURE 5.36 Finding a resultant vector

You can easily find $\theta = 110°$, but you really need an angle inside the shaded triangle. Use the property from geometry that adjacent angles in a parallelogram are supplementary; that is, they add up to $180°$. This tells you that $\phi = 70°$. Thus, you know SAS, so use the Law of Cosines to find the magnitude, $|\mathbf{v}|$:

$$|\mathbf{v}|^2 = 3^2 + 7^2 - 2(3)(7)\cos 70°$$

$$|\mathbf{v}| \approx 6.60569103$$

The direction of $\mathbf{v}$ can be found using the Law of Sines to find α:

$$\frac{\sin 70°}{|\mathbf{v}|} = \frac{\sin \alpha}{7}$$

$$\sin \alpha = \frac{7}{|\mathbf{v}|}\sin 70° \approx .9957850459$$

Thus, $\alpha \approx 84.737565°$.

Since $20° + 85° = 105°$, you can see that the direction of $\mathbf{v}$ should be measured from the south. Thus, the magnitude of $\mathbf{v}$ is 6.6 and the direction is S75°W.

EXAMPLE 3　**Resolving a vector**

Suppose a vector, $\mathbf{v}$, has a magnitude of 5.00 and a direction given by $\theta = 30.0°$, where θ is the angle the vector makes with the positive x-axis. Resolve this vector into horizontal and vertical components.

SOLUTION Let $\mathbf{v}_x$ be the horizontal component and $\mathbf{v}_y$ the vertical component, as shown in Figure 5.37. Then

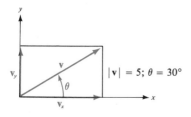

FIGURE 5.37 Resolving a vector

$$\cos\theta = \frac{|\mathbf{v}_x|}{|\mathbf{v}|} \qquad\qquad \sin\theta = \frac{|\mathbf{v}_y|}{|\mathbf{v}|}$$

$$|\mathbf{v}_x| = |\mathbf{v}|\cos\theta \qquad\quad |\mathbf{v}_y| = |\mathbf{v}|\sin\theta$$

$$\quad = 5\cos 30° \qquad\qquad = 5\sin 30°$$

$$\quad = \frac{5}{2}\sqrt{3} \qquad\qquad\quad = \frac{5}{2}$$

$$\quad \approx 4.33 \qquad\qquad\qquad = 2.50$$

Thus, $\mathbf{v}_x$ is a horizontal vector with magnitude 4.33 and $\mathbf{v}_y$ is a vertical vector with magnitude 2.50.

Inclined Plane

One application involves forces acting on an object that is resting on an inclined plane. If the object is resting on a horizontal plane, the weight of the object has the same magnitude as the force pressing against the plane, as shown in Figure 5.38a.

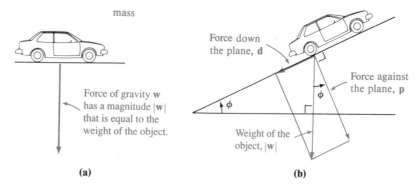

FIGURE 5.38 Inclined plane application

However, if the object is on an inclined plane, the weight, $|\mathbf{w}|$, of the object is the resultant of two forces—the downward pull, $\mathbf{d}$, along the inclined plane and the force pressing against the plane, $\mathbf{p}$, as shown in Figure 5.38b. Since $\mathbf{w}$ and $\mathbf{p}$ are perpendicular to the sides making up the angle ϕ, it follows that the angle of inclination, ϕ, is equal to the angle between the vectors representing the weight of the object and the force against the plane. You can use this information as shown in Example 4.

EXAMPLE 4   Inclined plane application

What is the smallest force necessary to keep a 3,420-lb car from sliding down a hill that makes an angle of 5.25° with the horizontal? Also, what is the force against the hill? Assume that friction is ignored.

SOLUTION The weight of the car, $|\mathbf{OC}|$, acts vertically downward. This vector can be resolved into two vectors: **OH**, the force against the hill, and **OD**, the force pushing down the hill, as shown in Figure 5.39.

FIGURE 5.39

$$\sin 5.25° = \frac{|\mathbf{CH}|}{|\mathbf{OC}|} \qquad\qquad \cos 5.25° = \frac{|\mathbf{OH}|}{|\mathbf{OC}|}$$

$$|\mathbf{CH}| = |\mathbf{OC}| \sin 5.25° \qquad\qquad |\mathbf{OH}| = |\mathbf{OC}| \cos 5.25°$$

$$= (3,420) \sin 5.25° \qquad\qquad = (3,420) \cos 5.25°$$

$$\approx 312.9 \qquad\qquad\qquad \approx 3,406$$

Since $|\mathbf{OD}| = |\mathbf{CH}|$, we see that thus a force of 312.9 lb is necessary to keep the car from sliding down the hill, and the force against the hill is 3,406 lb.

5.5 Problem Set

A *Find the resultant vector in Problems 1–3.*

1. **v** with direction west and magnitude 3.0 and **w** with direction north and magnitude 4.0

2. **v** with direction south and magnitude 4.0 and **w** with direction east and magnitude 3.0

3. **v** with direction east and magnitude 14 and **w** with direction north and magnitude 18

Resolve the vectors given in Problems 4–6 into horizontal and vertical components.

4. $|\mathbf{v}| = 10, \theta = 30°$

5. $|\mathbf{v}| = 23, \theta = 18°$

6. $|\mathbf{v}| = 120, \theta = 27°$

7. NAVIGATION A woman sets out in a rowboat on a due-west heading and rows at 4.8 mph. The current is carrying the boat due south at 12 mph. What is the true course of the rowboat, and how fast is the boat traveling relative to the ground?

8. AVIATION An airplane is headed due west at 240 mph. The wind is blowing due south at 43 mph. What is the true course of the plane, and how fast is it traveling across the ground?

9. PHYSICS A cannon (see Figure 5.40) is fired at an angle of 10.0° above the horizontal, with the initial speed of the cannonball 2,120 fps. Find the magnitude of the vertical and horizontal components of the initial velocity.

FIGURE 5.40

10. PHYSICS Answer Problem 9 for an angle of 17.8°.

11. AVIATION An airplane is heading 215° with a velocity of 723 mph. How far south has it traveled in one hour?

12. AVIATION An airplane is heading 43.0° with a velocity of 248 mph. How far east has it traveled in two hours?

B

13. NAVIGATION A boat (see Figure 5.41) moving at 18.0 knots heads in the direction of S38.2°W across a large river whose current is traveling at 5.00 knots in the direction of S13.1°W. Give the true course of the boat. How fast is it traveling as a result of these forces?

FIGURE 5.41

14. NAVIGATION A boat moving at 25.0 knots heads in the direction of N68.0°W across a river whose current is traveling at 12.0 knots in the direction of N16.0°W. Give the true course of the boat. How fast is it traveling as a result of these forces?

15. PHYSICS Two forces of 220 lb and 180 lb are acting on the same point in directions that differ by 52°. What is the magnitude of the resultant, and what is the angle that it makes with each of the given forces?

16. GEOMETRY If two sides of a parallelogram are 50 cm and 70 cm, and the diagonal connecting the ends of those sides is 40 cm, find the angles of the parallelogram.

17. PHYSICS If the resultant of two forces of 30 pounds and 50 pounds is a force of 60 pounds, find the angle the resultant makes with each component force.

18. NAVIGATION Two boats leave a dock at the same time; one travels S15°W at 8 mph, and the other travels N28°W at 12 mph. How far apart are they in three hours?

19. AVIATION Two airplanes leave an airport at the same time; one travels at 180 mph with a course of 280°, while the other travels at 260 mph with a course of 35°. How far apart are the planes in one hour?

20. NAVIGATION A boat traveling at a constant rate in the direction N25°W passes a 3-mile marker from a lighthouse whose direction from the marker is due west. Twenty minutes after the boat passes the marker, it is N58°E from the lighthouse. How fast is the boat traveling?

21. NAVIGATION A boat leaves a dock at noon and travels N53°W at 10 kph. When will the boat be 12 km from a lighthouse located 15 km due west of the dock?

22. AVIATION At noon a plane leaves an airport and flies with a course of 61° at 200 mph. When will the plane be 35 miles from a point located 50 miles due east of the airport?

23. NAVIGATION Two boats leave port at the same time. The first travels 24.1 mph in the direction of S58.5°W, while the other travels 9.80 mph in the direction of N42.1°E. How far apart are the boats after two hours?

24. NAVIGATION A sailboat (see Figure 5.42) is in a 5.30-mph current in the direction of S43.2°W and is being blown by a 3.20-mph wind in a direction of S25.3°W. Find the course and speed of the sailboat.

FIGURE 5.42

25. PHYSICS The world's steepest standard-gauge railroad grade measures 5.2° and is located between the Samala River Bridge and Zunil in Guatemala. What is the minimum force necessary to keep a boxcar weighing 52.0 tons from sliding down this incline? What is the force against the hill? Assume that friction is ignored.

26. PHYSICS What is the minimum force necessary to keep a 250-lb barrel from sliding down an inclined plane making an angle of 12° with the horizontal? What is the force against the plane? Assume that friction is ignored.

27. PHYSICS A force of 253 lb is necessary to keep a weight of exactly 400 lb from sliding down an inclined plane. What is the angle of inclination of the plane? Assume that friction is ignored.

28. PHYSICS A force of 486 lb is necessary to keep a weight of exactly 800 lb from sliding down an inclined plane. What is the angle of inclination of the plane? Assume that friction is ignored.

29. PHYSICS A cable that can withstand 5,250 lb is used to pull cargo up an inclined ramp for storage. If the inclination of the ramp is 25.5°, find the heaviest piece of cargo that can be pulled up the ramp. Assume that friction is ignored.

30. PHYSICS Answer the question posed in Problem 29 for a ramp whose angle of inclination is 18.2°.

31. SPACE SCIENCE The weight of astronauts on the moon is about $\frac{1}{6}$ of their weight on earth. This fact has a marked effect on such simple acts as walking, running, and jumping. To study these effects and to train astronauts for working under lunar-gravity conditions, scientists at NASA's Langley Research Center have designed an inclined-plane apparatus to simulate reduced gravity (see Figure 5.43). The apparatus consists of a sling that holds the astronaut in a position perpendicular to an inclined plane. The sling is attached to one end of a long cable that runs parallel to the inclined plane. The other end of the cable is attached to a trolley that runs along an overhead track. This device allows the astronaut to move freely in a plane perpendicular to the inclined plane. Let W be the weight of the astronaut and θ the angle between the inclined plane and the ground. Make a vector diagram showing the tension in the cable and the force exerted by the inclined plane against the feet of the astronaut.

FIGURE 5.43 Apparatus to simulate reduced gravity

32. SPACE SCIENCE From the point of view of the astronaut in Problem 31, the inclined plane is the ground and the astronaut's simulated mass (that is, the downward force against the inclined plane) is $W \cos \theta$. What value of θ is required to simulate lunar gravity?

5.6 Vector Operations

The process of resolving a vector can be simplified with a definition of scalar multiplication and two very special vectors. **Scalar multiplication** is the multiplication of a vector by a real number. It is called scalar multiplication because real numbers are sometimes called **scalars**. If c is a positive real number and **v** is a vector, then the vector c**v** is a vector representing the scalar multiplication of c and **v**. It is defined geometrically as a vector in the same direction as **v** but with a magnitude c times the original magnitude of **v**, as shown in Figure 5.44. If c is a negative real number, then the scalar multiplication results in a vector in exactly the opposite direction as **v** with length c times as long, as shown in Figure 5.44. If $c = 0$, then c**v** is the 0 vector.

FIGURE 5.44 Examples of scalar multiplication

Scalar multiplication allows us to define **subtraction** for vectors:

$$\mathbf{v} - \mathbf{w} = \mathbf{v} + (-\mathbf{w})$$

FIGURE 5.45 Vector subtraction

Geometrically, subtraction is shown in Figure 5.45.

Algebraic Representation of a Vector

Two very special vectors help us to treat vectors algebraically as well as geometrically.

THE i AND j VECTORS

i is the vector of unit length in the direction of the positive x-axis.

j is the vector of unit length in the direction of the positive y-axis.

Any vector **v** can be written as

$$\mathbf{v} = a\mathbf{i} + b\mathbf{j}$$

FIGURE 5.46
Algebraic representation of a vector

where a and b are the magnitude of the horizontal and vertical components, respectively. This is called the **algebraic representation of a vector** (see Figure 5.46).

EXAMPLE 1   **Finding the algebraic representation for a vector given magnitude and direction**

Find the algebraic representation for a vector **v** with magnitude 10 making an angle of 60° with the positive x-axis.

SOLUTION   Figure 5.47 shows the general procedure for writing the algebraic representation of a vector when given the magnitude and direction of that vector:

FIGURE 5.47

$$|b\mathbf{j}| = b$$
$$|a\mathbf{i}| = a$$
$$\cos \theta = \frac{a}{|\mathbf{v}|} \qquad a = |\mathbf{v}| \cos \theta$$
$$\sin \theta = \frac{b}{|\mathbf{v}|} \qquad b = |\mathbf{v}| \sin \theta$$
$$\mathbf{v} = |\mathbf{v}| \cos \theta \mathbf{i} + |\mathbf{v}| \sin \theta \mathbf{j}$$

From Figure 5.47,

$$a = |\mathbf{v}| \cos \theta \qquad b = |\mathbf{v}| \sin \theta$$
$$= 10 \cos 60° \qquad = 10 \sin 60°$$
$$= 5 \qquad\qquad \approx 8.7$$

Therefore,

$$\mathbf{v} = 5\mathbf{i} + 8.7\mathbf{j}$$

EXAMPLE 2   **Finding the algebraic representation for a vector given two points**

Find the algebraic representation for a vector **v** with initial point $(4, \, -3)$ and endpoint $(-2, 4)$.

SOLUTION   Figure 5.48 shows the general procedure for writing the algebraic representation of a vector when given the endpoints of that vector.

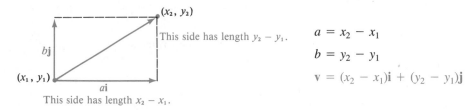

FIGURE 5.48

From Figure 5.48,

$$a = x_2 - x_1 \qquad b = y_2 - y_1$$
$$= -2 - 4 \qquad = 4 - (-3)$$
$$= -6 \qquad = 7$$

Thus,

$$\mathbf{v} = -6\mathbf{i} + 7\mathbf{j}$$

Magnitude of a Vector

EXAMPLE 3   **Finding the magnitude of a vector**

Find the magnitude of the vector in Example 2.

SOLUTION $\mathbf{v} = -6\mathbf{i} + 7\mathbf{j}$. Thus,

$$|\mathbf{v}| = \sqrt{(-6)^2 + (7)^2} \qquad \text{Pythagorean theroem}$$
$$= \sqrt{36 + 49}$$
$$= \sqrt{85}$$

Example 3 leads to the following general result.

MAGNITUDE OF A VECTOR

The **magnitude** of a vector $\mathbf{v} = a\mathbf{i} + b\mathbf{j}$ is given by
$$|\mathbf{v}| = \sqrt{a^2 + b^2}$$

The operations of addition, subtraction, and scalar multiplication can also be stated algebraically. Let $\mathbf{v} = a\mathbf{i} + b\mathbf{j}$ and $\mathbf{w} = c\mathbf{i} + d\mathbf{j}$. Then

$$\mathbf{v} + \mathbf{w} = (a + c)\mathbf{i} + (b + d)\mathbf{j}$$
$$\mathbf{v} - \mathbf{w} = (a - c)\mathbf{i} + (b - d)\mathbf{j}$$
$$c\mathbf{v} = ca\mathbf{i} + cb\mathbf{j}$$

EXAMPLE 4   Vector operations

Let $\mathbf{v} = 6\mathbf{i} + 4\mathbf{j}$ and $\mathbf{w} = -2\mathbf{i} + 3\mathbf{j}$.
Find: **a.** $|\mathbf{v}|$ **b.** $|\mathbf{w}|$ **c.** $\mathbf{v} + \mathbf{w}$ **d.** $\mathbf{v} - \mathbf{w}$ **e.** $-\mathbf{v}$ **f.** $-2\mathbf{w}$

SOLUTION   **a.** $|\mathbf{v}| = \sqrt{6^2 + 4^2}$ **b.** $|\mathbf{w}| = \sqrt{(-2)^2 + 3^2}$
$= \sqrt{36 + 16}$ $= \sqrt{4 + 9}$
$= 2\sqrt{13}$ $= \sqrt{13}$

c. $\mathbf{v} + \mathbf{w} = (6 - 2)\mathbf{i} + (4 + 3)\mathbf{j}$ **d.** $\mathbf{v} - \mathbf{w} = (6 + 2)\mathbf{i} + (4 - 3)\mathbf{j}$
$= 4\mathbf{i} + 7\mathbf{j}$ $= 8\mathbf{i} + \mathbf{j}$

e. $-\mathbf{v} = (-1)6\mathbf{i} + (-1)4\mathbf{j}$ **f.** $-2\mathbf{w} = (-2)(-2)\mathbf{i} + (-2)(3)\mathbf{j}$
$= -6\mathbf{i} - 4\mathbf{j}$ $= -4\mathbf{i} - 6\mathbf{j}$ ∎

You can see from Example 4 that the algebraic representation of a vector makes it easy to handle vectors and their operations. Another advantage of the algebraic representation is that it specifies a direction and a magnitude, but not a particular location. The directed line segment in Example 2 defined a given vector. However, there are infinitely many other vectors represented by the form $-6\mathbf{i} + 7\mathbf{j}$. Two of these are shown in Figure 5.49.

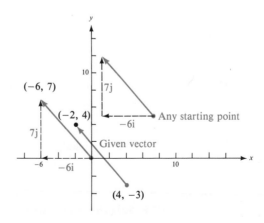

FIGURE 5.49 Some vectors represented by $\mathbf{v} = -6\mathbf{i} + 7\mathbf{j}$

Scalar Product

In arithmetic you are used to using the words *multiplication* and *product* to mean the same thing. When working with vectors these words are used to denote different ideas. Recall that scalar multiplication does not tell us how to multiply two vectors; instead, it tells us how to multiply a scalar and a vector. Now we define an operation called **scalar product** in order to multiply two vectors to obtain a number. It is called *scalar* product because we obtain a number, or scalar, as an answer. Sometimes scalar product is called *dot product* or *inner product*. In subsequent courses you will define another vector multiplication, called *vector product*, in which you obtain a vector as an answer.

SCALAR PRODUCT
Let $\mathbf{v} = a\mathbf{i} + b\mathbf{j}$ and $\mathbf{w} = c\mathbf{i} + d\mathbf{j}$. Then the **scalar product**, written $\mathbf{v} \cdot \mathbf{w}$, is defined by

$$\mathbf{v} \cdot \mathbf{w} = ac + bd$$

EXAMPLE 5   Finding the scalar product

Find the scalar product of the given vectors.

a. If $\mathbf{v} = 2\mathbf{i} + 5\mathbf{j}$ and $\mathbf{w} = 6\mathbf{i} - 3\mathbf{j}$, then

$$\begin{aligned} \mathbf{v} \cdot \mathbf{w} &= 2(6) + 5(-3) \\ &= 12 - 15 \\ &= -3 \end{aligned}$$

b. If $\mathbf{v} = \cos 30° \, \mathbf{i} + \sin 30° \, \mathbf{j}$ and $\mathbf{w} = \cos 60° \, \mathbf{i} - \sin 60° \, \mathbf{j}$, then

$$\begin{aligned} \mathbf{v} \cdot \mathbf{w} &= \cos 30° \cos 60° - \sin 30° \sin 60° \\ &= \cos(30° + 60°) \\ &= 0 \end{aligned}$$

c. If $\mathbf{v} = -\sqrt{3}\mathbf{i} + \sqrt{2}\mathbf{j}$ and $\mathbf{w} = 3\sqrt{3}\mathbf{i} + 5\sqrt{2}\mathbf{j}$, then

$$\begin{aligned} \mathbf{v} \cdot \mathbf{w} &= (-\sqrt{3})(3\sqrt{3}) + \sqrt{2}(5\sqrt{2}) \\ &= -9 + 10 \\ &= 1 \end{aligned}$$

d. If $\mathbf{v} = 2\mathbf{i} - 3\mathbf{j}$ and $\mathbf{w} = 4\mathbf{i} + a\mathbf{j}$, then

$$\mathbf{v} \cdot \mathbf{w} = 8 - 3a$$

Angle Between Vectors

There is a very useful geometric property for scalar product that is apparent if we find an expression for the angle between two vectors.

ANGLE BETWEEN VECTORS

The angle θ between nonzero vectors $\mathbf{v}$ and $\mathbf{w}$ is found by

$$\cos \theta = \frac{\mathbf{v} \cdot \mathbf{w}}{|\mathbf{v}| \, |\mathbf{w}|}$$

PROOF   Let $\mathbf{v} = a\mathbf{i} + b\mathbf{j}$ and $\mathbf{w} = c\mathbf{i} + d\mathbf{j}$ be drawn with their bases at the origin, as shown in Figure 5.50. Let x be the distance between the endpoints of the vectors. Then, by the Law of Cosines,

$$
\begin{aligned}
\cos \theta &= \frac{|\mathbf{v}|^2 + |\mathbf{w}|^2 - x^2}{2|\mathbf{v}| \, |\mathbf{w}|} \\[2mm]
&= \frac{\left(\sqrt{a^2 + b^2}\right)^2 + \left(\sqrt{c^2 + d^2}\right)^2 - \left(\sqrt{(a - c)^2 + (b - d)^2}\right)^2}{2|\mathbf{v}| \, |\mathbf{w}|} \\[2mm]
&= \frac{a^2 + b^2 + c^2 + d^2 - (a^2 - 2ac + c^2 + b^2 - 2bd + d^2)}{2|\mathbf{v}| \, |\mathbf{w}|} \\[2mm]
&= \frac{2ac + 2bd}{2|\mathbf{v}| \, |\mathbf{w}|} \\[2mm]
&= \frac{ac + bd}{|\mathbf{v}| \, |\mathbf{w}|} \\[2mm]
&= \frac{\mathbf{v} \cdot \mathbf{w}}{|\mathbf{v}| \, |\mathbf{w}|}
\end{aligned}
$$

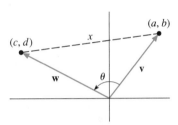

FIGURE 5.50 Finding the angle between two vectors

Orthogonal Vectors

There is a useful geometric property of vectors whose directions differ by 90°. If you are dealing with lines, they are called **perpendicular lines**, but if they are vectors forming a 90° angle, they are called **orthogonal**. Notice, if $\theta = 90°$, then $\cos 90° = 0$; therefore, from the angle between vectors formula,

$$0 = \frac{\mathbf{v} \cdot \mathbf{w}}{|\mathbf{v}| \, |\mathbf{w}|}$$

If you multiply both sides by $|\mathbf{v}| \, |\mathbf{w}|$, the result is the following important condition.

ORTHOGONAL VECTORS

Nonzero vectors $\mathbf{v}$ and $\mathbf{w}$ are **orthogonal** if and only if $\mathbf{v} \cdot \mathbf{w} = 0$.

EXAMPLE 6   Showing that given vectors are orthogonal

Show that $\mathbf{v} = 3\mathbf{i} - 2\mathbf{j}$ and $\mathbf{w} = 6\mathbf{i} + 9\mathbf{j}$ are orthogonal.

SOLUTION

$$\mathbf{v} \cdot \mathbf{w} = 3(6) + (-2)(9)$$
$$= 18 - 18 = 0$$

Since the scalar product is 0, the vectors are orthogonal.

EXAMPLE 7   **Finding a constant so that orthogonal vectors result**

Find a so that $\mathbf{v} = 3\mathbf{i} + a\mathbf{j}$ and $\mathbf{w} = \mathbf{i} - 2\mathbf{j}$ are orthogonal.

SOLUTION

$$\mathbf{v} \cdot \mathbf{w} = 3 - 2a$$

If they are orthogonal, then

$$3 - 2a = 0$$
$$a = \frac{3}{2}$$

5.6   Problem Set

A *Find the algebraic representation for each vector given in Problems 1–6. Approximate answers should be rounded to four decimal places. $|\mathbf{v}|$ is the magnitude of the vector $\mathbf{v}$, and θ is the angle the vector makes with the positive x-axis. A and B are the endpoints of the vector $\mathbf{v}$, and A is the base point. Draw each vector.*

1. $|\mathbf{v}| = 12$, $\theta = 60°$ **2.** $|\mathbf{v}| = 8$, $\theta = 30°$
3. $|\mathbf{v}| = 4$, $\theta = 112°$ **4.** $|\mathbf{v}| = 10$, $\theta = 214°$
5. $A(4, 1)$, $B(2, 3)$; $\overrightarrow{AB}$ **6.** $A(-1, -3)$, $B(4, 5)$; $\overrightarrow{AB}$

Find the magnitude of each of the vectors given in Problems 7–12.

7. $\mathbf{v} = 3\mathbf{i} + 4\mathbf{j}$ **8.** $\mathbf{v} = 5\mathbf{i} - 12\mathbf{j}$
9. $\mathbf{v} = 6\mathbf{i} - 7\mathbf{j}$ **10.** $\mathbf{v} = -3\mathbf{i} + 5\mathbf{j}$
11. $\mathbf{v} = -2\mathbf{i} + 2\mathbf{j}$ **12.** $\mathbf{v} = 5\mathbf{i} - 8\mathbf{j}$

State whether the given pairs of vectors in Problems 13–16 are orthogonal.

13. $\mathbf{v} = 3\mathbf{i} - 2\mathbf{j}$ **14.** $\mathbf{v} = 2\mathbf{i} + 3\mathbf{j}$
    $\mathbf{w} = 6\mathbf{i} + 9\mathbf{j}$ $\mathbf{w} = 6\mathbf{i} - 9\mathbf{j}$

15. $\mathbf{v} = 4\mathbf{i} - 5\mathbf{j}$
    $\mathbf{w} = 8\mathbf{i} + 10\mathbf{j}$

16. $\mathbf{v} = 5\mathbf{i} + 4\mathbf{j}$
    $\mathbf{w} = 8\mathbf{i} - 10\mathbf{j}$

B *In Problems 17–25 find $\mathbf{v} \cdot \mathbf{w}$, $|\mathbf{v}|$, $|\mathbf{w}|$, and $\cos \theta$, where θ is the angle between $\mathbf{v}$ and $\mathbf{w}$.*

17. $\mathbf{v} = 3\mathbf{i} + 4\mathbf{j}$ **18.** $\mathbf{v} = 8\mathbf{i} - 6\mathbf{j}$
    $\mathbf{w} = 5\mathbf{i} + 12\mathbf{j}$ $\mathbf{w} = -5\mathbf{i} + 12\mathbf{j}$

19. $\mathbf{v} = 2\mathbf{i} + \sqrt{5}\mathbf{j}$ **20.** $\mathbf{v} = 7\mathbf{i} - \sqrt{15}\mathbf{j}$
    $\mathbf{w} = 3\sqrt{5}\mathbf{i} - 3\mathbf{j}$ $\mathbf{w} = 2\sqrt{15}\mathbf{i} + 14\mathbf{j}$

21. $\mathbf{v} = 2\mathbf{i} + 3\mathbf{j}$ **22.** $\mathbf{v} = 3\mathbf{i} + 9\mathbf{j}$
    $\mathbf{w} = 6\mathbf{i} + 5\mathbf{j}$ $\mathbf{w} = 2\mathbf{i} - 5\mathbf{j}$

23. $\mathbf{v} = \mathbf{i}$ **24.** $\mathbf{v} = \mathbf{j}$ **25.** $\mathbf{v} = \mathbf{i}$
    $\mathbf{w} = \mathbf{i}$ $\mathbf{w} = \mathbf{j}$ $\mathbf{w} = -\mathbf{j}$

In Problems 26–31 find the angle θ to the nearest degree, $0° \leq \theta \leq 180°$, between the vectors $\mathbf{v}$ and $\mathbf{w}$.

26. $\mathbf{v} = \frac{1}{2}\mathbf{i} + \frac{\sqrt{3}}{2}\mathbf{j}$ **27.** $\mathbf{v} = \sqrt{2}\mathbf{i} - \sqrt{2}\mathbf{j}$
    $\mathbf{w} = \frac{1}{2}\mathbf{i} + \frac{1}{2}\mathbf{j}$ $\mathbf{w} = \frac{\sqrt{3}}{2}\mathbf{i} + \frac{1}{2}\mathbf{j}$

28. $\mathbf{v} = \mathbf{j}$ **29.** $\mathbf{v} = -\mathbf{i}$
    $\mathbf{w} = \frac{1}{2}\mathbf{i} - \frac{\sqrt{3}}{2}\mathbf{j}$ $\mathbf{w} = -2\sqrt{2}\mathbf{i} + 2\sqrt{2}\mathbf{j}$

30. $\mathbf{v} = 2\mathbf{i} + 3\mathbf{j}$ **31.** $\mathbf{v} = -3\mathbf{i} + 2\mathbf{j}$
    $\mathbf{w} = -\mathbf{i} + 4\mathbf{j}$ $\mathbf{w} = 6\mathbf{i} + 9\mathbf{j}$

In Problems 32–34 find a number a so that the given vectors are orthogonal.

32. $v = 2i + 3j$ **33.** $v = 4i - aj$ **34.** $v = ai + 5j$ **35.** If $a = |v|$ and $b = |w|$, show that $aw + bv$ and $bv - aw$
      $w = 5i + aj$ $w = -2i + 5j$ $w = ai - 15j$ are orthogonal.

5.7   Summary and Review

Objectives	*Pages / Examples / Problems*

5.1   Right Triangles

1. State the right-triangle definition of the trigonometric functions.   p. 180, Examples 1–2.

2. Round triangle descriptions to the correct number of significant digits.   p. 181, Examples 3; (all triangle solutions).

3. Solve right triangles.   pp. 182–184; Examples 4–5; Problems 1–18.

4. Solve applied right-triangle problems.   pp. 184–186; Examples 6–8; Problems 19–38.

5.2   Law of Cosines

5. State the Law of Cosines and tell when it is used.   p. 190.

6. Solve SSS triangles using the Law of Cosines.   pp. 190–191; Example 1; Problems 1–6, 18, 21, 26, 27.

7. Solve SAS triangles using the Law of Cosines.   p. 192; Example 2; Problems 7–12, 23, 24, 25.

8. Solve SSA triangles using the Law of Cosines.   pp. 192–195; Examples 3–6; Problems 13–17, 19, 22, 28.

9. Solve applied problems using the Law of Cosines.   p. 200; Example 3 (Section 5.3); Problems 29–36.

5.3   Law of Sines

10. State the Law of Sines and tell when it is used.   p. 198.

11. Solve ASA or AAS triangles using the Law of Sines.   p. 199; Example 1; Problems 1–24.

12. Solve applied problems using the Law of Sines.   pp. 200–201; Example 2; Problems 25–34; 38, 47.

13. Solve triangles by deciding the method of solution.   p. 200; Example 3; Problems 1–34.

5.4   Areas and Volumes

14. Find the area of a triangle given   pp. 205–206; Examples 1–3; Problems 1–24, 28.
   a. two sides and an included angle (SAS);
   b. two angles and an included side (ASA);
   c. three sides (SSS).

15. Find the area of a sector of a circle.   p. 207; Examples 4–5; Problems 25–27.

16. Find the volume of a cone when the height is not known.   pp. 207–208; Example 6; Problems 29–31.

5.5 Vector Triangles

17. Given two nonzero vectors, find the resultant vector.

pp. 211–213; Example 2; Problems 1–3.

18. Solve applied problems by finding the resultant vector.

pp. 213–214; Example 2; Problems 7–24.

19. Resolve a vector into horizontal and vertical components.

pp. 212–213; Example 3; Problems 4–6.

20. Solve applied problems involving inclined planes.

pp. 213–214; Example 4; Problems 25–31.

5.6 Vector Operations

21. Write the algebraic representation of a vector.

pp. 217–218; Examples 1–2; Problems 1–6.

22. Find the sum, difference, and magnitude of a vector.

pp. 218–219; Examples 3–4; Problems 7–12.

23. Find the scalar product of two vectors.

p. 220; Example 5; Problems 17–25.

24. Find the angle between vectors.

p. 221; Problems 26–31.

25. Determine whether two vectors are orthogonal.

pp. 221–222; Examples 6–7; Problems 13–16, 32–34.

TERMS

AAA [5.2]
AAS [5.3]
Accuracy [5.1]
Addition of vectors [5.5]
Adjacent side [5.1]
Algebraic representation [5.6]
Ambiguous case [5.2]
Angle between vectors [5.6]
Area [5.4]
ASA [5.3]
Bearing [5.1]
Components [5.5]

Depression, angle of [5.1]
Direction of a vector [5.5]
Elevation, angle of [5.1]
Heading [5.5]
Heron's formula [5.4]
Hypotenuse [5.1]
i, [5.6]
Inclination, angle of [5.1]
Inclined plane [5.5]
j, [5.6]
Law of Cosines [5.2]

Law of Sines [5.3]
Magnitude [5.5, 5.6]
Oblique triangles [5.2]
Opposite side [5.1]
Orthogonal [5.6]
Perpendicular lines [5.6]
Rectangular components [5.5]
Resolve [5.5]
Resultant vector [5.5]
SAS [5.2]
Scalar multiplication [5.6]

Scalar product [5.6]
Sector [5.4]
Significant digits [5.1]
Solved triangle [5.1]
SSA [5.2]
SSS [5.2]
Subtraction of vectors [5.6]
Sum of vectors [5.5]
Triangle [5.1]
Vector [5.5]
Volume of a cone [5.4]

CHAPTER 5 REVIEW OF OBJECTIVES

The problem numbers correspond to the objectives listed in Section 5.7.

1. Fill in the blanks by referring to the right triangle in Figure 5.51.

a. $\cos \beta =$ _____

b. $\sin \beta =$ _____

c. $\tan \beta =$ _____

FIGURE 5.51

2. Round each set of triangle descriptions to the least accurate of the measurements.

a. $a = 59.8$ $\alpha = 30.72298°$

$b = 77.9$ $\beta = 41.72231°$

$c = 111.6$ $\gamma = 107.5547°$

b. $a = 3.488649$ $\alpha = 79°$

$b = .8917898$ $\beta = 87.3°$

$c = 3.55$ $\gamma = 13.7°$

c. $a = .5976337$ $\alpha = 3.77°$
   $b = 7.7$ $\beta = 58°$
   $c = 8.0$ $\gamma = 118.226°$

3. Solve the triangle shown in Figure 5.51 if

$$a = 5.30 \quad \text{and} \quad b = 12.2$$

4. a. The tallest human-built structure in the world is the TV transmitter tower of KTHI-TV in North Dakota. Find its height if the angle of elevation at 501.0 ft away is 76.35°.

b. To measure the distance across the Rainbow Bridge in Utah, a surveyor selected two points P and Q on either end of the bridge. From point Q the surveyor measured 500 ft in the direction N38.4°E to point R. Point P was then determined to be in the direction S67.5°W. What is the distance across the Rainbow Bridge if all the preceding measurements are in the same plane and angle PQR is a right angle?

5. Fill in the blanks by using the Law of Cosines.

a. $a^2 = $ _____   **b.** $\cos \beta = $ _____

6. Solve the triangle where $a = 14$, $b = 27$, and $c = 19$.

7. Solve the triangle where $b = 7.2$, $c = 15$, and $\alpha = 113°$.

8. Ferndale is 7.0 miles N50°W of Fortuna. If I leave Fortuna at noon and travel due west at 2.0 mph, when will I be exactly 6.0 miles from Ferndale?

9. ENGINEERING A mine shaft is dug into the side of a sloping hill. The shaft is dug horizontally for 485 feet. Next, a turn is made so that the angle of elevation of the second shaft is 58.0°, thus forming a 58° angle between the shafts. The shaft is then continued for 382 feet before exiting, as shown in Figure 5.52. How far is it along a straight line from the entrance to the exit, assuming that all tunnels are in a single plane?

FIGURE 5.52 Determining the exit of a mine shaft

10. State the Law of Sines.

11. Solve the triangle where $a = 92.6$, $\alpha = 18.3°$, $\beta = 112.4°$.

12. A triangularly shaped garden has two angles measuring 46.5° and 105.8°, with the side opposite the 46.5° angle measuring 38.0 m. How much fence is needed to enclose the garden?

13. Fill in the blanks for the triangle ABC.

TO FIND	KNOWN	PROCEDURE	SOLUTION
Example:		Solution:	
β	a, b, α	$\dfrac{\sin \alpha}{a} = \dfrac{\sin \beta}{b}$	$\beta = \sin^{-1}\left(\dfrac{b \sin \alpha}{a}\right)$
α	a, b, c	**a.**	**b.**
α	a, b, β	**c.**	**d.**
α	a, β, γ	**e.**	**f.**
b	a, α, β	**g.**	**h.**
b	a, c, β	**i.**	**j.**

14. Find the area of each of the following triangles.

a. $b = 16$, $c = 43$, $\alpha = 113°$

b. $\alpha = 40.0°$, $\beta = 51.8°$, $c = 14.3$

c. $a = 121$, $b = 46$, $c = 92$

15. A farmer plows a circular field with a radius of 1,000 ft into separate equal sections. If one section has a central angle of 15° ($\frac{\pi}{12}$ radians), what is the area of one section?

16. If a conical pile of sawdust has a radius of 135 ft and an angle of elevation of 35.0°, what is the volume of the pile?

17. PHYSICS Consider two forces, one with a magnitude 5.0 in a S30°E direction and another with a magnitude of 12.0 in a N40°W direction. Find the resultant vector.

18. PHYSICS An object is hurled from a catapult due east with a velocity of 38 feet per second (fps). If the wind is blowing due south at 15 mph (22 fps), what are the true bearing and the velocity of the object?

19. Resolve the vector with magnitude 4.5 and $\theta = 51°$ into horizontal and vertical components.

20. PHYSICS What is the smallest force necessary to keep a 3-ton boxcar from sliding down a hill that makes an angle of 4° with the horizontal? Assume that friction is ignored.

21. Find the algebraic representation of the given vectors.

a. $|\mathbf{v}| = \sqrt{2}$, $\theta = 45°$

b. $|\mathbf{v}| = 9.3$, $\theta = 118°$

c. $A(1, -2), B(-5, -7); \overrightarrow{AB}$

22. Find the magnitude of $\mathbf{v} = 3\mathbf{i} - 2\mathbf{j}$ and $\mathbf{w} = \mathbf{i} - \mathbf{j}$.

23. Find $\mathbf{v} \cdot \mathbf{w}$ for the vectors in Problem 22.

24. Find the angle (to the nearest degree) between the vectors given in Problem 22.

25. If $\mathbf{t} = \mathbf{i} - a\mathbf{j}$, find a so that $\mathbf{v}$ (from Problem 22) and $\mathbf{t}$ are orthogonal.

DISCUSSION PROBLEMS

1. In your own words, describe what is meant by "the ambiguous case."

2. Outline your "method of attack" for solving triangles.

3. Describe a process for finding the area of a triangle.

4. Write a computer program for solving triangles.

5. Write a 500-word paper on the importance and uses of vectors.

MISCELLANEOUS PROBLEMS

Solve the triangles given in Problems 1–30.

1. $a = 24; c = 61; \beta = 58°$

2. $b = 34; c = 21; \gamma = 16°$

3. $a = 30; \beta = 50°; \gamma = 100°$

4. $b = 23; \alpha = 25°; \beta = 110°$

5. $b = 40; \alpha = 50°; \gamma = 60°$

6. $b = 90; \beta = 85°; \gamma = 25°$

7. $a = 15; b = 8.0; \gamma = 38°$

8. $a = 18; c = 11; \beta = 63°$

9. $b = 14; c = 12; \alpha = 82°$

10. $b = 21; c = 35; \alpha = 125°$

11. $a = 7.0; b = 8.0; c = 2.0$

12. $a = 10; b = 4.0; c = 8.0$

13. $a = 6.80; b = 12.2; c = 21.5$

14. $b = 4\,6; \alpha = 108°; \gamma = 38°$

15. $a = 14.5; b = 17.2; \alpha = 35.5°$

16. $\alpha = 48.0°; b = 25.5; c = 48.5$

17. $a = 121; b = 315; \gamma = 50.0°$

18. $a = 6.50; b = 8.30; c = 12.6$

19. $a = 11; b = 9.0; c = 8.0$

20. $a = 12; b = 6.0; c = 15$

21. $a = 12.4; b = 32.5; c = 35.1$

22. $a = 2.72; c = 3.57; \alpha = 43.7°$

23. $b = 68.2; \alpha = 145°; \beta = 52.4°$

24. $a = 151; b = 235; c = 412$

25. $\beta = 15.0°; \gamma = 18.0°; b = 23.5$

26. $b = 45.7; \alpha = 82.3°; \beta = 61.5°$

27. $b = 82.5; c = 52.2; \gamma = 23.1°$

28. $a = 352; b = 230; c = 418$

29. $a = 10.2; b = 11.8; \alpha = 47.0°$

30. $a = 8.62; b = 11.8; \alpha = 37.0°$

31. SURVEYING A surveyor needs to estimate the distance of a tunnel through a hill. The distances to either end are 1,000 ft and 1,236 ft with an angle of 50.6° between these sides, as shown in Figure 5.53. What is the length of the tunnel?

Proposed tunnel

FIGURE 5.53

32. SURVEYING What is the area of a trianglar lot whose sides measure 275 ft, 690 ft, and 735 ft?

33. AVIATION A pilot leaves Santa Rosa and flies 250 miles with a heading of 128° to the Nut Tree. He then leaves the Nut Tree at noon and flies north at 160 mph. When he is exactly 200 miles from Santa Rosa after leaving the Nut Tree, he must log his time. What time (to the nearest minute) should his log show?

34. AVIATION San Francisco and Los Angeles are approximately 450 miles apart. A pilot flying from San Francisco finds that she is 15.0° off course after 150 miles. How far is she from Los Angeles at this time?

35. A fire is simultaneously spotted from two observation towers that are located 10,520 ft apart, as shown in Figure 5.54. The first observation post reports the fire at N54.8°E and the second post reports the fire at N48.2°W. How far is the fire from each observation location? Assume that all the mea-

surements are in the same plane and that the towers are on an east–west line.

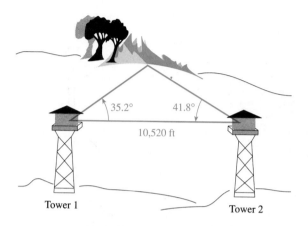

FIGURE 5.54 Triangulation for the location of a fire

36. A large advertising sign stands on top of a tall building. The angle of elevation from a point on level ground 275 ft from the base of the building up to the bottom of the sign is 18.5°, and to the top of the sign it is 21.5°. How tall is the sign?

37. A drive for mounting a camera for time-exposure astrophotography used in astronomy is illustrated in Figure 5.55.* Let $\beta = b/c$. Assume

a. $\theta = 2 \arcsin\left(\dfrac{a}{2r}\right)$ for the single-arm drive.

b. $\phi = \theta \arcsin\left(\dfrac{\sin \theta}{\beta}\right)$ for the Type-1 drive.

c. $\phi = \theta \arctan\left(\dfrac{\beta \sin \theta}{\beta \cos \theta + 1}\right)$ for the Type-2 drive.

d. $\phi = \theta \arctan\left(\dfrac{\beta \sin \theta}{\beta \cos \theta - 1}\right)$ for the Type-3 drive.

e. $\phi = \theta + \arcsin\left(\dfrac{\sin \theta}{\beta}\right)$ for the Type-4 drive.

If you were to construct a Type-3 mounting so that $r = 13.519$, $\beta = 6.464$, $a = 2.000$, and $b = 12.928$, what is ϕ, measured in degrees?

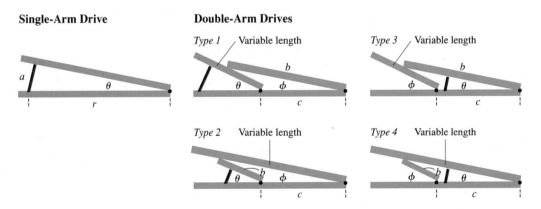

FIGURE 5.55 Dave Trott's double-arm drive

---

*From "Two Arms Are Better Than One," by Roger W. Sinnott, *Sky & Telescope*, April 1989, p. 437.

▎ Parking Lot Design

Courtesy of Wide World Photos, Inc.

$c =$ curb length per space

FIGURE 5.56 Street parking

A city engineer must design the parking along a certain street. She can use parallel or angle parking. Angle parking requires greater street width, as shown in Figure 5.56, but it allows more spaces along the curb. Let us assume that city code calls for a 10-ft width per parking space. This means that 100 ft of curb could accommodate ten parking spaces if $\theta = 90°$.

EXAMPLE 1 Finding the curb length

Find the curb length required for ten spaces if $\theta = 45°$.

SOLUTION From Figure 5.56, $\sin \theta = \dfrac{10}{c}$, so $c = \dfrac{10}{\sin \theta}$. For $\theta = 45°$,

$$c = \frac{10}{\sin 45°} \approx 14.14213562$$

This means that at 45° the curb length required for ten spaces is a little less than 142 ft. This is an increase in curb space of about 42% over the amount necessary for $\theta = 90°$. ▎

For this example, you can see that changing θ from 90° to 45° increased the amount of curb space needed by 42%. Now consider the change in the amount of street space needed. Would you guess it is larger or smaller? Suppose that city code calls for parking spaces to be 21 ft long, as shown in Figure 5.57.

The parking-space length cannot be measured from P, so the 21-ft requirement is measured from a point Q, where Q is drawn so that PQ is perpendicular to QR. We see from Figure 5.7 that, for $\theta = 90°$, $w = 21$. Let $\alpha = 90° - \theta$. Then

$$\cos \alpha = \frac{w}{21 + |PQ|}$$
$$w = (21 + |PQ|) \cos \alpha$$

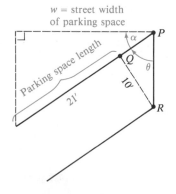

$w =$ street width of parking space

Parking space length

21'

FIGURE 5.57 Parking-space angle

Notice that $\tan \theta = \dfrac{10}{|PQ|}$, so $|PQ| = \dfrac{10}{\tan \theta} = 10 \cot \theta$, which can be substituted into the equation for w:

$$w = (21 + 10 \cot \theta) \cos \alpha$$
$$ = (21 + 10 \cot \theta) \sin \theta \qquad \text{Cofunctions of complementary angles are equal.}$$
$$w = 21 \sin \theta + 10 \cot \theta \sin \theta$$
$$ = 21 \sin \theta + 10 \cos \theta$$

For $\theta = 45°$,

$$w = 21 \sin 45° + 10 \cos 45°$$
$$ \approx 21.92$$

This result means that for an angle of 45° the width that the parking space extends into the street is *increased* by about 4%. Does this match your guess?

EXAMPLE 2   **Finding the street width as a function of parking-space angle**

How does the street width change if θ is changed from 90° to 30°?

SOLUTION $w = 21 \sin 30° + 10 \cos 30°$
$$\approx 19.16$$

This result is a decrease of about 1.8 ft, or about 9%, in the street width of the parking space.   ∎

PROBLEMS FOR FURTHER STUDY: PARKING-LOT DESIGN

1. Complete the table at the right for the parking-lot design problem presented in this section.

2. **Engineering** You can see from the table in Problem 1 that there appear to be some physical limitations. For example, if c exceeds 21 ft, parallel parking appears to be more advantageous (why?). Find the angle θ for which $c = 21$, and then specify what you think would be a reasonable domain for θ for angle parking.

3. **Engineering** Suppose you measured the parking width in Figure 5.57 from P instead of Q, as shown in Figure 5.58. For example, if $\theta = 50°$, then $\alpha = 40°$ and

$$\cos \alpha = \frac{w}{21}$$
$$w = 21 \cos 40°$$
$$\approx 16 \text{ ft}$$

But this would make the usable length of the space 21 ft $- |QP|$ or

$$21 - 10 \cot \theta = 21 - 10 \cot 50°$$
$$\approx 13 \text{ ft}$$

θ	c	w
10°		
20°		
30°		
40°		
50°		
60°		
70°		
80°		
90°		

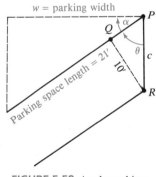

FIGURE 5.58 Angle parking

Complete a table similar to the one shown in Problem 1 for usable length of various parking spaces. Calculate values of θ for 30°, 40°, 50°, . . . , 80°, and 90°.

4. **Engineering** If street size were of primary importance, what type of parking would you recommend, based on your observations from Problems 1–3?

5. **Engineering** If curb size were of primary importance, what type of parking would you recommend, based on your observations from Problems 1–3?

▌ Abraham De Moivre
(1667–1754)

The naming of results in the history of mathematics has often been a strange process. In this chapter we introduce De Moivre's formula, named after Abraham De Moivre, who used his famous formula in *Philosophical Transactions* (1707) and in *Miscellanea analytica* (1730). He was an outstanding mathematician who worked extensively with probability. However, De Moivre's formula was really first used by Roger Cotes (1682–1716). On the other hand, De Moivre was the inventor of a result called Stirling's formula.

The great mathematician Karl Gauss used the concept of a complex number so that he could prove that all the roots of *any* algebraic equation have the form $a + bi$. In his work with fluid mechanics and the theory of gravitation of its particles, the mathematician Henri Poincaré (1854–1912) displayed a comprehensive grasp of the concept of a complex number. Poincaré has been described as the last universalist in mathematics in the sense that he considered all of mathematics within his province.

De Moivre was unable to receive a professorship, so he supported himself by making a London pub his headquarters and solving problems for anyone who came to him. Howard Eves told the following story about De Moivre's death: De Moivre noticed that each day he required a quarter of an hour more sleep than on the preceding day. De Moivre said he would die when the arithmetic progression reached 24 hours. Finally, after sleeping longer and longer every night, he went to bed and slept for more than 24 hours and then died in his sleep.

6

COMPLEX NUMBERS

Contents

6.1 The Imaginary Unit and Complex Numbers

6.2 Trigonometric Form of Complex Numbers

6.3 De Moivre's Formula

6.4 Summary and Review

Application for Further Study
   Electric Generator

From *Calculus, Second Edition* by James Stewart, (Pacific Grove: Brooks/Cole Publishing Company, 1991), p. A48.

Preview

In mathematics, several applications give rise to equations such as $x^2 + 1 = 0$, which cannot be solved in the set of real numbers. A set of **complex numbers** has been defined that includes not only the real numbers, but also square roots of negative numbers. The number i, defined in Section 6.1, is called the **imaginary unit,** although it is no more "imaginary" than the number 5. When working with complex and imaginary numbers, you must keep in mind that *all* numbers are concepts and that the number i is just as much a number as any of the "real" numbers. The trigonometric form of a complex number, introduced in Section 6.2, is a very useful form you will use not only in this course, but also in your future mathematical work.

Perspective

Complex numbers are not used extensively in beginning calculus, but instead are saved for a branch of calculus called **complex analysis.** There is, however, a very important use of complex numbers in a topic of calculus called Taylor series. This formula is called **Euler's formula** and is one of the most elegant numbers in all of mathematics:

$$e^{i\pi} = -1$$

Notice that this number is derived in Example 8 in the following illustration as a result of Euler's formula in an excerpt from a current calculus text.

Here we have used the Taylor series for cos y and sin y (Equations 10.50 and 10.49). The result is a famous formula called **Euler's formula:**

(6)
$$e^{iy} = \cos y + i \sin y$$

Combining Euler's formula with Equation 5, we get

(7)
$$e^{x+iy} = e^x e^{iy} = e^x(\cos y + i \sin y)$$

Example 8   Evaluate: (a) $e^{i\pi}$ (b) $e^{-1+i\pi/2}$

Solution
   (a) From Euler's equation (6) we have

$$e^{i\pi} = \cos \pi + i \sin \pi = -1 + i(0) = -1$$

6.1   The Imaginary Unit and Complex Numbers*

Definition of Complex Numbers

To find the roots of certain equations, you must sometimes consider the square roots of negative numbers. Since the set of real numbers does not allow for such a possibility, a number that is *not a real number* is defined. This number is denoted by the symbol i.

THE IMAGINARY UNIT

> The number i, called the **imaginary unit,** is defined as a number with the following properties:
>
> $$i^2 = -1 \quad \text{and if } a > 0, \quad \sqrt{-a} = i\sqrt{a}$$

With this number, you can write the square roots of any negative numbers as the product of a real number and the number i. Thus, $\sqrt{-9} = i\sqrt{9} = 3i$. For any positive real number b, $\sqrt{-b} = \sqrt{b}i$. We now consider a new set of numbers.

COMPLEX NUMBERS

> The set of all numbers that can be written in the form
>
> $$a + bi$$
>
> with real numbers a and b and the imaginary unit i is called the set of **complex numbers.**

If $b = 0$, then $a + bi = a + 0i = a$, which is a **real number**; thus, the real numbers form a subset of the complex numbers. If $a = 0$ and $b \neq 0$, then $a + bi = 0 + bi = bi$ is called a **pure imaginary number.** If $a = 0$ and $b = 1$, then $a + bi = 0 + 1 \cdot i = i$, which is the imaginary unit. Sometimes a single symbol, usually z, is used to represent a complex number $a + bi$ (see Problems 32–40).

Operations with Complex Numbers

A complex number is **simplified** if it is written in the form $a + bi$, where a and b are real numbers in simplified form, and i is the imaginary unit. To work with complex numbers, you will need definitions of equality, along with the usual arithmetic operations. Let $a + bi$ and $c + di$ be complex numbers. Values that could cause division by zero are excluded.

---

*If you have recently had an intermediate algebra course, you can skip or briefly review this section.

OPERATIONS WITH COMPLEX NUMBERS

> **Equality:** $a + bi = c + di$ if and only if $a = c$ and $b = d$
>
> **Addition:** $(a + bi) + (c + di) = (a + c) + (b + d)i$
>
> **Subtraction:** $(a + bi) - (c + di) = (a - c) + (b - d)i$
>
> **Multiplication:** $(a + bi)(c + di) = (ac - bd) + (ad + bc)i$
>
> **Division:** $\dfrac{a + bi}{c + di} = \dfrac{(ac + bd) + (bc - ad)i}{c^2 + d^2} = \dfrac{ac + bd}{c^2 + d^2} + \dfrac{bc - ad}{c^2 + d^2}i$

It is not necessary to memorize these definitions because you can deal with two complex numbers as you would any binomials as long as you remember that $i^2 = -1$.

EXAMPLE 1   Operations with complex numbers

Simplify each expression.

a. $(4 + 5i) + (3 + 4i) = (4 + 3) + (5 + 4)i = 7 + 9i$

b. $(2 - i) - (3 - 5i) = (2 - 3) + (-1 + 5)i = -1 + 4i$

c. $(5 - 2i) + (3 + 2i) = 8 + 0i$ or 8

d. $(4 + 3i) - (4 + 2i) = 0 + i$ or i

e. $(2 + 3i)(4 + 2i) = 8 + 16i + 6i^2$

$$= 8 + 16i - 6$$

$$= 2 + 16i$$

f. $i^{95} = i^{92+3}$

$$= (i^4)^{23}i^3$$

$$= (1)^{23}i^2i \quad \text{Since } i^4 = 1$$

$$= (1)(-1)i$$

$$= -i$$

EXAMPLE 2   Verifying the definition of multiplication of complex numbers

Verify that multiplying $(a + bi)(c + di)$ in the usual algebraic way gives the same result as that shown in the definition of multiplication of complex numbers.

SOLUTION

$$(a + bi)(c + di) = ac + adi + bci + bdi^2$$

$$= ac + (ad + bc)i - bd$$

$$= (ac - bd) + (ad + bc)i$$

EXAMPLE 3   Multiplying complex numbers

Simplify $(4 - 3i)(4 + 3i)$.

SOLUTION

$$(4 - 3i)(4 + 3i) = 16 - 9i^2$$
$$= 16 + 9$$
$$= 25$$

Example 3 gives a clue for dividing complex numbers. The definition would be difficult to remember, so instead we use the idea of *conjugates*. The complex numbers $a + bi$ and $a - bi$ are called **complex conjugates**, and each is the conjugate of the other:

$$(a + bi)(a - bi) = a^2 - b^2i^2$$
$$= a^2 + b^2$$

which is a real number. Thus, simplify a quotient by using the conjugate of the denominator, as illustrated by Examples 4–6. This process is similar to a previous technique used for simplifying trigonometric identities.

EXAMPLE 4   Dividing complex numbers

Simplify $\dfrac{15 - 5i}{2 - i}$.

Multiply by 1.

$$\frac{15 - 5i}{2 - i} = \frac{15 - 5i}{2 - i} \cdot \frac{2 + i}{2 + i}$$

Conjugates

$$= \frac{30 + 5i - 5i^2}{4 - i^2}$$

$$= \frac{35 + 5i}{5}$$

$$= 7 + i$$

You can check this division by multiplying $(2 - i)(7 + i)$:

$$(2 - i)(7 + i) = 14 - 5i - i^2$$
$$= 15 - 5i$$

EXAMPLE 5   Dividing complex numbers

$$\frac{6 + 5i}{2 + 3i} = \frac{6 + 5i}{2 + 3i} \cdot \frac{2 - 3i}{2 - 3i}$$

$$= \frac{12 - 8i - 15i^2}{4 - 9i^2}$$

$$= \frac{27}{13} - \frac{8}{13}i$$

EXAMPLE 6   Verifying the definition of division of complex numbers

Verify that the conjugate method gives the same result as the definition of division of complex numbers.

SOLUTION

$$\frac{a + bi}{c + di} = \frac{a + bi}{c + di} \cdot \frac{c - di}{c - di}$$

$$= \frac{ac - adi + bci - bdi^2}{c^2 - d^2i^2}$$

$$= \frac{(ac + bd) + (bc - ad)i}{c^2 + d^2}$$

$$= \frac{ac + bd}{c^2 + d^2} + \frac{bc - ad}{c^2 + d^2}i$$

6.1 Problem Set

A *Simplify the expressions in Problems 1–14.*

1. a. $(3 + 3i) + (5 + 4i)$ **b.** $(6 - 2i) + (5 + 3i)$
c. $(4 - 2i) - (3 + 4i)$
2. a. $5 - (2 - 3i)$ **b.** $5i - (5 + 5i)$
c. $(5 - 3i) - (5 + 2i)$
3. a. $(3 + 4i) - (7 + 4i)$ **b.** $-2(-4 + 5i)$
c. $4(2 - i) - 3(-1 - i)$
4. a. $6(3 + 2i) + 4(-2 - 3i)$ **b.** $i(2 + 3i)$
c. $i(5 - 2i)$
5. a. $(3 - i)(2 + i)$ **b.** $(4 - i)(2 + i)$
c. $(5 - 2i)(5 + 2i)$
6. a. $(8 - 5i)(8 + 5i)$ **b.** $(3 - 5i)(3 + 5i)$
c. $(7 - 9i)(7 + 9i)$
7. a. i^2 **b.** $-i^2$ **c.** $(-i)^2$
8. a. i^3 **b.** $-i^3$ **c.** $(-i)^3$

9. a. i^4 **b.** $-i^4$ **c.** $(-i)^4$
10. a. i^5 **b.** $-i^5$ **c.** $(-i)^5$
11. a. i^6 **b.** i^8 **c.** i^{11}
12. a. i^{26} **b.** i^{25} **c.** i^{27}
13. a. i^{144} **b.** i^{917} **c.** i^{236}
14. a. i^{2001} **b.** $-i^{1992}$ **c.** i^{1994}

B *Simplify the expressions in Problems 15–29.*

15. $(6 - 2i)^2$ **16.** $(3 + 3i)^2$ **17.** $(4 + 5i)^2$

18. $(3 - 5i)^3$ **19.** $\dfrac{-3}{1 + i}$ **20.** $\dfrac{5}{4 - i}$

21. $\dfrac{2}{i}$ **22.** $\dfrac{5}{i}$ **23.** $\dfrac{3i}{5 - 2i}$

24. $\dfrac{-2i}{3+i}$ **25.** $\dfrac{5+3i}{4-i}$ **26.** $\dfrac{4-2i}{3+i}$

27. $\dfrac{1-6i}{1+6i}$ **28.** $\dfrac{2+7i}{2-7i}$ **29.** $\dfrac{3-2i}{5+i}$

C

30. Simplify $(1.9319+.5176i)(2.5981+1.5i)$.

31. Simplify $\dfrac{-3.2253+8.4022i}{3.4985+1.9392i}$

Let $z_1=(1+\sqrt{3})+(2+\sqrt{3})i$ and $z_2=(2+\sqrt{12})+\sqrt{3}i$. Perform the indicated operation in Problems 32–37.

32. z_1+z_2 **33.** z_1-z_2 **34.** z_1z_2

35. $\dfrac{z_1}{z_2}$ **36.** $(z_1)^2$ **37.** $(z_2)^2$

For each of Problems 38–40, let $z_1=a+bi$, $z_2=c+di$, and $z_3=e+fi$.

38. Prove the commutative laws for complex numbers. That is, prove that

$$z_1+z_2=z_2+z_1$$
$$z_1z_2=z_2z_1$$

39. Prove the associative laws for complex numbers. That is, prove that

$$z_1+(z_2+z_3)=(z_1+z_2)+z_3$$
$$z_1(z_2z_3)=(z_1z_2)z_3$$

40. Prove the distributive law for complex numbers. That is, prove that

$$z_1(z_2+z_3)=z_1z_2+z_1z_3$$

6.2   Trigonometric Form of Complex Numbers

Plotting Complex Numbers

Consider a graphical representation of a complex number. To give a graphical representation of complex numbers such as

$$2+3i, \quad -i, \quad -3-4i, \quad 3i, \quad -2+\sqrt{2}i, \quad \frac{3}{2}-\frac{5}{2}i$$

a two-dimensional coordinate system is used: the horizontal axis represents the **real axis** and the vertical axis is the **imaginary axis,** so that $a+bi$ is represented by the ordered pair (a, b). Remember that a and b represent real numbers, so we plot (a, b) in the usual manner, as shown in Figure 6.1. The coordinate system in Figure 6.1 is called the **complex plane,** or the **Gaussian plane,** in honor of Karl Friedrich Gauss (1777–1855).

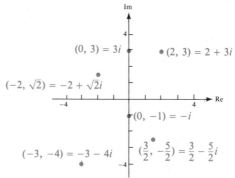

FIGURE 6.1 Complex plane

The **absolute value** of a complex number z is, graphically, the distance between z and the origin (just as it is for real numbers). The absolute value of a complex number is also called the **modulus**. The distance formula leads to the following definition.

ABSOLUTE VALUE, OR MODULUS, OF A COMPLEX NUMBER

If $z = a + bi$, then the *absolute value*, or *modulus*, of z is denoted by $|z|$ and defined by

$$|z| = \sqrt{a^2 + b^2}$$

EXAMPLE 1   Finding the absolute value of complex numbers

Find the absolute value.

a. $3 + 4i$
absolute value:
$$|3 + 4i| = \sqrt{3^2 + 4^2}$$
$$= \sqrt{25}$$
$$= 5$$

b. $-2 + \sqrt{2}i$
absolute value:
$$|-2 + \sqrt{2}i| = \sqrt{4 + 2}$$
$$= \sqrt{6}$$

c. -3
absolute value:
$$|-3 + 0i| = \sqrt{9 + 0}$$
$$= 3$$

This example shows that the definition of absolute value for complex number is consistent with the definition of absolute value given for real numbers.

d. $4i$
absolute value:
$$|4i| = \sqrt{0^2 + 4^2}$$
$$= 4$$

Trigonometric Form of a Complex Number

The form $a + bi$ is called the **rectangular form** of a complex number, but another very useful representation uses trigonometry. Consider the graphical representation of a complex number $a + bi$, as shown in Figure 6.2.

Let r be the distance from the origin to (a, b) and let θ be the angle the segment makes with the real axis. Then

$$r = \sqrt{a^2 + b^2}$$

and θ, called the **argument**, is chosen so that it is the smallest nonnegative angle the terminal side makes with the positive real axis. From the definition of the trigonometric functions,

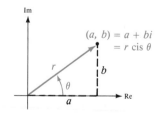

FIGURE 6.2 Trigonometric form and rectangular form of a complex number

$$\cos \theta = \frac{a}{r} \qquad \sin \theta = \frac{b}{r} \qquad \tan \theta = \frac{b}{a}$$

$$a = r \cos \theta \qquad b = r \sin \theta$$

Therefore,
$$a + bi = r \cos \theta + ir \sin \theta$$
$$= r(\cos \theta + i \sin \theta)$$

Sometimes, $r(\cos \theta + i \sin \theta)$ is abbreviated by

$$r(\cos \theta + i \sin \theta)$$

$$r(c \qquad i\ s \qquad \theta) \qquad \text{or } r \operatorname{cis} \theta \quad \text{(see Figure 6.2)}$$

TRIGONOMETRIC FORM OF A COMPLEX NUMBER

> The **trigonometric form** of a complex number $z = a + bi$ is
>
> $$r(\cos \theta + i \sin \theta) = r \operatorname{cis} \theta$$
>
> where $r = \sqrt{a^2 + b^2}$; $\tan \theta = \frac{b}{a}$ if $a \neq 0$; $a = r \cos \theta$; $b = r \sin \theta$. This representation is unique for $0 \leq \theta < 360°$ for all z except $0 + 0i$.

The placement of θ in the proper quadrant is an important consideration because there are two values of $0 \leq \theta < 360°$ that will satisfy the relationship

$$\tan \theta = \frac{b}{a}$$

For example, compare the following:

1. $-1 + i;\quad a = -1, b = 1;\qquad \tan \theta = \frac{1}{-1} \quad \text{or} \quad \tan \theta = -1$

2. $1 - i;\qquad a = 1, b = -1;\qquad \tan \theta = \frac{-1}{1} \quad \text{or} \quad \tan \theta = -1$

Notice the same trigonometric equation for both complex numbers, even though $-1 + i$ is in Quadrant II and $1 - i$ is in Quadrant IV. This consideration of quadrants is even more important when you are doing the problem on a calculator, since the sequence of steps is

$$\boxed{1}\ \boxed{+/-}\ \boxed{\text{INV}}\ \boxed{\tan} \quad \text{or} \quad \boxed{\tan^{-1}}\ \boxed{-1}$$

giving the result $-45°$, which is not true for either example since $0 \leq \theta < 360°$. The entire process can be taken care of quite simply if you let θ' be the reference angle for θ. Then find the reference angle

$$\theta' = \tan^{-1} \left| \frac{b}{a} \right|$$

After you know the reference angle and the quadrant, it is easy to find θ. For these examples,

$$\theta' = \tan^{-1}|-1|$$
$$= 45°$$

For Quadrant II, $\theta = 135°$; for Quadrant IV, $\theta = 315°$.

EXAMPLE 2   Changing from rectangular to trigonometric form

Change the complex numbers to trigonometric form.

a. $1 - \sqrt{3}i$

$a = 1$ and $b = -\sqrt{3}$; the number is in Quadrant IV.

$$r = \sqrt{1^2 + (-\sqrt{3})^2} \qquad \theta' = \tan^{-1}\left|\frac{-\sqrt{3}}{1}\right|$$

$$= \sqrt{4} \qquad\qquad\qquad = 60°$$

$$= 2 \qquad\qquad\qquad \text{The reference angle is } 60°;$$
$$\text{in Quadrant IV: } \theta = 300°.$$

Thus, $1 - \sqrt{3}i = 2 \operatorname{cis} 300°$.

b. $6i$

$a = 0$ and $b = 6$; notice that $\tan \theta$ is not defined for $\theta = 90°$. By inspection, $6i = 6 \operatorname{cis} 90°$.

CALCULATOR COMMENT

Many calculators have a built-in conversion from rectangular to trigonometric form. Most calculators call this **rectangular–polar** conversion $(x, y) \Leftrightarrow (r, \theta)$ where $x + yi \Leftrightarrow r \operatorname{cis} \theta$. For example, on the TI-81 these conversions are found when you press the $\boxed{\text{MATH}}$ key; they are called R→P (rectangular to trig form) and P→R (trig to rectangular form)

c. $4.310 + 5.516i$

$a = 4.310$ and $b = 5.516$; the number is in Quadrant I.

$$r = \sqrt{(4.310)^2 + (5.516)^2} \qquad \theta = \tan^{-1}\left|\frac{5.516}{4.310}\right|$$

$$\approx \sqrt{49} \qquad\qquad\qquad \approx \tan^{-1}(1.2798)$$

$$= 7 \qquad\qquad\qquad \approx 52°$$

Thus, $4.310 + 5.516i \approx 7 \operatorname{cis} 52°$.

EXAMPLE 3   Changing from trigonometric to rectangular form

Change the complex numbers to rectangular form.

a. 4 cis 150°

$r = 4$ and $\theta = 150°$

$a = 4 \cos 150°$ $b = 4 \sin 150°$

$$= 4\left(-\frac{\sqrt{3}}{2}\right) \qquad = 4\left(\frac{1}{2}\right)$$

$$= -2\sqrt{3} \qquad\qquad = 2$$

Thus, 4 cis 150° $= -2\sqrt{3} + 2i$.

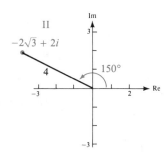

b. $5(\cos 218° + i \sin 218°)$

$r = 5$ and $\theta = 218°$

$a = 5 \cos 218°$ $b = 5 \sin 218°$

≈ -3.94 ≈ -3.08

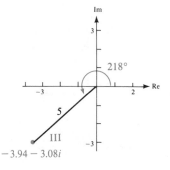

Thus, $5(\cos 218° + i \sin 218°) \approx -3.94 - 3.08i$.

6.2 Problem Set

A *Plot each pair of numbers in Problems 1–9, and find the absolute value of each number.*

1. $3 + i, 5 + 5i$

2. $7 - i, 2 - 5i$

3. $3 + 2i, -2 + 5i$

4. $-1 + 3i, 4 - i$

5. $2 + 2i, -1 + i$

6. $3 + 5i, -3 + i$

7. $2, -4i$

8. $-5, -i$

9. $-2i, 5$

B *Plot and then change each of the numbers in Problems 10–21 to trigonometric form. Write the angle to the nearest degree.*

10. $1 + i$

11. $1 - i$

12. $\sqrt{3} - i$

13. $\sqrt{3} + i$

14. $-1 - \sqrt{3}i$

15. 1

16. 5

17. $-i$

18. $4i$

19. $1.7207 + 2.4575i$

20. $5.7956 - 1.5529i$

21. $-.6946 + 3.9392i$

Plot and then change each of the numbers in Problems 22–31 to rectangular form. Use exact values whenever possible; otherwise, give answers with accuracy to four decimal places.

22. $2(\cos 45° + i \sin 45°)$

23. $3(\cos 60° + i \sin 60°)$

24. $5(\cos \frac{4\pi}{3} + i \sin \frac{4\pi}{3})$

25. $\cos \frac{5\pi}{6} + i \sin \frac{5\pi}{6}$

26. cis 0

27. 4 cis 30°

28. 7 cis 135°

29. 6 cis 247°

30. 9 cis 190°

31. 10 cis 371°

C

32. If $z = a + bi$ and $\bar{z} = a - bi$, show that $|z| = \sqrt{z \cdot \bar{z}}$.

33. If $z_1 = a + bi$ and $z_2 = c + di$, show that $|z_1 + z_2| \leq |z_1| + |z_2|$. This is called the triangle inequality.

34. In calculus it is shown (using series) that

$$e^{i\theta} = \cos\theta + i\sin\theta$$

which is one of the most remarkable formulas in all of mathematics. This means, among other things, that $re^{i\theta} = r \operatorname{cis}\theta$.

Plot the following complex numbers.

a. $e^{i\pi/4}$ **b.** $2e^{-i\pi/6}$ **c.** $-e^{-i\pi}$

d. $2e^{2\pi i}$ **e.** $\sqrt{2}e^{5i\pi/4}$

6.3 De Moivre's Formula

Products and Quotients of Complex Numbers

In Section 6.1, the operations involving complex numbers were fairly simple, so you encountered little difficulty when carrying them out. For example,

$$(1 + 2i)(2 + 3i) = 2 + 7i + 6i^2$$
$$= 2 + 7i + 6(-1)$$
$$= -4 + 7i$$

But consider

$$(1.9319 + .5176i)(2.5981 + 1.5i) = 5.0193 + 2.89785i + 1.34478i + .7764i^2$$
$$= 4.2426 + 4.2427i$$

You almost need a calculator to carry out the operations. If these numbers are changed to trigonometric form, are the calculations easier?

$$1.9319 + .5176i \approx 2 \operatorname{cis} 15°$$
$$2.5981 + 1.5i \approx 3 \operatorname{cis} 30°$$

The product, changed to trigonometric form, is

$$4.2426 + 4.2427i \approx 6 \operatorname{cis} 45° \qquad r = \sqrt{4.2426^2 + 4.2427^2} \approx 6.000013171 \approx 6$$

$$\theta = \tan^{-1}\left(\frac{4.2426}{4.2427}\right) \approx 44.99932476° \approx 45°$$

If we did several problems of this type, we might notice (by working in trigonometric form)

$$(2 \operatorname{cis} 15°)(3 \operatorname{cis} 30°) = 6 \operatorname{cis} 45°$$

Product of 2 and 3 is 6.

45 is the sum of 15 and 30.

A similar result holds for division. For division, consider the number

$$10(\cos 150° + i\sin 150°) \div 2(\cos 120° + i\sin 120°) = \left(-5\sqrt{3} + 5i\right) \div \left(-1 + \sqrt{3}i\right)$$

In algebraic (rectangular) form:

$$\frac{-5\sqrt{3} + 5i}{-1 + \sqrt{3}i} = \frac{-5\sqrt{3} + 5i}{-1 + \sqrt{3}i} \cdot \frac{-1 - \sqrt{3}i}{-1 - \sqrt{3}i}$$

$$= \frac{5\sqrt{3} + 15i - 5i - 5\sqrt{3}i^2}{1 - 3i^2}$$

$$= \frac{10\sqrt{3} + 10i}{4}$$

$$= \frac{5}{2}\sqrt{3} + \frac{5}{2}i$$

$$= 5 \text{ cis } 30° \qquad \text{Can you fill in the details of changing to trigonometric form?}$$

But from the trigonometric forms,

$$\frac{10 \text{ cis } 150°}{2 \text{ cis } 120°} = 5 \text{ cis } 30°$$

would seem to indicate that

$$\frac{10 \text{ cis } 150°}{2 \text{ cis } 120°} = \frac{10}{2} \text{ cis}(150° \overset{\downarrow}{-} 120°) \qquad \text{Difference of arguments}$$

$$\uparrow \underline{\hspace{3cm}} \text{Quotient of 10 and 2}$$

which is true not only for this example but also in general.

These multiplication and division rules are summarized by the following theorem.

PRODUCTS AND QUOTIENTS OF COMPLEX NUMBERS IN TRIGONOMETRIC FORM

Let $z_1 = r_1 \text{ cis } \theta_1$ and $z_2 = r_2 \text{ cis } \theta_2$ be nonzero complex numbers. Then

$$z_1 z_2 = r_1 r_2 \text{ cis } (\theta_1 + \theta_2) \qquad \frac{z_1}{z_2} = \frac{r_1}{r_2} \text{ cis}(\theta_1 - \theta_2)$$

PROOF OF PRODUCT FORM:

$$z_1 z_2 = (r_1 \text{ cis } \theta_1)(r_2 \text{ cis } \theta_2)$$

$$= [r_1(\cos \theta_1 + i \sin \theta_1)][r_2(\cos \theta_2 + i \sin \theta_2)]$$

$$= r_1 r_2(\cos \theta_1 + i \sin \theta_1)(\cos \theta_2 + i \sin \theta_2)$$

$$= r_1 r_2(\cos \theta_1 \cos \theta_2 + i \cos \theta_1 \sin \theta_2 + i \sin \theta_1 \cos \theta_2 - \sin \theta_1 \sin \theta_2)$$

$$= r_1 r_2[(\cos \theta_1 \cos \theta_2 - \sin \theta_1 \sin \theta_2) + i(\cos \theta_1 \sin \theta_2 + \sin \theta_1 \cos \theta_2)]$$

$$= r_1 r_2[\cos(\theta_1 + \theta_2) + i \sin (\theta_1 + \theta_2)]$$

$$= r_1 r_2 \text{ cis}(\theta_1 + \theta_2)$$

The proof of the quotient form is similar and is left as a problem.

EXAMPLE 1   Products of complex numbers in trigonometric form

Simplify.

a. $5 \text{ cis } 38° \cdot 4 \text{ cis } 75° = 5 \cdot 4 \text{ cis}(38° + 75°)$

$\qquad\qquad\qquad\qquad\quad = 20 \text{ cis } 113°$

b. $\sqrt{2} \text{ cis } 188° \cdot 2\sqrt{2} \text{ cis } 310° = 4 \text{ cis } 498°$

$\qquad\qquad\qquad\qquad\qquad\quad = 4 \text{ cis } 138°$

c. $(2 \text{ cis } 48°)^3 = (2 \text{ cis } 48°)(2 \text{ cis } 48°)^2$ Notice that this result is the same as

$\qquad\qquad\quad\ = (2 \text{ cis } 48°)(4 \text{ cis } 96°)$ $(2 \text{ cis } 48°)^3 = 2^3 \text{ cis}(3 \cdot 48°)$

$\qquad\qquad\quad\ = 8 \text{ cis } 144°$ $= 8 \text{ cis } 144°$

EXAMPLE 2   Finding the quotient of complex numbers in trigonometric form

Find $\dfrac{15(\cos 48° + i \sin 48°)}{5(\cos 125° + i \sin 125°)}$.

SOLUTION $\dfrac{15 \text{ cis } 48°}{5 \text{ cis } 125°} = 3 \text{ cis}(48° - 125°)$

$\qquad\qquad\qquad\qquad = 3 \text{ cis}(-77°)$

$\qquad\qquad\qquad\qquad = 3 \text{ cis } 283°$ Keep the arguments between 0° and 360°.

De Moivre's Formula

As you can see from Example 1c, multiplication in trigonometric form extends quite nicely to any positive integral power in a result called **De Moivre's formula.**

DE MOIVRE'S FORMULA

If n is a natural number, then

$$(r \text{ cis } \theta)^n = r^n \text{ cis } n\theta$$

for a complex number $r \text{ cis } \theta = r(\cos \theta + i \sin \theta)$.

EXAMPLE 3   Using De Moivre's formula to find a power of a complex number

Simplify $\left(1 - \sqrt{3}i\right)^5$.

SOLUTION First change to trigonometric form: $a = 1; b = -\sqrt{3}$; Quadrant IV.

$$r = \sqrt{1 + 3} \qquad \theta' = \tan^{-1}\left|-\frac{\sqrt{3}}{1}\right|$$

$$= 2 \qquad\qquad\qquad = 60°$$

$$\theta = 300° \qquad \text{Quadrant IV}$$

$$\left(1 - \sqrt{3}i\right)^5 = (2 \text{ cis } 300°)^5$$
$$= 2^5 \text{ cis}(5 \cdot 300°)$$
$$= 32 \text{ cis } 1{,}500°$$
$$= 32 \text{ cis } 60°$$

If you want the answer in rectangular form, you can now change back:

$$a = 32 \cos 60° \qquad b = 32 \sin 60°$$
$$= 32\left(\frac{1}{2}\right) \qquad\qquad = 32\left(\frac{1}{2}\sqrt{3}\right)$$
$$= 16 \qquad\qquad\qquad = 16\sqrt{3}$$

Thus, $\left(1 - \sqrt{3}i\right)^5 = 16 + 16\sqrt{3}i$. ▌

Although De Moivre's formula is useful for powers, as illustrated by Example 3, its main usefulness is in finding the complex roots of numbers. Recall from algebra that $\sqrt[n]{r} = r^{1/n}$ is used to denote the principal nth root of r. However, $r^{1/n}$ is only *one* of the nth roots of r. How do you find *all* nth roots of r? You can use a calculator or logarithms to find the principal root, and then use the following theorem to find the other roots.

*n*TH ROOT THEOREM

If n is any positive integer, then the nth roots of r cis θ are given by

$$\sqrt[n]{r} \text{ cis}\left(\frac{\theta + 360°k}{n}\right) \quad \text{or} \quad \sqrt[n]{r} \text{ cis}\left(\frac{\theta + 2\pi k}{n}\right)$$

for $k = 0, 1, 2, 3, \ldots, n - 1$.

The proof of this theorem follows directly from De Moivre's formula and is left as a problem.

EXAMPLE 4   Finding the square roots of a complex number

Find the square roots of $-\frac{9}{2} + \frac{9}{2}\sqrt{3}i$.

SOLUTION   First change to trigonometric form:

$$r = \sqrt{\left(-\frac{9}{2}\right)^2 + \left(\frac{9}{2}\sqrt{3}\right)^2} \qquad \theta' = \tan^{-1}\left|\frac{\frac{9}{2}\sqrt{3}}{-\frac{9}{2}}\right|$$
$$= \sqrt{\frac{81}{4} + \frac{81 \cdot 3}{4}} \qquad\qquad = \tan^{-1}\left(\sqrt{3}\right)$$
$$= 60°$$

$$= \sqrt{81\left(\frac{1}{4} + \frac{3}{4}\right)} \qquad\qquad \theta = 120° \quad \text{Quadrant II}$$

$$= 9$$

By the nth root theorem, the square roots of 9 cis 120° are

$$9^{1/2} \text{ cis}\left(\frac{120° + 360°k}{2}\right) = 3 \text{ cis}(60° + 180°k)$$

$$k = 0: \quad 3 \text{ cis } 60° = \frac{3}{2} + \frac{3}{2}\sqrt{3}i$$

$$k = 1: \quad 3 \text{ cis } 240° = -\frac{3}{2} - \frac{3}{2}\sqrt{3}i$$

All other integral values of k repeat one of the previously found roots. For example,

$$k = 2: \quad 3 \text{ cis } 420° = \frac{3}{2} + \frac{3}{2}\sqrt{3}i$$

Check:
$$\left(\frac{3}{2} + \frac{3}{2}\sqrt{3}i\right)^2 = \frac{9}{4} + \frac{9}{2}\sqrt{3}i + \frac{9}{4} \cdot 3i^2$$

$$= -\frac{9}{2} + \frac{9}{2}\sqrt{3}i$$

$$\left(-\frac{3}{2} - \frac{3}{2}\sqrt{3}i\right)^2 = \frac{9}{4} + \frac{9}{2}\sqrt{3}i + \frac{9}{4} \cdot 3i^2$$

$$= -\frac{9}{2} + \frac{9}{2}\sqrt{3}i$$

EXAMPLE 5   Finding the fifth roots of a complex number

Find the fifth roots of 32.

SOLUTION   You might ask "Why not find these on a calculator?" The reason you cannot is that it will give you only one of the roots—not all five.

Begin by writing 32 in trigonometric form: 32 = 32 cis 0°. The fifth roots are found by

$$32^{1/5} \text{ cis}\left(\frac{0° + 360°k}{5}\right) = 2 \text{ cis } 72°k$$

$k = 0: \quad 2 \text{ cis } 0° \quad = 2 \longleftarrow$ —————— The first root, which is located so that its

$k = 1: \quad 2 \text{ cis } 72° = .6180 + 1.9021i$ argument is θ/n, is called the **principal nth

$k = 2: \quad 2 \text{ cis } 144° = -1.6180 + 1.1756i$ **root**. This is the root you can find on a calcu-

$k = 3: \quad 2 \text{ cis } 216° = -1.6180 - 1.1756i$ lator if the number whose root you are finding

$k = 4: \quad 2 \text{ cis } 288° = .6180 - 1.9021i$ is real and positive.

All other integral values for k repeat those listed here. |

When all the fifth roots of 32 are represented graphically, as shown in Figure 6.3, they all lie on a circle of radius 2 and are equally spaced.

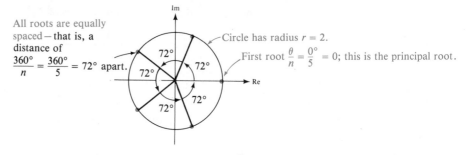

FIGURE 6.3 Graphical representation of the fifth roots of 32

If n is a positive integer, then the nth roots of a complex number $a + bi = r \operatorname{cis} \theta$ are equally spaced on the circle of radius r centered at the origin.

EXAMPLE 6   **Solving a cubic equation using De Moivre's formula**

Solve $x^3 + 1 = 0$.

SOLUTION $x^3 = -1$, so we are looking for the cube roots of -1.

Since $-1 = \operatorname{cis} 180°$ we have $1^{1/3} \operatorname{cis} \dfrac{180° + 360°k}{3} = \operatorname{cis}(60° + 120°k)$

$$k = 0: \quad \operatorname{cis} 60° = \frac{1}{2} + \frac{\sqrt{3}}{2}i$$

$$k = 1: \quad \operatorname{cis} 180° = -1$$

$$k = 2: \quad \operatorname{cis} 300° = \frac{1}{2} - \frac{\sqrt{3}}{2}i$$

Thus, the roots of $x^3 + 1 = 0$ are $\frac{1}{2} \pm \frac{\sqrt{3}}{2}i, -1$. |

6.3 Problem Set

A *Perform the indicated operations in Problems 1–11. Leave your answer in a form that matches the form in the question.*

1. $2 \operatorname{cis} 60° \cdot 3 \operatorname{cis} 150°$

2. $4(\cos 65° + i \sin 65°) \cdot 12(\cos 87° + i \sin 87°)$

3. $\dfrac{5(\cos 315° + i \sin 315°)}{2(\cos 48° + i \sin 48°)}$

4. $\dfrac{8 \operatorname{cis} 30°}{4 \operatorname{cis} 15°}$

5. $\dfrac{12 \operatorname{cis} 120°}{4 \operatorname{cis} 250°}$

6. $(2 \text{ cis } 50°)^3$

7. $(3 \text{ cis } 60°)^4$

8. $(\cos 210° + i \sin 210°)^5$

9. $(2 - 2i)^4$

10. $(1 + i)^6$

11. $(\sqrt{3} - i)^8$

B *Find the indicated roots of the numbers in Problems 12–21. Leave your answers in trigonometric form.*

12. Square roots of 16 cis 100°

13. Cube roots of 8 cis 240°

14. Fourth roots of 81 cis 88°

15. Fifth roots of 32 cis 200°

16. Cube roots of 64 cis 216°

17. Fifth roots of 32 cis 160°

18. Cube roots of -1

19. Cube roots of 27

20. Fourth roots of $1 + i$

21. Fourth roots of $-1 - i$

Find the indicated roots of the numbers in Problems 22–27. Leave your answers in trigonometric form.

22. Sixth roots of -64

23. Sixth roots of $64i$

24. Ninth roots of 1

25. Ninth roots of $-1 + i$

26. Tenth roots of i

27. Tenth roots of 1

Find the indicated roots of the numbers in Problems 28–33. Leave your answers in rectangular form. Also, show the roots graphically. Use exact values whenever possible.

28. Cube roots of 1

29. Fourth roots of 16

30. Cube roots of -8

31. Cube roots of $4\sqrt{3} - 4i$

32. Square roots of $\dfrac{25}{2} - \dfrac{25\sqrt{3}}{2}i$

33. Fourth roots of $12.2567 + 10.2846i$

Solve the equations in Problems 34–41 for all complex roots. Leave your answers in rectangular form and give approximate answers to four decimal places.

34. $x^3 - 1 = 0$

35. $x^5 - 1 = 0$

36. $x^5 + 1 = 0$

37. $x^6 - 1 = 0$

38. $x^4 + 16 = 0$

39. $x^4 - 16 = 0$

40. $x^4 + x^3 + x^2 + x + 1 = 0$
 [*Hint*: Multiply both sides by $(x - 1)$ and compare with Problem 35.]

41. $x^5 + x^4 + x^3 + x^2 + x + 1 = 0$
 [*Hint*: Multiply both sides by $(x - 1)$ and compare with Problem 37.]

C *For Problems 42–45, leave your answers in trigonometric form.*

42. Find the cube roots of $\left(4\sqrt{2} + 4\sqrt{2}i\right)^2$.

43. Find the fifth roots of $\left(-16 + 16\sqrt{3}i\right)^3$.

44. Prove that

$$\frac{r_1 \text{ cis } \theta_1}{r_2 \text{ cis } \theta_2} = \frac{r_1}{r_2} \text{ cis } (\theta_1 - \theta_2)$$

45. Prove that

$$\left[\sqrt[n]{r} \text{ cis } \left(\frac{\theta + 350°k}{n} \right) \right]^n = r \text{ cis } \theta$$

46. In the previous section (Problem 34) we noted $re^{i\theta} = r \text{ cis } \theta$. Use this fact and the laws of exponents to prove De Moivre's formula.

6.4   Summary and Review

Objectives	*Pages / Examples / Problems*

6.1   The Imaginary Unit and Complex Numbers

1. Add and subtract complex numbers.

p. 233; Example 1; Problems 1–4.

2. Multiply complex numbers.

pp. 233–234; Examples 1–3; Problems 4–18, 30.

3. Divide complex numbers.

pp. 234–235; Examples 4–6; Problems 19–29, 29.

6.2   Trigonometric Form of Complex Numbers

4. Find the absolute value of a complex number and plot that number in the Gaussian plane.

p. 236; Figure 6.1 and Example 1; Problems 1–9.

5. Write complex numbers in trigonometric form.

pp. 237–239; Example 2; Problems 10–21.

6. Write complex numbers in rectangular form.

p. 240; Example 3; Problems 22–31.

6.3   De Moivre's Formula

7. Multiply and divide complex numbers in trigonometric form.

pp. 241–243; Examples 1 and 2; Problems 1–5.

8. Raise complex numbers to integral powers.

pp. 243–244; Example 3; Problems 6–11.

9. Find the nth roots of a complex number.

pp. 244–246; Examples 4 and 5; Problems 12–33.

10. Solve equations for all complex roots.

p. 246; Example 6; Problems 34–41.

TERMS

Absolute value [6.2]
Argument [6.2]
Complex conjugates [6.1]
Complex number [6.1]
Complex plane [6.2]
De Moivre's formula [6.3]

Gaussian plane [6.2]
i [6.1]
Imaginary axis [6.2]
Imaginary unit [6.1]
Modulus [6.2]
nth root theorem [6.3]

Operations with complex
  numbers [6.1]
Principal nth root [6.3]
Pure imaginary number [6.1]
Real axis [6.2]
Real number [6.1]

Rectangular form [6.2]
Simplify complex
  numbers [6.1]
Trigonometric form [6.2]

CHAPTER 6 REVIEW OF OBJECTIVES

The problem numbers correspond to the objectives listed in Section 6.4. *Simplify the expressions in Problems* 1–3.

1. a. $(6 + 3i) + (-4 + 2i)$ **b.** $(2 - 5i) - (3 - 2i)$

2. a. $2i(-1 + i)(-2 + 2i)$ **b.** $(2 - 5i)^2$

3. a. $\dfrac{2 + 3i}{1 - i}$ **b.** $\dfrac{-4}{i}$

4. Plot each number and find its absolute value.

  a. $4 + 2i$ **b.** $1 + 5i$

  c. $-4 - i$ **d.** $4 - 2i$

5. Change the given complex numbers to trigonometric form.

  a. $7 - 7i$ **b.** $-3i$

6. Change the given complex numbers to rectangular form.

  a. $2 \text{ cis } 150°$ **b.** $4\left(\cos \frac{7\pi}{4} + i \sin \frac{7\pi}{4}\right)$

7. Simplify $\dfrac{2 \text{ cis } 158° \cdot 4 \text{ cis } 212°}{(2 \text{ cis } 312°)^3}$ and leave your answer in trigonometric form.

8. Simplify $\left(\sqrt{12} - 2i\right)^4$ and leave your answer in rectangular form.

9. Find the indicated roots and represent them in trigonometric form, in rectangular form, and graphically.

  a. Square roots of $\frac{7}{2}\sqrt{3} - \frac{7}{2}i$ **b.** Fourth roots of 1

10. Solve $x^3 - 27 = 0$.

DISCUSSION PROBLEMS

1. What is a number? Discuss different sets of numbers used in mathematics.

2. What is a complex number? Find some examples of complex numbers used in disciplines other than mathematics.

3. Why is De Moivre's formula important?

4. Show that the set of complex numbers forms a field.

5. Write a 500-word paper on trigonometry and astronomy.

MISCELLANEOUS PROBLEMS

Change the rectangular-form complex numbers to trigonometric form in Problems 1–8.

1. $9 - 9i$ **2.** -6

3. $-\frac{7}{2}\sqrt{3} + \frac{7}{2}i$ **4.** $2 + 3i$

5. 9 **6.** $-1.5321 - 1.2856i$

7. $16i$ **8.** $-2 + 3i$

Change the trigonometric-form complex numbers to rectangular form in Problems 9–14.

9. $9\left(\cos\frac{3\pi}{4} + i\sin\frac{3\pi}{4}\right)$ **10.** $2(\cos 3 + i\sin 3)$

11. $3\operatorname{cis} 240°$ **12.** $5\operatorname{cis} 270°$

13. $5\operatorname{cis} 25°$ **14.** $3\operatorname{cis}\frac{11\pi}{6}$

Simplify the given expressions and leave your answer in the form indicated in Problems 15–18.

15. $\left(\sqrt{12} - 2i\right)^4$; rectangular

16. $\dfrac{(3 + 3i)\left(\sqrt{3} - 1\right)}{1 + i}$; rectangular

17. $\dfrac{2\operatorname{cis} 185° \cdot 4\operatorname{cis} 223°}{(2\operatorname{cis} 300°)^3}$; trigonometric

18. $2i(-1 + i)(-2 + 2i)$; trigonometric

Find the indicated roots in Problems 19–30. Leave your answers in trigonometric form unless otherwise indicated.

19. Cube roots of i **20.** Sixth roots of -64

21. Sixth roots of $64i$ **22.** Ninth roots of 1

23. Ninth roots of $-1 + i$ **24.** Tenth roots of i

25. Tenth roots of 1 **26.** Fourth roots of -16

27. Cube roots of $-16 + 16\sqrt{3}i$

28. Cube roots of $4\operatorname{cis}\frac{5\pi}{4}$

29. Square roots of $-\frac{7}{2}\sqrt{3} + \frac{7}{2}i$ (rectangular form)

30. Cube roots of $4\sqrt{2} + 4\sqrt{2}i$ (rectangular form)

▌ Electric Generator

Courtesy of P. G. & E.

FIGURE 6.4 An electric generator

In 1831 Michael Faraday (1791–1867) discovered induced electric currents by performing an experiment with a coil of wire and a magnet. He reasoned that if current through a wire would produce a magnetic field, then why not the reverse, in which a magnet is made to produce current in a wire?

Consider the electric generator illustrated in Figure 6.4. The current will flow first in one direction and then in the other, according to the turning of the armature, because each wire moves up across the field one instant and down at the next. (See Figure 6.5.)

FIGURE 6.5 Alternating current from an electric generator

The graph of the current leads us to suspect that the equation for the current is trigonometric, which it is. The current, I, can be found from an equation of the form

$$I = k \sin \omega t$$

for constants k and ω. Although it is beyond the scope of this course, we might mention that the complex numbers introduced in this chapter play an important role in the study of electricity.

For alternating currents, the current sometimes lags behind the voltage, V, which means that the power supplied to any alternating-current circuit is given by

$$P = VI \cos \theta$$

The quantity $\cos \theta$ is called the power factor, and it is desirable to have θ as close to 0 as possible because VI should be as close to P as possible. Notice that, since $-1 \le \cos \theta \le 1$ has a maximum value, $\cos \theta = 1$, $VI \le P$ unless $\cos \theta = 1$. A low power factor should be avoided because, for a given supply voltage V, a large current I would be needed to transmit appreciable amounts of electrical energy. (See Figure 6.6.)

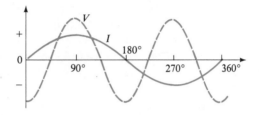

FIGURE 6.6 The current is lagging 45° behind the voltage

EXAMPLE 1   **Finding the power factor**

If $\theta = 15°$, find the power factor.

SOLUTION $P = VI \cos 15° \approx VI(.9659)$; the power factor is about 97%.

EXAMPLE 2   **Finding the maximum current lag**

If you need to have a power factor of at least 90%, what is the maximum current lag (to the nearest degree)?

SOLUTION $$\cos \theta = .90$$
$$\theta = 25.8°$$

Since $\cos \theta$ is decreasing in the first quadrant, if $\cos \theta \ge .90$, then $\theta \le 25.8°$. Thus, the maximum current lag to the nearest degree is 25°.

PROBLEMS FOR FURTHER STUDY: ELECTRIC GENERATOR

1. Graph the current if $k = 15$ and $\omega = 120\pi$.

2. Engineering The voltage is given by $V = k \sin \lambda(t - p)$

where $k = 10$, $\lambda = 60\pi$, and $p = \frac{1}{2}$. Graph the voltage.

3. Engineering If you need to have a power factor of at least 96%, what will the maximum current lag be?

■ Charlotte Angas Scott
(1858–1931)

■ The science of figures is most glorious and beautiful.

N. FRISCHLINUS
Dialog 1.

■ Indeed, mathematics, the indispensable tool of the sciences, defying the senses to follow its splendid flights, is demonstrating today, as it never has been demonstrated before, the supremacy of the pure reason.

NICHOLAS BUTLER
The Meaning of Education and Other Essays and Addresses

The first prominent woman mathematician in America was Charlotte Angas Scott. She is best known because of her lifelong work as an educator. She studied in England, but soon after receiving her Ph.D., she came to America and developed the mathematics program at Bryn Mawr College. Today, it is difficult for us to understand the problems encountered by a woman in Scott's day who wished to receive a college education. Scott, for example, was permitted to take her final undergraduate examinations at Cambridge only informally. According to Karen Rappaport,*

"Scott tied for eighth place in the mathematics exam. Mathematics was an unprecedented area for a woman to excel in and the achievement attracted public attention, especially when her name was not mentioned at the official ceremony. When the official name was read, loud shouts of "Scott of Girton" could be heard in the gallery. The public also responded to the slight. The February 7, 1880 issue of Punch stated:

'But when the academy doors are reopened to the Ladies let them be opened to their full worth. Let us not hear of any restrictions or exclusions from this or that function or privilege . . . .' As a result of this support and a public petition, women were formally admitted to the Tripos Exams the following year but they were still not permitted Cambridge degrees. (That event did not occur until 1948.)"

Mathematically, Scott was concerned with the study of specific algebraic curves of degree higher than two. She published thirty papers in the field of algebraic geometry, as well as a text, *Modern Analytic Geometry,* in 1894. In 1922, members of the American Mathematical Society and many former students organized a dinner to honor Scott. Alfred North Whitehead came from England to give the main address, in which he stated: "A life's work such as that of Professor Charlotte Angas Scott is worth more to the world than many anxious efforts of diplomatists."

---

*Karen D. Rappaport, "Two American Women Mathematicians," *MATYC Journal,* Fall (1980), p. 203.

7

POLAR CURVES

Contents

7.1 Polar Coordinates

7.2 Graphing Polar-Form Equations

7.3 Special Polar-Form Curves

7.4 Parametric Equations

7.5 Intersection of Polar-Form Equations

7.6 Summary and Review

Application for Further Study
    Communication of Bees

Cumulative Review for Chapters 5–7

From *Calculus, Fourth Edition*, by Stanley I. Grossman, copyright © 1988 by Harcourt Brace Jovanovich, Inc., reprinted by permission of the publisher.

Preview

Up to now, the only coordinate system you have studied in mathematics is a rectangular coordinate system. You have used this system to plot pairs of real numbers (Cartesian coordinate system), as well as to plot single complex numbers (Gaussian plane). This chapter introduces another method of representing a point in a plane: a *polar coordinate system*. It is reported that polar coordinates were used to plot global locations on the floor of the Paris observatory (see Problems 13 and 14 of Problem Set 7.1).

Perspective

Polar coordinates open up avenues of investigation that would be very difficult if we were limited to rectangular coordinates. The page illustrated here, from a leading calculus book, shows a curve called a *cardioid* (Example 1), a circle, and a curve called a limaçon (Example 2). Their respective equations are given in polar coordinates as

$$r = 1 + \sin \theta \qquad r = 5 \sin \theta \qquad r = 2 + \sin \theta$$

We will introduce these curves in Section 7.3.

EXAMPLE 2

Calculate the area enclosed by the cardioid $r = 1 + \sin \theta$.

Solution. The cardioid is sketched in Figure 3. We have

$$\text{Area} = \int_0^{2\pi} \int_0^{1+\sin\theta} r\,dr\,d\theta = \int_0^{2\pi} \left\{ \frac{r^2}{2} \Big|_0^{1+\sin\theta} \right\} d\theta$$

$$= \frac{1}{2} \int_0^{2\pi} (1 + 2\sin\theta + \sin^2\theta)\,d\theta$$

$$= \frac{1}{2} \int_0^{2\pi} \left(1 + 2\sin\theta + \frac{1 - \cos 2\theta}{2} \right) d\theta$$

$$= \frac{1}{2} \left(\frac{3\theta}{2} - 2\cos\theta - \frac{\sin 2\theta}{4} \right) \Big|_0^{2\pi} = \frac{3\pi}{2}$$

$r = 1 + \sin\theta$

Figure 3

7.1   Polar Coordinates

Plotting Points

Up to this point in the book, we have used a rectangular coordinate system. Now, consider a different system, called the **polar coordinate system.** In this system, fix a point O, called the *origin,* or **pole,** and represent a point in the plane by an ordered pair $P(r, \theta)$, where θ measures the angle from the positive x-axis and r represents the directed distance from the pole to the point P. Both r and θ can be any real numbers. When plotting points, consider a **polar axis,** fixed at the pole and coinciding with the x-axis. Now rotate the polar axis through an angle θ as shown in Figure 7.1. You might find it helpful to rotate your pencil as the axis—the tip points in the positive direction and the eraser in the negative direction. If θ is positive, the angle is measured in a counterclockwise direction, and if θ is negative, it is measured in a clockwise direction.

FIGURE 7.1 Polar-form points

Next, plot r on the polar axis (pencil). Notice that any real number can be plotted on this real number line. Plotting points seems easy, but it is necessary that you completely understand this process. Study each part of Example 1 and make sure you understand how each point is plotted.

EXAMPLE 1   Plotting points given in polar-form

Plot each of the following polar-form points: $A\left(4, \frac{\pi}{3}\right)$, $B\left(-4, \frac{\pi}{3}\right)$, $C\left(3, -\frac{\pi}{6}\right)$, $D\left(-3, -\frac{\pi}{6}\right)$, $E(-3, 3)$, $F(-3, -3)$, $G(-4, -2)$, $H\left(5, \frac{3\pi}{2}\right)$, $I\left(-5, \frac{\pi}{2}\right)$, $J\left(5, -\frac{\pi}{2}\right)$.

SOLUTION   Points A–D illustrate the basic ideas of plotting polar-form points.

Plot $A\left(4, \frac{\pi}{3}\right)$; positve θ	Plot $B\left(-4, \frac{\pi}{3}\right)$; positive θ	Plot $C\left(3, -\frac{\pi}{6}\right)$; negative θ	Plot $D\left(-3, -\frac{\pi}{6}\right)$; negative θ
(1) Rotate x-axis through an angle $\theta = \frac{\pi}{3}$	(1) Rotate $\theta = \frac{\pi}{3}$	(1) Rotate $\theta = -\frac{\pi}{6}$	(1) Rotate $\theta = -\frac{\pi}{6}$
(2) Plot the point positive 4 (point-end of pencil)	(2) Plot the point negative 4 (eraser end of pencil)	(2) Plot the point positive 3 (point end)	(2) Plot the point negative 3 (eraser end)

Points *E–H* illustrate common situations that can sometimes be confusing. Make sure you understand each example.

Plot $E(-3, 3)$.

(1) Angle is 3
(2) Plot -3 (eraser end)

Plot $F(-3, -3)$.

(1) Angle is -3
(2) Plot -3 (eraser end)

Plot $G(-4, -2)$.

(1) Angle is -2
(2) Plot -4 (eraser end)

Plot $H\left(5, \frac{3\pi}{2}\right)$.

Coordinates $I\left(-5, \frac{\pi}{2}\right), J\left(5, -\frac{\pi}{2}\right)$, and $H\left(5, \frac{3\pi}{2}\right)$ all represent the same point.

Primary Representations of a Point

One thing you will notice from Example 1 is that ordered pairs in polar form are not associated in a one-to-one fashion with points in the plane. Indeed, given any point in the plane, there are infinitely many ordered pairs of polar coordinates associated with that point. If you are given a point (r, θ) other than the pole in polar coordinates, then $(-r, \theta + \pi)$ also represents the same point. In addition, there are infinitely many others, all of which have the same first component as one of these and have second components that are multiples of 2π added to these angles. We call (r, θ) and $(-r, \theta + \pi)$ the **primary representations of the point** if the angles θ and $\theta + \pi$ are between 0 and 2π.

⊘ Remember to simplify so the second component represents a nonnegative angle less than one revolution. ⊘

PRIMARY REPRESENTATIONS OF A POINT IN POLAR FORM

Every point in polar-form has two primary representations:

(r, θ), where $0 \le \theta < 2\pi$ and $(-r, \pi + \theta)$, where $0 \le \pi + \theta < 2\pi$

EXAMPLE 2   Finding the primary representations of polar-form points

Give both primary representations for each of the given points.

a. $\left(3, \frac{\pi}{4}\right)$ has primary representations

$\left(3, \frac{\pi}{4}\right)$ and $\left(-3, \frac{5\pi}{4}\right)$.

$\frac{\pi}{4} + \pi = \frac{5\pi}{4}$

b. $\left(5, \frac{5\pi}{4}\right)$ has primary representations

$\left(5, \frac{5\pi}{4}\right)$ and $\left(-5, \frac{\pi}{4}\right)$.

$\frac{5\pi}{4} + \pi = \frac{9\pi}{4}$, but $\left(-5, \frac{9\pi}{4}\right)$ is not a primary representation of the point $\left(5, \frac{5\pi}{4}\right)$ since $\frac{9\pi}{4} > 2\pi$.

c. $\left(-6, -\frac{2\pi}{3}\right)$ has primary representa-

tions $\left(-6, \frac{4\pi}{3}\right)$ and $\left(6, \frac{\pi}{3}\right)$.

θ should be positive.

d. $(9, 5)$ has primary representations $(9, 5)$ and $(-9, 5 - \pi)$; a point such as $(-9, 5 - \pi)$ is usually approximated by writing $(-9, 1.86)$.

Notice that $(-9, 5 + \pi)$ is not a primary representation, since $5 + \pi > 2\pi$.

e. $(9, 7)$ has primary representations $(9, 7 - 2\pi)$ or $(9, .72)$ and $(-9, 7 - \pi)$ or $(-9, 3.86)$.

Add or subtract from θ **whatever** multiple of 2π will force the new angle to be between 0 and approximately 6.28 (2π).

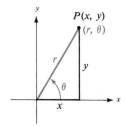

FIGURE 7.2 Relationship between rectangular and polar coordinates

Relationship Between Polar and Rectangular Coordinates

The relationship between the two coordinate systems can easily be found by using the definition of the trigonometric functions (see Figure 7.2).

RELATIONSHIP BETWEEN RECTANGULAR AND POLAR COORDINATES

1. To change **from polar to rectangular**:

$$x = r \cos \theta \qquad y = r \sin \theta$$

2. To change **from rectangular to polar**:

$$r = \sqrt{x^2 + y^2} \qquad \theta' = \tan^{-1}\left|\frac{y}{x}\right|, \quad x \neq 0$$

where θ' is the reference angle for θ. Place θ in the proper quadrant by noting the signs of x and y. If $x = 0$, then $\theta' = \frac{\pi}{2}$.

EXAMPLE 3   Changing from polar form to rectangular form

Change the polar coordinates $\left(-3, \frac{5\pi}{4}\right)$ to rectangular coordinates.

SOLUTION $x = -3\cos\dfrac{5\pi}{4} = -3\left(-\dfrac{\sqrt{2}}{2}\right) = \dfrac{3\sqrt{2}}{2}$ $y = -3\sin\dfrac{5\pi}{4} = -3\left(-\dfrac{\sqrt{2}}{2}\right) = \dfrac{3\sqrt{2}}{2}$

$$\underbrace{\left(-3, \frac{5\pi}{4}\right)}_{\text{Polar form}}; \underbrace{\left(\frac{3\sqrt{2}}{2}, \frac{3\sqrt{2}}{2}\right)}_{\substack{\text{Rectangular} \\ \text{form}}}$$

∎

EXAMPLE 4   Changing from rectangular form to polar form

Write both primary representations of the polar-form coordinates for the point $\left(\frac{5\sqrt{3}}{2}, -\frac{5}{2}\right)$.

SOLUTION $r = \sqrt{\left(\dfrac{5\sqrt{3}}{2}\right)^2 + \left(-\dfrac{5}{2}\right)^2}$ $\theta' = \tan^{-1}\left|\dfrac{-\dfrac{5}{2}}{\dfrac{5\sqrt{3}}{2}}\right| = \tan^{-1}\left(\dfrac{1}{\sqrt{3}}\right)$

$$= \sqrt{\frac{75}{4} + \frac{25}{4}}$$ $= \dfrac{\pi}{6}$ $\theta = \dfrac{11\pi}{6}$ (Quadrant IV)

$$= 5$$

$$\underbrace{\left(\frac{5\sqrt{3}}{2}, -\frac{5}{2}\right)}_{\text{Rectangular form}}; \quad \underbrace{\left(5, \frac{11\pi}{6}\right); \left(-5, \frac{5\pi}{6}\right)}_{\text{Polar form}} \qquad \frac{11\pi}{6} + \pi = \frac{17\pi}{6}, \text{ and } \frac{17\pi}{6} \text{ is coterminal with } \frac{5\pi}{6}.$$

∎

Points on a Polar Curve

If r and θ are related by an equation, we can speak of the *graph of the equation*. For example, $r = 5$ is the equation of a circle with center at the origin and radius 5. Also $\theta = \frac{\pi}{3}$ is the equation of the line passing through the pole as drawn in Figure 7.3.

As with other equations we have graphed in this book, we will begin graphing polar-form curves by plotting some points. However, you must first be able to recognize whether a point in polar form satisfies a given equation.

FIGURE 7.3

Graphs of $r = 5$ and $\theta = \frac{\pi}{3}$

EXAMPLE 5   Showing that a point lies on a graph

Show that each of the given points lies on the graph of $r = \dfrac{2}{1 - \cos\theta}$.

a. $\left(2, \frac{\pi}{2}\right)$ **b.** $\left(-2, \frac{3\pi}{2}\right)$ **c.** $(-1, 2\pi)$

SOLUTION   Begin by substituting the given coordinates into the equation.

a. $2 \overset{?}{=} \dfrac{2}{1 - \cos \frac{\pi}{2}} = \dfrac{2}{1 - 0} = 2$

Thus, the point $\left(2, \frac{\pi}{2}\right)$ is on the curve, since it satisfies the equation.

b. $-2 \overset{?}{=} \dfrac{2}{1 - \cos \frac{3\pi}{2}} = \dfrac{2}{1 - 0} = 2$ This equation is **not** true.

The equation is not satisifed, but we *cannot* say that the point is not on the curve.

Indeed, we see from part **a** that it is on the curve, since $\left(-2, \frac{3\pi}{2}\right)$ and $\left(2, \frac{\pi}{2}\right)$ name the same point! So even if one primary representation of a point does not satisfy the equation, we must still check the other primary representation of the point.

c. $-1 \overset{?}{=} \dfrac{2}{1 - \cos 2\pi} = \dfrac{2}{1 - 1}$,   which is undefined.

Check the other representation of the same point—namely, $(1, \pi)$:

$$1 \overset{?}{=} \dfrac{2}{1 - \cos \pi} = \dfrac{2}{1 - (-1)} = 1$$

Thus, the point is on the curve.   ∎

Example 5 leads us to the following definition of what is necessary for a polar-form point to **satisfy** an equation.

A POINT SATISFYING AN EQUATION

> An ordered pair representing a polar-form point (other than the pole) **satisfies an equation** involving a function of $n\theta$ (for an integer n) if and only if at least one of its primary representations satisfies the given equation.

We turn our attention to graphing polar-form curves in the next section.

7.1 Problem Set

A *In Problems 1–6 plot each of the given polar-form points. Give both primary representations, and give the rectangular coordinates of the point.*

1. $\left(4, \frac{\pi}{4}\right)$ **2.** $\left(6, \frac{\pi}{3}\right)$ **3.** $\left(5, \frac{2\pi}{3}\right)$

4. $\left(3, -\frac{\pi}{6}\right)$ **5.** $\left(\frac{3}{2}, -\frac{5\pi}{6}\right)$ **6.** $(-4, 4)$

In each of Problems 7–12, plot the given rectangular-form points and give both primary representations in polar form.

7. $(5, 5)$ **8.** $\left(-1, \sqrt{3}\right)$

9. $\left(2, -2\sqrt{3}\right)$ **10.** $(-2, -2)$

11. $(3, -3)$ **12.** $(3, 7)$

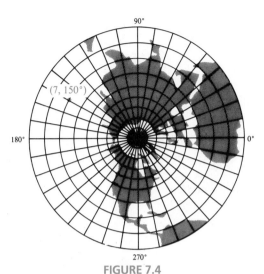

90°

(7, 150°)

180°   0°

270°

FIGURE 7.4
This is reproduced from the floor of the Paris
Observatory

13. GEOGRAPHY Name the country in Figure 7.4 in which each of the named points is located.

a. $(5.5, 260°)$

b. $(7.5, 78°)$

c. $(-2, 140°)$

d. $(-4, 80°)$

14. GEOGRAPHY Name possible coordinates for each of the following cities (see Figure 7.4).

a. Miami, Florida

b. Los Angeles, California

c. Mexico City, Mexico

d. London, England

B *In Problems 15–19 tell whether each of the given points lies on the curve* $r = \dfrac{5}{1 - \sin \theta}$.

15. $\left(10, \frac{\pi}{6}\right)$ **16.** $\left(5, \frac{\pi}{2}\right)$ **17.** $\left(-10, \frac{5\pi}{6}\right)$

18. $\left(-\frac{10}{3}, \frac{5\pi}{6}\right)$

19. $\left(20 + 10\sqrt{3}, \frac{\pi}{3}\right)$

In Problems 20–24 tell whether each of the given points lies on the curve $r = 2(1 - \cos \theta)$.

20. $\left(1, \frac{\pi}{3}\right)$

21. $\left(1, -\frac{\pi}{3}\right)$

22. $\left(-1, \frac{\pi}{3}\right)$

23. $\left(-2, \frac{\pi}{2}\right)$

24. $\left(-2 - \sqrt{2}, \frac{\pi}{4}\right)$

Find three ordered pairs satisfying each of the equations in Problems 25–36. Give both primary representations for each point.

25. $r^2 = 9 \cos \theta$

26. $r^2 = 9 \cos 2\theta$

27. $r = 3\theta$

28. $r = 5\theta$

29. $r\theta = 4$

30. $r = 2 - 3 \sin \theta$

31. $r = 2(1 + \cos \theta)$

32. $r = 2 \cos \theta$

33. $r = 6 \cos 3\theta$

34. $r = \tan \theta$

35. $\dfrac{r}{1 - \sin \theta} = 2$

36. $r = \dfrac{8}{1 - 2 \cos \theta}$

C

37. Derive the equations for changing from polar coordinates to rectangular coordinates.

38. Derive the equations for changing from rectangular coordinates to polar coordinates.

39. What is the distance between $\left(3, \frac{\pi}{3}\right)$ and $\left(7, \frac{\pi}{4}\right)$? Explain why you cannot use the distance formula for these ordered pairs.

40. What is the distance between polar-form points (r, θ) and (c, α)?

7.2 **Graphing Polar-Form Equations**

The general process for graphing a new type of curve is to plot points, make generalizations, and then graph future similar curves using those generalizations along with plotting only a few key points.

EXAMPLE 1   Graphing by plotting points

Graph $r - \theta = 0$ for $\theta \geq 0$.

SOLUTION   Set up a table of values.

θ	r
0	0
1	1
2	2
3	3
4	4
5	5
6	6

Choose a θ, and then find a corresponding r so that (r, θ) satisfies the equation.

Plot each of these points and connect them as shown in Figure 7.5.

Notice that as θ increases, r must also increase.

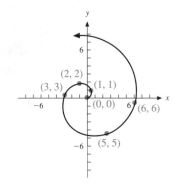

FIGURE 7.5 Graph of $r - \theta = 0$*

EXAMPLE 2   Graphing a circle by plotting points

Graph $r = \cos \theta$ by plotting points.

	DISPLAY
If $\theta = 0$, then $r = 1$; plot $(1, 0)$.	1.
If $\theta = 1$, then $r \approx .54$; plot $(.54, 1)$.	.5403023059
If $\theta = 2$, then $r \approx -.42$; plot $(-.42, 2)$.	-.4161468365
If $\theta = 3$, then $r \approx -.99$; plot $(-.99, 3)$.	-.9899924966
If $\theta = 4$, then $r \approx -.65$; plot $(-.65, 4)$.	-.6536436209
If $\theta = 5$, then $r \approx .28$; plot $(.28, 5)$.	.2836621855
If $\theta = 6$, then $r \approx .96$; plot $(.96, 6)$.	.9601702867
If $\theta = 7$, then $r \approx .75$; plot $(.75, 7)$.	.7539022543
If $\theta = 8$, then $r \approx -.15$; plot $(-.15, 8)$.	-.1455000338

FIGURE 7.6 Graph of $r = \cos \theta$

Plot these points as shown in Figure 7.6. Notice that, if these points are connected, they form a circle with center at $(0, .5)$ and radius $.5$.

*Many books use what is called **polar graph paper**, but such paper is not really necessary. It also obscures the fact that polar curves and rectangular curves are both plotted on a Cartesian coordinate system only with a different meaning attached to the ordered pairs. You can estimate the angles as necessary without polar graph paper. Remember, just as when graphing rectangular curves, the key is not in plotting many points, but in recognizing the curve and then plotting a few key points.

Cardioids

EXAMPLE 3   **Graphing a new type of curve by constructing a table of values**

Graph $r = 2(1 - \cos \theta)$ by plotting points.

SOLUTION   Construct a table of values by choosing values for θ and approximating the corresponding values for r.

θ	0	$\frac{\pi}{6}$	$\frac{\pi}{3}$	$\frac{\pi}{2}$	$\frac{2\pi}{3}$	$\frac{5\pi}{6}$	π	$\frac{7\pi}{6}$	$\frac{4\pi}{3}$	$\frac{3\pi}{2}$	$\frac{5\pi}{3}$	$\frac{11\pi}{6}$
r (APPROX. VALUE)	0	.27	1	2	3	3.7	4	3.7	3	2	1	.27

These points are plotted and then connected, as shown in Figure 7.7.

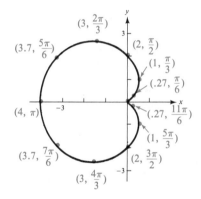

FIGURE 7.7 Graph of $r = 2(1 - \cos \theta)$

Polar coordinates were introduced by Jacob Bernoulli in 1691 but were not used to any extent until the close of the 18th century. Jacob was a member of one of the most distinguished families in mathematics. He was one of 12 related Bernoullis, all of whom achieved some fame and recognition as mathematicians. In regard to polar coordinates, Jacob was so moved by the way the equiangular spiral reproduces itself that he asked that it be engraved on his tombstone with the inscription

Eadem mutata resurgo
("I shall arise the same, though changed").

The curve in Example 3 is called a **cardioid** because it is heart-shaped. Compare the curve graphed in Example 3 with the general curve $r = a(1 - \cos \theta)$, which is called the **standard-position cardioid.** Consider the following table of values:

θ	0	$\frac{\pi}{2}$	π	$\frac{3\pi}{2}$
r	0	a	$2a$	a

← These values for θ should be included whenever you are making a graph in polar coordinates and are using the method of plotting points.

NOTE: This table presents the key points for a cardioid.

These reference points are all that is necessary to plot other standard-position cardioids, because they will all have the same shape as the one shown in Figure 7.7.

Rotations of Polar-Form Curves

What about cardioids that are not in standard position? In Chapter 2 we translated the graphs to new locations, but the translation of a polar-form curve is not so easy because points are not labeled in a rectangular fashion. We can, however, easily rotate the polar-form curve.

ROTATION OF POLAR-FORM GRAPHS

> The polar graph of $r = f(\theta - \alpha)$ is the same as the polar graph of $r = f(\theta)$ that has been rotated through an angle α.

EXAMPLE 4   Graphing a rotated cardioid

Graph $r = 3 - 3 \cos\left(\theta - \frac{\pi}{6}\right)$.

SOLUTION   Recognize this as a cardioid with $a = 3$ and a rotation of $\frac{\pi}{6}$. Plot the four points shown in Figure 7.8 and draw the cardioid.

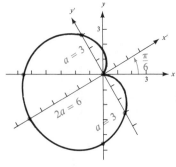

FIGURE 7.8

Graph of $r = 3 - 3 \cos\left(\theta - \frac{\pi}{6}\right)$

If the rotation is 180°, the equation simplifies considerably. Consider

$$r = 3 - 3 \cos(\theta - \pi) \qquad \text{Cardioid with a 180° rotation}$$
$$= 3 - 3\,[\cos\theta \cos\pi + \sin\theta \sin\pi] \qquad \text{Identity 16}$$
$$= 3 - 3\,[\cos\theta(-1) + \sin\theta(0)] \qquad \text{Exact values}$$
$$= 3 - 3(-1)\cos\theta$$
$$= 3 + 3\cos\theta$$

Compare this with Example 4 and you will see that the only difference is a 180° rotation instead of a 30° rotation. This means that, whenever you graph an equation of the form

$$r = a(1 + \cos\theta) \qquad \text{Note the plus sign.}$$

instead of $r = a(1 - \cos\theta)$, it is a standard-form cardioid with a 180° rotation.

EXAMPLE 5   Graphing a cardioid by using an identity

Graph $r = 4(1 - \sin\theta)$.

SOLUTION   Notice that

$$\sin\theta = \cos\left(\frac{\pi}{2} - \theta\right) = \cos\left(\theta - \frac{\pi}{2}\right)$$

This is a cardioid with $a = 4$ that has been rotated $\frac{\pi}{2}$. The graph is shown in Figure 7.9. A curve of the form $r = a(1 - \sin\theta)$ is a cardioid that has been rotated $\frac{\pi}{2}$.

FIGURE 7.9

Graph of $r = 4(1 - \sin\theta)$

The cardioid is only one of the interesting polar-form curves. It is a special case of a curve called a **limaçon**, which is developed in Problem 34 of Problem Set 7.2. Polar-form curves with multiple angles are discussed in the next section.

7.2   Problem Set

A *Sketch each of the equations in Problems 1–18 by relating it to the standard-position cardioid. Use a rotation if necessary.*

1. $r = 2 - 2 \cos \theta$

2. $r = 3(1 - \cos \theta)$

3. $r = 4 - 4 \cos \theta$

4. $r - 3 = 3 \cos \theta$

5. $r - 5 = 5 \cos \theta$

6. $r - 4 = 4 \cos \theta$

7. $r = 2 - 2 \sin \theta$

8. $r = 1 - \sin \theta$

9. $4r = 1 - \sin \theta$

10. $r = 2 + 2 \sin \theta$

11. $r - 3 = 3 \sin \theta$

12. $r - 4 \sin \theta = 4$

13. $r = 3 + 3 \cos\left(\theta - \frac{\pi}{3}\right)$

14. $r = 2 - 2 \cos\left(\theta - \frac{\pi}{4}\right)$

15. $r = 2 + 2 \cos\left(\theta - \frac{\pi}{6}\right)$

16. $r = 3 + 3 \sin\left(\theta - \frac{\pi}{6}\right)$

17. $r = 2 - 2 \sin\left(\theta - \frac{\pi}{3}\right)$

18. $r = 2 - 2 \sin\left(\theta - \frac{\pi}{6}\right)$

B *Sketch each of the equations in Problems 19–33 by plotting polar-form points. For your convenience, plot positive values of θ only.*

19. $r = \theta$

20. $r - 2\theta = 0$

21. $r = 3\theta$

22. $r = 5 \sin \frac{\pi}{6}$

23. $r = 4 \cos \frac{\pi}{3}$

24. $r = 3 \tan \frac{\pi}{4}$

25. $\theta = 2$

26. $\theta = 3$

27. $\theta = \frac{1}{2}$

28. $r = 2 \sin \theta$

29. $r = 3 \cos \theta$

30. $r = \tan \theta$

31. $r = \sec \theta$

32. $r = \csc \theta$

33. $r = \cot \theta$

C

*34. The limaçon is a curve of the form $r = b \pm a \cos \theta$ or $r = b \pm a \sin \theta$, where $a > 0$ and $b > 0$. There are four types of limaçons:

a. $\frac{b}{a} < 1$ (limaçon with inner loop); graph $r = 2 - 3 \cos \theta$ by plotting points.

b. $\frac{b}{a} = 1$ (cardioid); graph $r = 3 - 3 \cos \theta$.

c. $1 < \frac{b}{a} < 2$ (limaçon with a dimple); graph $r = 3 - 2 \cos \theta$ by plotting points.

d. $\frac{b}{a} \geq 2$ (convex limaçon); graph $r = 3 - \cos \theta$ by plotting points.

*35. Graph the following limaçons (see Problem 34).

a. $r = 1 - 2 \cos \theta$ **b.** $r = 2 - \cos \theta$

c. $r = 2 + 3 \cos \theta$ **d.** $r = 2 - 3 \sin \theta$

e. $r = 2 + 3 \sin \theta$ **f.** $r = 3 - 2 \sin \theta$

36. *Spirals* are interesting mathematical curves. There are three general types of spirals:

a. A spiral of Archimedes has the form $r = a\theta$; graph $r = 2\theta$ by plotting points (let $\theta > 0$).

b. A hyperbolic spiral has the form $r\theta = a$; graph $r\theta = 2$ by plotting points (let $\theta > 0$).

c. A logarithmic spiral has the form $r = a^{k\theta}$; graph $r = 2^{\theta}$ by plotting points (let $\theta > 0$).

37. Identify and graph the following spirals (see Problem 36).

a. $r = \theta$ **b.** $r = -\theta$ **c.** $r\theta = 1$

d. $r\theta = -1$ **e.** $r = 2^{2\theta}$ **f.** $r = 3^{\theta}$

*Problems 34 and 35 are required for Section 7.5.

7.3   Special Polar-Form Curves

Rose Curves

The cardioid discussed in the last section involved a cosine or a sine of a single angle. The first example of this section is a function of a double angle and is called a **rose curve.**

EXAMPLE 1   Graphing a new type of curve by constructing a table of values

Graph $r = 4 \cos 2\theta$.

SOLUTION   When presented with a polar-form curve that you do not recognize, you will graph the curve by plotting points. You can use tables, exact values, or a calculator. A calculator was used to form the following table.

θ	r (APPROX. VALUE)	θ	r (APPROX. VALUE)	θ	r (APPROX. VALUE)	θ	r (APPROX. VALUE)
0	4	$\frac{\pi}{2}$	-4	π	4	$\frac{3\pi}{2}$	-4
$\frac{\pi}{12}$	3.5	$\frac{7\pi}{12}$	-3.5	$\frac{13\pi}{12}$	3.5	$\frac{19\pi}{12}$	-3.5
$\frac{\pi}{6}$	2	$\frac{2\pi}{3}$	-2	$\frac{7\pi}{6}$	2	$\frac{5\pi}{3}$	-2
$\frac{\pi}{4}$	0	$\frac{3\pi}{4}$	0	$\frac{5\pi}{4}$	0	$\frac{7\pi}{4}$	0
$\frac{\pi}{3}$	-2	$\frac{5\pi}{6}$	2	$\frac{4\pi}{3}$	-2	$\frac{11\pi}{6}$	2
$\frac{5\pi}{12}$	-3.5	$\frac{11\pi}{12}$	3.5	$\frac{17\pi}{12}$	-3.5	$\frac{23\pi}{12}$	3.5

For each θ you choose, press:

Algebraic: | θ | $\times$ | 2 | $=$ | $\cos$ | $\times$ | 4 | $=$ |　　DISPLAY: ∃.464101615

↑——— For example, if $\theta = \frac{\pi}{12}$, press | π | $\div$ | 12 | here.

RPN: | θ | ENTER | 2 | $\times$ | $\cos$ | 4 | $\times$ |　　DISPLAY: ∃.464101616

The graph is shown in Figure 7.10. Notice that as θ increases from the starting point 0 to $\frac{\pi}{4}$, r decreases from 4 to 0. As θ increases from $\frac{\pi}{4}$ to $\frac{\pi}{2}$, you can see that r is negative and decreases from 0 to -4. Study the table at the left as it relates to the graph in Figure 7.10.

θ	r	SECTION OF CURVE
0 to $\frac{\pi}{4}$	4 to 0	1
$\frac{\pi}{4}$ to $\frac{\pi}{2}$	0 to -4	2
$\frac{\pi}{2}$ to $\frac{3\pi}{4}$	-4 to 0	3
$\frac{3\pi}{4}$ to π	0 to 4	4
π to $\frac{5\pi}{4}$	4 to 0	5
$\frac{5\pi}{4}$ to $\frac{3\pi}{2}$	0 to -4	6
$\frac{3\pi}{2}$ to $\frac{7\pi}{4}$	-4 to 0	7
$\frac{7\pi}{4}$ to 2π	0 to 4	8

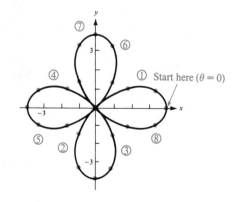

FIGURE 7.10 Graph of $r = 4\cos 2\theta$

The equation

$$r = a \cos n\theta$$

is a four-leaved rose if $n = 2$, and the length of the leaves is a. In general, if n is an even number, the curve has $2n$ leaves; if n is odd, the number of leaves is n. These leaves are equally spaced on a radius of a.

EXAMPLE 2   Graphing a rotated rose curve

Graph $r = 4 \cos 2(\theta - \frac{\pi}{4})$.

SOLUTION   This is a rose curve with four leaves of length 4 equally spaced on a circle. However, this curve has been rotated $\frac{\pi}{4}$, as shown in Figure 7.11.

Other leaves equally spaced from first leaf. First leaf rotated $\frac{\pi}{4}$.

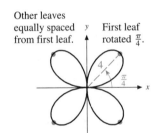

FIGURE 7.11

Graph of $r = 4 \cos 2(\theta - \frac{\pi}{4})$

Notice that Example 2 can be rewritten as a sine curve:

$$r = 4 \cos 2\left(\theta - \frac{\pi}{4}\right)$$

$$= 4 \cos\left(2\theta - \frac{\pi}{2}\right)$$

$$= 4 \cos\left(\frac{\pi}{2} - 2\theta\right)$$

$$= 4 \sin 2\theta$$

These steps can be reversed to graph a rose curve written in terms of a sine function.

EXAMPLE 3   Graphing a rose curve by using an identity

Graph $r = 5 \sin 4\theta$.

SOLUTION

$$r = 5 \sin 4\theta$$

$$= 5 \cos\left(\frac{\pi}{2} - 4\theta\right)$$

$$= 5 \cos\left(4\theta - \frac{\pi}{2}\right)$$

$$= 5 \cos 4\left(\theta - \frac{\pi}{8}\right)$$

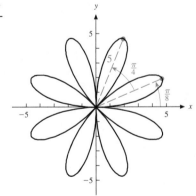

FIGURE 7.12

Graph of $r = 5 \sin 4\theta$

Recognize this as a rose curve rotated $\frac{\pi}{8}$. There are eight leaves of length 5. The leaves are a distance of $\frac{\pi}{4}$ apart, as shown in Figure 7.12.

CALCULATOR COMMENT

Converge, which has been mentioned before, graphs curves of the form $r = f(\theta)$. For example, we might use a computer program to answer a question that many students ask, but which must wait until calculus for an answer—namely, what is the width of a rose leaf?

Consider, for example, the graph of $r = \cos n\theta$. It has either n or $2n$ leaves. Let's consider n an odd number. If $n = 3$, then the rose has three leaves of the same size. As n increases in size, the number of leaves increases and the width of each leaf decreases.

If we let $n = 3$ and wish to change the width of a three-leaved rose, we can vary the exponent, m, on $r = [\cos 3\theta]^m$. Consider the examples for $r = [\cos 3\theta]^{1/8}$ (wider leaves) and $r = [\cos 3\theta]^8$, $r > 0$ (narrower leaves).

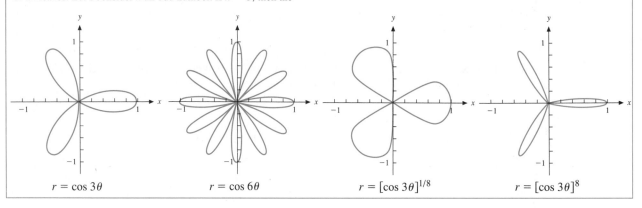

$r = \cos 3\theta$ $r = \cos 6\theta$ $r = [\cos 3\theta]^{1/8}$ $r = [\cos 3\theta]^8$

Lemniscates

The third and last general type of polar-form curve we will consider is called a **lemniscate** and has the general form

$$r^2 = a^2 \cos 2\theta$$

EXAMPLE 4   Graphing a lemniscate

Graph $r^2 = 9 \cos 2\theta$

SOLUTION As before, when graphing a curve for the first time, begin by plotting points. For this example, be sure to obtain two values for r when solving this quadratic equation. For example, if $\theta = 0$, then $\cos 2\theta = 1$ and $r^2 = 9$, so $r = 3$ or -3.

θ	0	$\frac{\pi}{12}$	$\frac{\pi}{6}$	$\frac{\pi}{4}$	$\frac{\pi}{4}$ to $\frac{3\pi}{4}$	$\frac{5\pi}{6}$	$\frac{11\pi}{12}$	π
r (APPROX. VALUE)	± 3	± 2.8	± 2.1	0	undefined	± 2.1	± 2.8	± 3

Notice that for $\frac{\pi}{4} < \theta < \frac{3\pi}{4}$ there are no values for r since $\cos 2\theta$ is negative. For $\pi \leq \theta \leq 2\pi$, the values repeat the sequence given above, so these points are plotted and then connected, as shown in Figure 7.13.

θ	r	SECTION OF CURVE
0 to $\frac{\pi}{4}$	3 to 0	1 (two values) Quadrants I and III
$\frac{\pi}{4}$ to $\frac{3\pi}{4}$	No value	
$\frac{3\pi}{4}$ to π	0 to 3	2 (two values) Quadrants II and IV
π to 2π	Repeat of values from 0 to π	

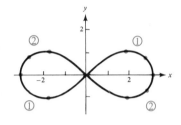

FIGURE 7.13 Graph of $r^2 = 9 \cos 2\theta$

The graphs of $r^2 = a^2 \cos 2\theta$ and $r^2 = a^2 \sin 2\theta$ are called *lemniscates*. There are always two leaves to a lemniscate and the length of the leaves is a. The procedure for graphing a lemniscate given in the form $r^2 = a^2 \sin 2\theta$ is illustrated by Example 5.

EXAMPLE 5   Graphing a rotated lemniscate

Graph $r^2 = 16 \sin 2\theta$.

SOLUTION
$$r^2 = 16 \sin 2\theta$$
$$= 16 \cos\left(\frac{\pi}{2} - 2\theta\right)$$
$$= 16 \cos\left(2\theta - \frac{\pi}{2}\right)$$
$$= 16 \cos 2\left(\theta - \frac{\pi}{4}\right)$$

This is a lemniscate whose leaf has length $\sqrt{16} = 4$, rotated $\frac{\pi}{4}$ as shown in Figure 7.14

FIGURE 7.14 Graph of $r^2 = 16 \sin 2\theta$

Summary of Polar Curves

We conclude this section by summarizing the special types of polar-form curves we have examined. There are many others, some of which are presented in the problems, but these three special types are the most common.

Cardioid: $r = a\,(1 \pm \cos\theta)$ **or** $r = a\,(1 \pm \sin\theta)$

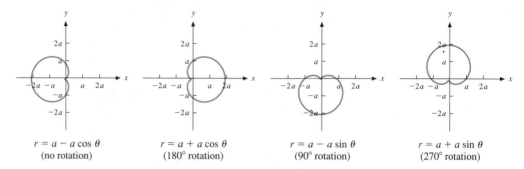

$r = a - a\cos\theta$
(no rotation)

$r = a + a\cos\theta$
(180° rotation)

$r = a - a\sin\theta$
(90° rotation)

$r = a + a\sin\theta$
(270° rotation)

Rose curve: $r = a\cos n\theta$ **or** $r = a\sin n\theta$ (n is a positive integer). Length of each leaf is a.

1. If n is odd, the rose is n-leaved.

One leaf: If $n = 1$, the rose is a curve with one leaf and is circular.

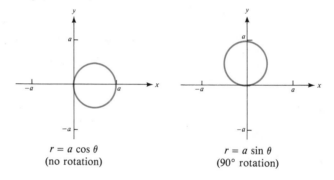

$r = a\cos\theta$
(no rotation)

$r = a\sin\theta$
(90° rotation)

2. If n is even, the rose is $2n$-leaved.

Two leaves: See the lemniscate described below.

Three leaves: $n = 3$ *Four leaves:* $n = 2$

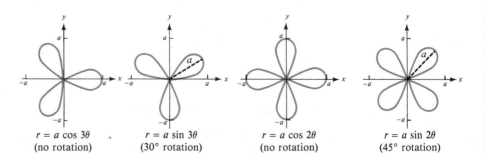

$r = a\cos 3\theta$
(no rotation)

$r = a\sin 3\theta$
(30° rotation)

$r = a\cos 2\theta$
(no rotation)

$r = a\sin 2\theta$
(45° rotation)

Lemniscate: $r^2 = a^2 \cos 2\theta$ **or** $r^2 = a^2 \sin 2\theta$

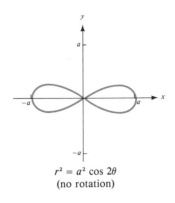

$r^2 = a^2 \cos 2\theta$
(no rotation)

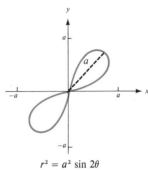

$r^2 = a^2 \sin 2\theta$
(45° rotation)

7.3   Problem Set

A *Identify each of the curves in Problems 1–15 as a cardioid, a rose curve (state number of leaves), a lemniscate, or none of the above.*

1. $r^2 = 9 \cos 2\theta$

2. $r = 2 \sin 2\theta$

3. $r = 3 \sin 3\theta$

4. $r^2 = 2 \cos 2\theta$

5. $r = 2 - 2 \cos \theta$

6. $r = 3 + 3 \sin \theta$

7. $r^2 = \sin 3\theta$

8. $r = 4 \sin 30°$

9. $r = 5 \cos 60°$

10. $r = 5 \sin 8\theta$

11. $r = 3\theta$

12. $r\theta = 3$

13. $\theta = \tan \frac{\pi}{4}$

14. $r = 5(1 - \sin \theta)$

15. $\cos \theta = 1 - r$

B *Sketch each of the equations in Problems 16–30 by relating it to a standard rose curve or lemniscate. Use a rotation if necessary.*

16. $r = 2 \cos 2\theta$

17. $r = 3 \cos 3\theta$

18. $r = 5 \sin 3\theta$

19. $r^2 = 9 \cos 2\theta$

20. $r^2 = 16 \cos 2\theta$

21. $r^2 = 9 \sin 2\theta$

22. $r = 3 \cos 3\left(\theta - \frac{\pi}{3}\right)$

23. $r = 2 \cos 2\left(\theta + \frac{\pi}{3}\right)$

24. $r = 5 \cos 3\left(\theta - \frac{\pi}{4}\right)$

25. $r = \sin 3\left(\theta + \frac{\pi}{6}\right)$

26. $r = \sin\left(2\theta + \frac{\pi}{3}\right)$

27. $r = \cos\left(2\theta + \frac{\pi}{3}\right)$

28. $r^2 = 16 \cos 2\left(\theta - \frac{\pi}{6}\right)$

29. $r^2 = 9 \cos\left(2\theta - \frac{\pi}{3}\right)$

30. $r^2 = 9 \cos\left(2\theta - \frac{\pi}{4}\right)$

C

31. The *strophoid* is a curve of the form $r = a \cos 2\theta \sec \theta$; graph this curve where $a = 2$ by plotting points.

32. The *bifolium* has the form $r = a \sin \theta \cos^2\theta$; graph this curve where $a = 1$ by plotting points.

33. The *folium of Descartes* has the form

$$r = \frac{3a \sin \theta \cos \theta}{\sin^3\theta + \cos^3\theta}$$

Graph this curve where $a = 2$ by plotting points.

7.4   Parametric Equations

It is sometimes convenient to graph curves that consist of points (x, y) where *each* of the variables x and y is defined as a function. For example, suppose

$$x = f(t) \quad \text{and} \quad y = g(t)$$

CALCULATOR COMMENT

Most graphing calculators have a parametric mode. For the example at the right, press

MODE PARAMETRIC

Y= $X_{1T} = 1 + 3T$
      $Y_{1T} = 2T$

You should obtain the following graph:

for functions f and g, where the domain of these functions is some interval I. If we let

$$f(t) = 1 + 3t \quad \text{and} \quad y = g(t) = 2t \qquad (\text{for } 0 \leq t \leq 5)$$

then, if $t = 1$,

$$x = f(1) = 1 + 3(1) = 4$$
$$y = g(1) = 2(1) = 2$$

Plot the point $(x, y) = (4, 2)$ as one point on the curve. Other values are shown in the following table and are plotted in Figure 7.15.

t	0	1	2	3	4	5
x	1	4	7	10	13	16
y	0	2	4	6	8	10

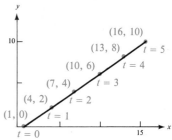

FIGURE 7.15 Graph of $x = 1 + 3t, y = 2t$

The variable t is called a **parameter**, and the equations $x = 1 + 3t$ and $y = 2t$ are called **parametric equations** for the line segment shown in Figure 7.15.

PARAMETER AND
PARAMETRIC EQUATIONS

> Let t be a number in an interval I. Consider the curve defined by the set of ordered pairs (x, y), where
>
> $$x = f(t) \quad \text{and} \quad y = g(t)$$
>
> for functions f and g defined on I. Then the variable t is called a **parameter** and the equations $x = f(t)$ and $y = g(t)$ are called the **parametric equations** for the curve defined by (x, y).

EXAMPLE 1   Graphing a circle using a parameter

Plot the curve represented by the parametric equations

$$x = \cos \theta$$
$$y = \sin \theta$$

SOLUTION   The parameter is θ and you can generate a table of values by using a calculator:

θ	0°	15°	30°	45°	60°	75°	90°	120°	$\cdots$
x	1.00	.97	.87	.71	.50	.26	.00	−.50	$\cdots$
y	.00	.26	.50	.71	.87	.97	1.00	.87	$\cdots$

If you use your calculator for a problem like Example 1, the graph will look strange:

If you change the scale to

Xmin = −1.2, Xmax = 1.2
Ymin = −1.2, Ymax = 1.2,

the graph will still look strange:

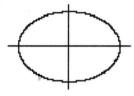

To make a circle look like a circle, you must fix the scale so the units are the same in both the x and y directions. On a TI-81 you can do this by pressing ZOOM and then 5: Square:

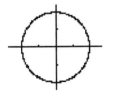

These points are plotted in Figure 7.16. If the plotted points are connected, you can see that the curve is a circle.

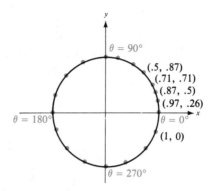

FIGURE 7.16 Graph of $x = \cos\theta, y = \sin\theta$

Converge, which has been mentioned before, does an excellent job of graphing parametric equations. For example, the curve shown below is the Converge plot of the curve with the following parametric equation:

$$x = 4\sin\theta + 3\cos\theta \quad y = 4\cos\theta + 3\sin\theta$$

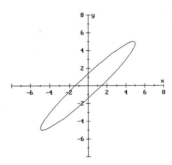

It is possible to recognize the parametric equations in Example 1 as a unit circle if you square both sides of the equation and add:

$$x^2 = \cos^2\theta$$
$$y^2 = \sin^2\theta$$
$$x^2 + y^2 = \cos^2\theta + \sin^2\theta$$

So

$$x^2 + y^2 = 1$$

This process is called **eliminating the parameter.**

   Suppose a light is attached to the edge of a bike wheel. The path of light is shown in Figure 7.17, and the curve traced by the light is called a **cycloid.** If the radius of the wheel is a, we can find the parametric equations for the cycloid. Let $P(x, y)$ be any point of the cycloid, and let θ be the amount of rotation of the circle, as shown in Figure 7.18.

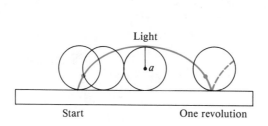

FIGURE 7.17 Graph of a cycloid   **FIGURE 7.18**

Then,

$$
\begin{aligned}
x &= |OA| \\
&= |OB| - |AB| \\
&= \overset{\frown}{PB} - |PQ| \quad \text{where } \overset{\frown}{PB} \text{ is the arc length.} \\
&\qquad\qquad\qquad\quad \text{Also notice } \overset{\frown}{PB} = a\theta \text{ by the} \\
&\qquad\qquad\qquad\quad \text{arc length formula.} \\
&= a\theta - a \sin \theta \quad \text{since } \sin \theta = \frac{|PQ|}{a}
\end{aligned}
$$

Also,

$$
\begin{aligned}
y &= |PA| \\
&= |QB| \\
&= |CB| - |CQ| \\
&= a - a \cos \theta \quad \text{since } \cos \theta = \frac{|PQ|}{a}
\end{aligned}
$$

Therefore,

$$
\begin{aligned}
x &= a\theta - a \sin \theta = a(\theta - \sin \theta) \\
y &= a - a \cos \theta = a(1 - \cos \theta)
\end{aligned}
$$

CALCULATOR COMMENT

Two special cases of cycloids involve the path traced by a fixed point P on a circle. If it rolls *inside* a larger circle, the trace is called a **hypocycloid** and as it rolls around the *outside* of a fixed circle, the trace is called an **epicycloid**. Without computer assistance, these interesting curves are beyond the scope of this course. However, if you have access to a computer you might look at an article by Florence and Sheldon Gordon entitled "Mathematics Discovery via Computer Graphics: Hypocycloids and Epicycloids," in *The Two-Year College Mathematics Journal*, November 1984, p. 441.

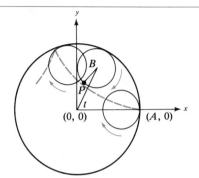

Hypocycloid

If R is the radius of the fixed circle and r is the radius of the rolling circle, then the parametric equations are:

Hypocycloid: $x = (R - r)\cos t + r\cos\left(\dfrac{R - r}{r}\right)t;$

$\qquad\qquad y = (R - r)\sin t - r\sin\left(\dfrac{R - r}{r}\right)t$

Epicycloid: $x = (R + r)\cos t - r\cos\left(\dfrac{R + r}{r}\right)t;$

$\qquad\qquad y = (R + r)\sin t - r\sin\left(\dfrac{R + r}{r}\right)t$

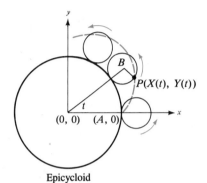

Epicycloid

Some variations of these curves, which were drawn by computer, are shown below.

Hypocycloids

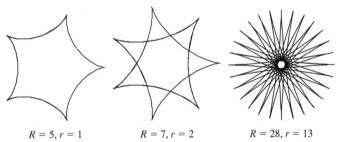

$R = 5, r = 1$ $\qquad\qquad$ $R = 7, r = 2$ $\qquad\qquad$ $R = 28, r = 13$

Epicycloids

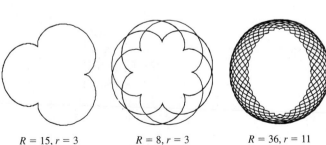

$R = 15, r = 3$ $\qquad\qquad$ $R = 8, r = 3$ $\qquad\qquad$ $R = 36, r = 11$

EXAMPLE 2   Graphing a cycloid using a parameter

Graph the cycloid for $a = 2$ and $0 \leq \theta \leq 2\pi$.

SOLUTION $x = 2(\theta - \sin \theta)$ $y = 2(1 - \cos \theta)$

θ	0	$\frac{\pi}{6}$	$\frac{\pi}{3}$	$\frac{\pi}{2}$	$\frac{2\pi}{3}$	$\frac{5\pi}{6}$	π	$\frac{7\pi}{6}$	$\frac{4\pi}{3}$	$\frac{3\pi}{2}$	$\frac{5\pi}{3}$	$\frac{11\pi}{6}$	2π
x	0	.05	.36	1.14	2.46	4.24	6.28	8.33	10.11	11.42	12.20	12.52	12.57
y	0	.27	1.00	2.00	3.00	3.73	4.00	3.73	3.00	2.00	1.00	.27	0

These points are plotted in Figure 7.19.

FIGURE 7.19 Graph of a cycloid

7.4 Problem Set

A *Graph the curves whose parametric equations are given in Problems 1–12.*

1. $x = 4t, y = -2t$ **2.** $x = t + 1, y = 2t$

3. $x = t, y = 2 + \frac{2}{3}(t - 1)$ **4.** $x = t, y = 3 - \frac{3}{5}(t + 2)$

5. $x = 2t, y = t^2 + t + 1$ **6.** $x = 3t, y = t^2 - t + 6$

7. $x = t, y = t^2 + 2t + 3$ **8.** $x = t, y = 2t^2 - 5t + 6$

9. $x = 3 \cos \theta, y = 3 \sin \theta$ **10.** $x = 2 \cos \theta, y = 2 \sin \theta$

11. $x = 4 \cos \theta, y = 3 \sin \theta$ **12.** $x = 5 \cos \theta, y = 2 \sin \theta$

B *Graph the curves whose parametric equations are given in Problems 13–26.*

13. $x = 10 \cos t, y = 10 \sin t$

14. $x = 8 \sin t, y = 8 \cos t$

15. $x = 5 \cos \theta, y = 3 \sin \theta$

16. $x = 4 \cos \theta, y = 2 \sin \theta$

17. $x = \theta - \sin \theta, y = 1 - \cos \theta$

18. $x = \theta + \sin \theta, y = 1 - \cos \theta$

19. $x = 4 \tan 2t, y = 3 \sec 2t$

20. $x = 2 \tan 2t, y = 4 \sec 2t$

21. $x = 1 + \cos t, y = 3 - \sin t$

22. $x = 2 - \sin t, y = -3 + \cos t$

23. $x = 3 \cos \theta + \cos 3\theta, y = 3 \sin \theta - \sin 3\theta$

24. $x = \cos t + t \sin t, y = \sin t - t \cos t, 0 \leq t \leq 2\pi$

25. $x = \sin t, y = \csc t, 0 \leq t \leq 2\pi$

26. $x = \cos \theta, y = \sec \theta, 0 \leq t \leq 2\pi$

*27. Graph the hypocycloid where $R = 4$ and $r = 1$.

*28. Graph the hypocycloid where $R = 7$ and $r = 1$.

*29. Graph the epicycloid where $R = 4$ and $r = 1$.

*30. Graph the epicycloid where $R = 6$ and $r = 1$.

*See Calculator Comment on page 271. You do not need a computer or graphing calculator to work these problems.

C

31. Suppose a string is wound around a circle of radius a. The string is then unwound in the plane of the circle while it is held tight, as shown in Figure 7.20. Find the equation for this curve, called the *involute of a circle*.

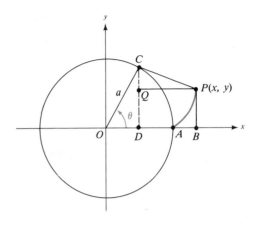

FIGURE 7.21

FIGURE 7.20 Graph of the involute of a circle

Hint for Problem 31:   Consider Figure 7.21 and find the coordinates of $P(x, y)$. Notice that

$$x = |OB| \qquad y = |PB|$$

Find x and y in terms of θ, the amount of rotation in radians.

*32. Derive the parametric equations for the hypocycloid. .

*33. Derive the parametric equations for the epicycloid.

34. COMPUTER APPLICATION Write a paper describing what happens if the radius of the inner rolling circle is larger than the radius of the outer fixed circle.

35. COMPUTER APPLICATION Write a paper describing what happens if the radius of the outer rolling circle is larger than the radius of the inner fixed circle.

7.5 ## Intersection of Polar-Form Curves

In order to find the points of intersection of graphs in rectangular form, you need only find the simultaneous solution of the equations that define those graphs. It is not always necessary to draw the graphs. This follows because in a rectangular coordinate system there is one-to-one correspondence between ordered pairs satisfying an equation and points on its graph.

However, as you saw in Section 7.1, this one-to-one property is lost when working with polar coordinates. This means that the simultaneous solution of two equations in polar-form may introduce extraneous points of intersection or may even fail to yield all points of intersection. For this reason, our method for finding the intersection of polar-form curves will include sketching the graphs.

---

*See Calculator Comment on page 271. You do not need a computer or graphing calculator to work these problems.

Graphical Solution

EXAMPLE 1   Simultaneous solution of a system of polar-form curves by graphing

Find the points of intersection of the curves $r = 2 \cos \theta$ and $r = 2 \sin \theta$.

SOLUTION   First, consider the simultaneous solution of the system of equations.

$$\begin{cases} r = 2 \cos \theta \\ r = 2 \sin \theta \end{cases}$$

By substitution,

$$2 \sin \theta = 2 \cos \theta$$

$$\sin \theta = \cos \theta$$

$$\theta = \frac{\pi}{4} + 2n\pi, \frac{5\pi}{4} + 2n\pi \qquad (n \text{ an integer})$$

Then find r using the primary representation for θ:

$$r = 2 \cos \frac{\pi}{4} \qquad r = 2 \cos \frac{5\pi}{4}$$

$$= 2\left(\frac{\sqrt{2}}{2}\right) \qquad = 2\left(\frac{-\sqrt{2}}{2}\right)$$

$$= \sqrt{2} \qquad\qquad = -\sqrt{2}$$

This gives the points $\left(\sqrt{2}, \frac{\pi}{4}\right)$ and $\left(-\sqrt{2}, \frac{5\pi}{4}\right)$. Writing the primary representations for these points, we see they represent the same point. Thus, the simultaneous solution yields one point of intersection. Next, consider the graphs of these curves, shown in Figure 7.22.

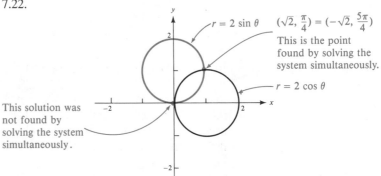

This is the point found by solving the system simultaneously.

This solution was not found by solving the system simultaneously.

FIGURE 7.22 Graphs of $r = 2 \cos \theta$ and $r = 2 \sin \theta$

It looks as though $(0, 0)$ is also a point of intersection. Check this point in each of the given equations:

$$r = 2 \cos \theta: \quad \text{If } r = 0, \text{ then } \theta = \frac{\pi}{2}.$$

$$r = 2 \sin \theta: \quad \text{If } r = 0, \text{ then } \theta = \pi.$$

At first it does not seem that $(0, 0)$ satisfies the equation since $r = 0$ gives $\left(0, \frac{\pi}{2}\right)$ and $(0, \pi)$, respectively. Notice that these coordinates are different and do not satisfy the equations simultaneously. But, if you plot these coordinates, you will see that $(0, 0)$, $\left(0, \frac{\pi}{2}\right)$, and $(0, \pi)$ are all the same point. Points of intersection for the given curves are $\left\{(0, 0), \left(\sqrt{2}, \frac{\pi}{4}\right)\right\}$.

The pole is often a solution for a system of equations, even though it may not satisfy the equations simultaneously. For this reason, it is necessary to check separately to see whether the pole lies on the given graph.

GRAPHICAL SOLUTION OF THE INTERSECTION OF POLAR CURVES

1. Find the simultaneous solution of the given system of equations.
2. Determine whether the pole lies on the two graphs.
3. Graph the curves to look for other points of intersection.

EXAMPLE 2   Points of intersection of polar-form curves by graphing

Find the points of intersection of the curves $r = 2 + 4 \cos \theta$ and $r = 6 \cos \theta$.

SOLUTION   STEP 1:   Solve the equations simultaneously:

$$\begin{cases} r = 2 + 4 \cos \theta \\ r = 6 \cos \theta \end{cases}$$

By substitution,

$$6 \cos \theta = 2 + 4 \cos \theta$$
$$2 \cos \theta = 2$$
$$\cos \theta = 1$$
$$\theta = 0$$

If $\theta = 0$, then $r = 6$, so an intersection point is $(6, 0)$.

STEP 2:   Determine whether the pole lies on the graphs.

a. If $r = 0$, then

$$0 = 2 + 4 \cos \theta$$
$$\cos \theta = -\frac{1}{2}$$
$$\theta = \frac{2\pi}{3}, \frac{4\pi}{3}$$

Thus, $\left(0, \frac{2\pi}{3}\right)$ and $\left(0, \frac{4\pi}{3}\right)$ satisfy the first equation, and the pole lies on this graph.

b. If $r = 0$, then

$$0 = 6 \cos \theta$$
$$\theta = \frac{\pi}{2}, \frac{3\pi}{2}$$

Thus, $\left(0, \frac{\pi}{2}\right)$ and $\left(0, \frac{3\pi}{2}\right)$ satisfy the second equation, so the pole also lies on this graph.

STEP 3:   Graph the curves to look for other points of intersection. The first curve is a limaçon, and the second is a circle as shown in Figure 7.23.

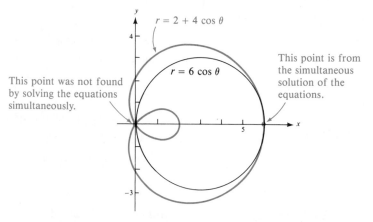

FIGURE 7.23 Graphs of $r = 2 + 4\cos\theta$ and $r = 6\cos\theta$

Notice that there are no other points of intersection, so the intersection points are $(6, 0)$ and $(0, 0)$.   ▌

EXAMPLE 3   **Points of intersection of polar-form curves by graphing**

Find the points of intersection of the curves $r = \frac{3}{2} - \cos\theta$ and $\theta = \frac{2\pi}{3}$.

SOLUTION   Solve $\begin{cases} r = \frac{3}{2} - \cos\theta \\ \theta = \frac{2\pi}{3} \end{cases}$

By substitution,

$$r = \frac{3}{2} - \cos\frac{2\pi}{3}$$
$$= \frac{3}{2} - \left(-\frac{1}{2}\right)$$
$$= 2$$

The solution of $\left(2, \frac{2\pi}{3}\right)$. Also, if $r = 0$, the first equation has no solution since

$$0 = \frac{3}{2} - \cos\theta$$
$$\cos\theta = \frac{3}{2}$$

and $\cos\theta$ cannot be larger than 1.

Now look at the graph, as shown in Figure 7.24.

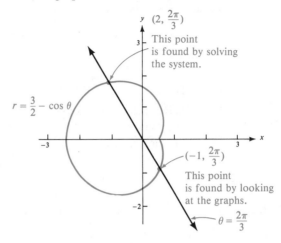

FIGURE 7.24 Graphs of $r = \frac{3}{2} - \cos\theta$ and $\theta = \frac{2\pi}{3}$

From the graph we see that $\left(-1, \frac{2\pi}{3}\right)$ looks like a point of intersection. It satisfies the equation $\theta = \frac{2\pi}{3}$, but what about $r = \frac{3}{2}\cos\theta$? Check $\left(-1, \frac{2\pi}{3}\right)$:

$$-1 \overset{?}{=} \frac{3}{2} - \cos\left(\frac{2\pi}{3}\right)$$

$$= \frac{3}{2} - \left(-\frac{1}{2}\right)$$

$$= 2 \qquad \text{Not satisfied.}$$

However, if you check the other representation $\left(-1, \frac{2\pi}{3}\right) = \left(1, \frac{5\pi}{3}\right)$:

$$1 \overset{?}{=} \frac{3}{2} - \cos\left(\frac{5\pi}{3}\right)$$

$$= \frac{3}{2} - \left(\frac{1}{2}\right)$$

$$= 1 \qquad \text{Satisfied.}$$

You would not have found this other point without checking the graph.   ▮

Analytic Solution

Now you might ask if there is a procedure you can use that will yield all of the points of intersection without relying on the graph. The answer is yes, if you realize that polar-form equations can have different representations just as we found for polar-form points. Recall, a polar-form point has two primary representations

$$(r, \theta) \quad \text{and} \quad (-r, \theta + \pi)$$

where the angle (θ or $\theta + \pi$, respectively) is between 0 and 2π. For every polar-form equation $r = f(\theta)$, there are equations

$$r = (-1)^n f(\theta + n\pi)$$

for any interger n that yield exactly the same curve.

<table>
<tr><td>ANALYTIC SOLUTION
FOR THE INTERSECTION
OF POLAR CURVES</td><td>1. Determine whether the pole lies on the two graphs.

2. Solve each equation of one graph simultaneously with each equation of the other graph. If $r = f(\theta)$ is the given equation, then the other equations of the same graph are

$$r = (-1)^n f(\theta + n\pi)$$</td></tr>
</table>

EXAMPLE 4   **Analytic solution of a simultaneous system of polar-form curves**

Find the points of intersection of the curves

$$r = 1 - 2\cos\theta \quad \text{and} \quad r = 1$$

SOLUTION   *Check pole:*   If $r = 0$, then the second equation ($r = 1$) is not satisfied. That is, the graph of the second equation does not pass through the origin.

Next, solve the equations simultaneously; write out the alternate forms of the equations:

$r = 1 - 2\cos\theta$: $r = (-1)^n[1 - 2\cos(\theta + n\pi)]$

If $n = 0$, $r = 1 - 2\cos\theta$

If $n = 1$, $r = -[1 - 2\cos(\theta + \pi)]$

$\qquad\qquad\qquad = -[1 + 2\cos\theta]$

$\qquad\qquad\qquad = -1 - 2\cos\theta$

For other integral values, one of these two equations is repeated:

$r = 1$: $r = (-1)^n \cdot 1$

If $n = 0$, $r = 1$

If $n = 1$, $r = -1$

For other integral values, one of these two equations is repeated. Solve the systems:

$\begin{cases} r = 1 - 2\cos\theta \\ r = 1 \end{cases}$ $\begin{cases} r = 1 - 2\cos\theta \\ r = -1 \end{cases}$ $\begin{cases} r = -1 - 2\cos\theta \\ r = 1 \end{cases}$ $\begin{cases} r = -1 - 2\cos\theta \\ r = -1 \end{cases}$

$\quad 1 = 1 - 2\cos\theta \qquad -1 = 1 - 2\cos\theta \qquad 1 = -1 - 2\cos\theta \qquad -1 = -1 - 2\cos\theta$

$\quad 0 = \cos\theta \qquad\qquad 1 = \cos\theta \qquad\qquad -1 = \cos\theta \qquad\qquad 0 = \cos\theta$

$\quad \theta = \dfrac{\pi}{2}, \dfrac{3\pi}{2} \qquad\qquad \theta = 0 \qquad\qquad\quad \theta = \pi \qquad\qquad\quad$ Same as first equation

Points: $\left(1, \frac{\pi}{2}\right), \left(1, \frac{3\pi}{2}\right), \left(1, \pi\right)$

The graphs of these curves are shown in Figure 7.25.

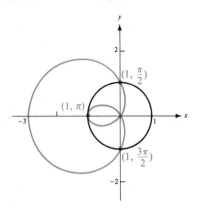

FIGURE 7.25 Graphs of $r = 1 - 2\cos\theta$ and $r = 1$

7.5   **Problem Set**

A *Find the points of intersection of the curves given by the equations in Problems 1–30. Sketch each of the graphs as a check of your work since you may have introduced extraneous solutions. You need to give only one primary representation for each point of intersection; give the one with a positive r or with $0 \le \theta < 2\pi$.*

B

1. $\begin{cases} r = 4\cos\theta \\ r = 4\sin\theta \end{cases}$

2. $\begin{cases} r = 8\cos\theta \\ r = 8\sin\theta \end{cases}$

3. $\begin{cases} r = 2\cos\theta \\ r = 1 \end{cases}$

4. $\begin{cases} r = 4\cos\theta \\ r = 2 \end{cases}$

5. $\begin{cases} r^2 = 9\cos 2\theta \\ r = 3 \end{cases}$

6. $\begin{cases} r^2 = 4\sin 2\theta \\ r = 2 \end{cases}$

7. $\begin{cases} r = 2(1 + \cos\theta) \\ r = 2(1 - \cos\theta) \end{cases}$

8. $\begin{cases} r = 2(1 + \sin\theta) \\ r = 2(1 - \sin\theta) \end{cases}$

9. $\begin{cases} r^2 = 4\cos 2\theta \\ r = 2 \end{cases}$

10. $\begin{cases} r^2 = 9\sin 2\theta \\ r = 3 \end{cases}$

11. $\begin{cases} r^2 = 4\sin 2\theta \\ r = 2\sqrt{2}\cos\theta \end{cases}$

12. $\begin{cases} r^2 = 9\cos 2\theta \\ r = 3\sqrt{2}\sin\theta \end{cases}$

13. $\begin{cases} r^2 = \cos 2(\theta - 90°) \\ r^2 = \cos 2\theta \end{cases}$

14. $\begin{cases} r = 1 + \sin\theta \\ r = 1 + \cos\theta \end{cases}$

15. $\begin{cases} r = a(1 + \sin\theta) \\ r = a(1 - \sin\theta) \end{cases}$

16. $\begin{cases} r = 2\theta \\ \theta = \dfrac{\pi}{6} \end{cases}$

17. $\begin{cases} r = 3\theta \\ \theta = \dfrac{\pi}{3} \end{cases}$

18. $\begin{cases} r^2 = \cos 2\theta \\ r = \sqrt{2}\sin\theta \end{cases}$

19. $\begin{cases} r = 2(1 - \cos\theta) \\ r = 4\sin\theta \end{cases}$

20. $\begin{cases} r = 2(1 + \cos\theta) \\ r = 4\sin\theta \end{cases}$

21. $\begin{cases} r = 2(1 - \sin\theta) \\ r = 4\cos\theta \end{cases}$

22. $\begin{cases} r = 2\cos\theta + 1 \\ r = \sin\theta \end{cases}$

23. $\begin{cases} r = 2\sin\theta + 1 \\ r = \cos\theta \end{cases}$

24. $\begin{cases} r = \dfrac{5}{3 - \cos\theta} \\ r = 2 \end{cases}$

25. $\begin{cases} r = \dfrac{2}{1 + \cos\theta} \\ r = 2 \end{cases}$

26. $\begin{cases} r = \dfrac{4}{1 - \cos\theta} \\ r = 2\cos\theta \end{cases}$

27. $\begin{cases} r = \dfrac{1}{1 + \cos\theta} \\ r = 2(1 - \cos\theta) \end{cases}$

28. $\begin{cases} r = 2\cos\theta \\ r = 2\sec\theta \end{cases}$

29. $\begin{cases} r = 2\sin\theta \\ r = 2\csc\theta \end{cases}$

30. $\begin{cases} r\sin\theta = 1 \\ r = 4\sin\theta \end{cases}$

7.6 Summary and Review

Objectives	*Pages / Examples / Problems*

7.1 Polar Coordinates

1. Plot polar-form points.

pp. 254–255; Example 1; Problems 1–6, 13–14.

2. Write the two primary representations in polar form, given a point in either rectangular form or polar form.

p. 256; Examples 2 and 4; Problems 1–12.

3. Write the rectangular form of a point, given the polar form.

p. 257; Example 3; Problems 1–6.

4. Determine whether a given point lies on a graph defined by a polar-form equation.

pp. 257–258; Example 5; Problems 15–36.

7.2 Graphing Polar-Form Equations

5. Graph polar-form curves by plotting points.

pp. 260–261; Examples 1–2; Problems 19–37.

6. Graph cardioids by rotating and comparing to a standard-position cardioid.

pp. 261–262; Examples 3–5; Problems 1–18.

7.3 Special Polar-Form Curves

7. Recognize and graph rose curves.

pp. 263–265; Examples 1–3; Problems 16–18, 22–27.

8. Recognize and graph lemniscates.

pp. 266–267; Examples 4–5; Problems 19–21, 28–30.

9. Classify and identify the following polar-form curves by inspection of their equations: cardioid, rose curve (number of leaves), and lemniscate. Also be able to determine the rotation, if any.

pp. 267–269; Problems 1–15.

7.4 Parametric Equations

10. Sketch a curve represented by parametric equations.

pp. 269–274; Examples 1–2; Problems 1–26.

7.5 Intersection of Polar-Form Curves

11. Find the intersection of polar-form curves.

pp. 275–281; Examples 1–4; Problems 1–30.

TERMS

Cardioid [7.2, 7.3]	Parameter [7.4]	Polar coordinate system [7.1]	Rose curve [7.3]
Lemniscate [7.3]	Parametric equations [7.4]	Pole [7.1]	Rotation [7.2]
Limaçon [7.2]	Polar axis [7.1]	Primary representations of a point [7.1]	Standard-position cardioid [7.2]

CHAPTER 7 REVIEW OF OBJECTIVES

The problem numbers correspond to the objectives listed in Section 7.6.

1. Plot the given polar-form points.

    a. $\left(3, -\frac{2\pi}{3}\right)$ **b.** $\left(-5, -\frac{\pi}{10}\right)$ **c.** $(-2, 2)$ **d.** $\left(5, \sqrt{75}\right)$

2. Give both primary representations in polar form for the given ponts.

    a. $\left(3, -\frac{2\pi}{3}\right)$ is in polar form.

    b. $\left(-5, -\frac{\pi}{10}\right)$ is in polar form.

    c. $\left(-3, 3\sqrt{3}\right)$ is in rectangular form.

    d. $(2, 5)$ is in rectangular form.

3. Give the rectangular form for the given points.

    a. $\left(3, -\frac{2\pi}{3}\right)$ **b.** $\left(-5, -\frac{\pi}{10}\right)$

4. Determine whether the given points lie on the curve defined by the equation

$$r = \frac{10}{2 + \cos \theta}$$

    a. $\left(4, \frac{\pi}{6}\right)$ **b.** $\left(-4, \frac{4\pi}{3}\right)$ **c.** $\left(5, \frac{3\pi}{2}\right)$ **d.** $(5, \pi)$

5. Sketch the given curves by plotting points.

    a. $r - 3\theta = 0, \theta > 0$

    b. $\theta - 2 = 0$

    c. $r - 2 \tan \frac{3\pi}{4} = 0$

Sketch the curves in Problems 6–8 by comparing them with standard forms.

6. $r + 3 \sin \theta = 3$

7. $r = 6 \cos 3\left(\theta - \frac{\pi}{10}\right)$

8. $r^2 = 4 \sin 2\theta$

9. Identify each of the given curves.

    a. $r^2 = 9 \cos 2\theta$ **b.** $r = 2 + 2 \sin \theta$

    c. $r = 4 \cos 5\theta$ **d.** $r = 5 \sin \frac{\pi}{4}$

10. Graph the curve given by the parametric equations

$$x = 5 \cos \theta \qquad y = 3 \sin \theta$$

11. Find the intersection of the curves $r = 4 - 4 \sin \theta$ and $r = 4 - 8 \sin \theta$.

DISCUSSION PROBLEMS

1. Describe the procedure you use to graph a polar-form curve.

2. In your own words, explain the concept of parametric equations.

3. Outline your "method of attack" for graphing polar-form curves.

4. In each chapter we considered cardioids, rose curves, and lemniscates. Research other polar-form curves and identify them by name, equation, and graph.

5. Write a 500-word paper extending the rectangular and polar coordinate system to three dimensions. Are there other three-dimensional coordinate systems?

MISCELLANEOUS PROBLEMS

Identify each of the curves in Problems 1–16.

1. $r = \cos 2\theta$ **2.** $r = 2 \cos \theta$

3. $r = \cos 2$ **4.** $r^2 = \cos 2\theta$

5. $r = 2 - 2 \cos \theta$ **6.** $r = 2 + 2 \cos \theta$

7. $r = 4 \sin 3\theta$ **8.** $r = 4 \sin 3$

9. $r^2 = 4 \sin 2\theta$ **10.** $r = 4 + 4 \sin \theta$

11. $r = \cos 3\left(\theta - \frac{\pi}{6}\right)$ **12.** $r^2 = \cos 2\left(\theta - \frac{\pi}{6}\right)$

13. $r = \sin 3\left(\theta - \frac{\pi}{6}\right)$ **14.** $r^2 = 3 \sin 2\left(\theta - \frac{\pi}{6}\right)$

15. $r - 2 = 2 \sin(\theta - 3)$ **16.** $r - 3 = 3 \cos(\theta - 2)$

Graph each of the curves in Problems 17–32 by comparing them with standard forms.

17. $r = \cos 2\theta$

18. $r = 2\cos\theta$

19. $r = \cos 2$

20. $r^2 = \cos 2\theta$

21. $r = 2 - 2\cos\theta$

22. $r = 2 + 2\cos\theta$

23. $r = 4\sin 3\theta$

24. $r = 4\sin 3$

25. $r^2 = 3\sin 2\theta$

26. $r = 4 + 4\sin\theta$

27. $r = 3\cos 3\left(\theta - \frac{\pi}{6}\right)$

28. $r^2 = \cos 2\left(\theta - \frac{\pi}{6}\right)$

29. $r = \sin 3\left(\theta - \frac{\pi}{6}\right)$

30. $r^2 = 3\sin 2\left(\theta - \frac{\pi}{6}\right)$

31. $r - 2 = 2\sin(3 - \theta)$

32. $r - 3 = 3\cos(2 - \theta)$

Graph the curves in Problems 33–40 by plotting points.

33. $r = \tan 2\theta, 0 \le \theta \le \pi$

34. $r - 2\theta = 0, 0 \le \theta \le \pi$

35. $r\theta = 1, 0 \le \theta \le 2\pi$

36. $r = \dfrac{8}{4 - 2\cos\theta}, 0 \le \theta \le 2\pi$

37. $x = 5\cos\theta, y = 3\sin\theta$

38. $x = 4\cos\theta, y = 4\sin\theta$

39. $x = 1 + 4\cos\theta, y = 2 - 4\sin\theta$

40. $x = 2 - 5\cos\theta, y = 1 + 5\sin\theta$

▌ Communication of Bees

Courtesy of David Cavagnaro/Dandelet Interlinks.

Polar coordinates can be applied to the manner in which bees communicate the distance and source of a food supply.* Scout bees looking for pollen keep track of their position relative to the hive by referencing it to the sun's position. When a scout finds a food source, the bee returns to the hive, gives away samples of the food, and then performs a dance inside the hive.

The dance is done on the vertical hanging honeycomb inside the hive (see Figure 7.26). The vertical direction represents the sun's direction, and the angle θ represents the direction of the food source.

FIGURE 7.26 Communication of bees. The bee runs straight ahead for a short distance (about a centimeter), returns to the starting point through a 360° turn to the left, again runs through the straight stretch returning through a 360° turn to the right, and so on in an alternating fashion. The straight part of the run is emphasized by lateral wagging of the abdomen.

---

*From Karl von Frisch, *The Dance Language and Orientation of Bees* (Cambridge, Mass.: Belknap Press, 1967). I obtained the idea for this application from Warren Page, New York Technical College.

The direction of the straight waggle-run gives an index of the distance, ρ (see Figure 7.27), from the hive, H, to the food source, F. The bee's estimate of ρ depends not on optical perception of the terrain, nor on the duration of flight time, but rather on the bee's expenditure of energy to get there. Notice that the bee is using polar coordinates to fix the location of the food source. See Problem 4 for a development of this idea.

FIGURE 7.27 Bee using polar coordinates

PROBLEMS FOR FURTHER STUDY: COMMUNICATION OF BEES

1. Entomology Suppose the equation $r = 1 - \sin\theta$ is used to describe the dance shown by the bee in Figure 7.26. Graph this equation. What is the direction of the food source relative to the sun?

2. Entomology Modify the equation in Problem 1 to indicate a food source that is in a direction of 30° away from the direction of the sun.

3. Entomology Suppose the equation

$$r = 2 - 2\sin(\theta - 50°)$$

is used to describe the dance shown by the bee in Figure 7.26. Graph this equation. Compare this food source to the one described in Problem 1.

4. Entomology In the bee's dance, a straight run of three seconds indicates a distance of about $\frac{1}{2}$ mile, and a straight run of five seconds means about 2 miles. If the pollen is 20° to the right of the sun, we could use the polar coordinates (ρ, θ) to indicate the location of the pollen as $(3, 20°)$ or $(5, 20°)$. What do you think would be the coordinates for a food source about 1 mile from the hive in the same direction? What do you suppose $(6, 20°)$ means in terms of the time of the straight portion of the dance and the distance from the hive?

| Chapters 5–7

Sample Test I: Objective Form

1. State whether you would use the Law of Cosines, Law of Sines, or a right-triangle solution for each part. If the solution involves the ambiguous case, so state. You do *not* need to solve these triangles.

	a	b	c	α	β	γ
a.	38	—	42	—	108°	—
b.	14.6	28.1	19.4	—	—	—
c.	—	52	49	—	—	63°
d.	83	52	—	121°	—	—
e.	418	—	385	—	218.0°	—
f.	281	318	412	—	—	—
g.	3.0	6.0	—	30°	—	—
h.	18	15	—	—	106°	—
i.	55	61	—	57°	—	—
j.	—	14.1	18.1	—	—	27.4°

2. Identify each of the given curves.

   a. $r = 3 \sin 2\theta$ b. $r^2 = 3 \sin 2\theta$ c. $r = 4 \cos 3\theta$ d. $r = 3(1 + \sin \theta)$

   e. $r = 2 \sin \frac{\pi}{2}$ f. $r - 5 = 5 \cos \theta$ g. $r^2 = 5 \cos 2\theta$ h. $\theta = 3 \sin \frac{\pi}{6}$

   i. $r = 6(1 - \sin \theta)$ j. $r = 2 + 2 \cos \theta$

3. Solve the triangle given in Problem 1d.
4. Solve the triangle given in Problem 1i.
5. A draftsman drew to scale (10 ft = 10 in.) a triangularly shaped lot with sides 45 ft, 82 ft, and 60 ft. On the plot plan, what are the measures of the angles?
6. For each given polar-form point, give both primary representations in polar form, and then change it to rectangular form.

   a. $\left(3, -\frac{\pi}{4}\right)$ b. $\left(-5, -\frac{\pi}{6}\right)$ c. $(-3, 9.4248)$

7. Write the cube roots (in trigonometric form) of $-\frac{5}{2}\sqrt{3} + \frac{5}{2}i$.
8. Write the square roots (in trigonometric form) of $-8i$.
9. Graph $r = 3 \cos 3\theta$.
10. Graph $r^2 = 4 \sin 2\theta$.

Sample Test II: Fill In and Multiple Choice

Fill in Problems 1–5 with the word or words to make the statements complete and correct.

1. The Law of Cosines for a correctly labeled triangle states $c^2 =$ _____ and
$\cos \beta =$ _____ .

2. The Law of Sines states _____ .

3. To change a complex number r cis θ to rectangular form we use $a =$ _____ and
$b =$ _____ .

4. If $z_1 = r_1$ cis θ_1 and $z_2 = r_2$ cis θ_2, then $z_1 z_2 =$ _____ .

5. The polar graph $r = f(\theta - \alpha)$ is the same as the polar graph of _____ that has
been _____ .

Choose the BEST answer from the choices given.

6. If you are given $a = 38$, $c = 42$, and $\beta = 108°$, the correct method of solution is

    A. Law of Sines—one solution   B. Law of Cosines—ambiguous case

    C. Law of Cosines—one solution   D. right-triangle solution

    E. no solution

7. If you are given $a = 55$, $b = 61$, and $\alpha = 57°$, the correct method of solution is

    A. Law of Sines—one solution   B. Law of Cosines—ambiguous case

    C. Law of Cosines—one solution   D. right-triangle solution

    E. no solution

8. If you are given $b = 26$, $\beta = 47°$, and $\gamma = 76°$, the correct method of solution is

    A. Law of Sines—one solution   B. Law of Cosines—ambiguous case

    C. Law of Cosines   D. right-triangle solution

    E. no solution

9. Which of the following pairs of vectors are orthogonal?

    A. $\mathbf{v} = \cos 20°\mathbf{i} + \sin 20°\mathbf{j}$,   B. $\mathbf{v} = \frac{1}{\sqrt{3}}\mathbf{i} - \frac{2}{\sqrt{5}}\mathbf{j}$,

        $\mathbf{w} = \cos 70°\mathbf{i} + \sin 70°\mathbf{j}$ $\mathbf{w} = 2\mathbf{i} - \frac{\sqrt{5}}{5\sqrt{3}}\mathbf{j}$

    C. $\mathbf{v} = \mathbf{i}$, $\mathbf{w} = \mathbf{j}$ D. all are orthogonal

    E. none is orthogonal

10. Robin, in the airport's control tower, observes Batman in the Batplane at a distance of
80 km on a bearing of 120° and the Joker in his Learjet 90 km from the airport on a
bearing of 140°. How far apart are Batman and the Joker?

    A. 31 km   B. 93 km   C. 147 km   D. 968 km

    E. none of these

11. $(3 + 5i)^2$ simplified is

    A. -16 B. $-16 + 15i$ C. $-16 + 30i$ D. $9 + 30i + 25i^2$ E. none of these

12. $\dfrac{(5 - 5i)(\sqrt{3} - i)}{1 - i}$ simplified is

    A. $5\sqrt{3} - 5i$ B. $10(2 - i)$

    C. $20 \text{ cis } 330°$ D. $5(1 - i)^2(\sqrt{3} - i)$

    E. none of these

13. $\dfrac{2 \text{ cis } 128° \cdot 9 \text{ cis } 285°}{(3 \text{ cis } 4°)^3}$ is

    A. $6 \text{ cis } 409°$ B. $6 \text{ cis } 49°$

    C. $6 \text{ cis } 401°$ D. $6 \text{ cis } 41°$

    E. none of these

14. One of the cube roots of $\frac{3}{2} - \frac{3}{2}\sqrt{3}\,i$ is

    A. $27 \text{ cis } 180°$ B. $\sqrt[3]{3} \text{ cis } 220°$

    C. $\sqrt[3]{3} \text{ cis } 20°$ D. $1.5 - 2.5981i$

    E. none of these

15. The other primary representation of $\left(3, \frac{\pi}{10}\right)$ is

    A. $\left(-3, \frac{\pi}{10}\right)$ B. $\left(-3, \frac{\pi}{10}\right)$ C. $\left(-3, \frac{9\pi}{10}\right)$ D. $\left(-3, \frac{11\pi}{10}\right)$

    E. none of these

16. The curve $\theta = \text{constant}$ is a

    A. cardioid   B. rose curve

    C. lemniscate   D. circle

    E. line

17. The graph of $r = 3 \sin 2\theta$ is a

    A. cardioid   B. rose curve

    C. lemniscate   D. circle

    E. line

18. What is the rotation of the curve $r^2 = \sin 2\theta$ when compared to the standard-position curve $r^2 = \cos 2\theta$?

    A. $90°$ B. $60°$ C. $45°$ D. $30°$

    E. none of these

19. The graph of $r = 3 \sin 60°$ is a

    A. cardioid   B. rose curve

    C. lemniscate   D. circle

    E. line

20. A point satisfying the equation $r = \dfrac{3}{1 + \sin \theta}$ is

    A. $\left(\frac{\pi}{3}, 12 - 6\sqrt{3}\right)$ B. $\left(\frac{1}{2}, \frac{\pi}{3}\right)$ C. $\left(-6, \frac{\pi}{6}\right)$ D. $\left(12 - 6\sqrt{3}, \frac{\pi}{6}\right)$

    E. none of these

▪ René Descartes
(1596–1650)

▌ It is impossible not to feel stirred at the thought of the emotions of men at certain historic moments of adventure and discovery—Columbus when he first saw the Western shore, Pizarro when he stared at the Pacific Ocean, Franklin when the electric spark came from the string of his kite, Galileo when he first turned his telescope to the heavens. Such moments are also granted to students in the abstract regions of thought, and high among them must be placed the morning when Descartes lay in bed and invented the method of coordinate geometry.

A. N. WHITEHEAD
An Introduction to Mathematics

In this text we have seen many ideas that revolutionized mathematics and the nature of thought. None is more profound than the powerful idea of associating ordered pairs of real numbers with points in a plane, thereby making possible a correspondence between curves in the plane and equations in two variables. This idea came from the French mathematician René Descartes.

Descartes was born in France on March 31, 1596. He came from a wealthy family, but was a weakling with frail health. Because of his delicate health he formed a lifelong habit of spending the mornings in bed, and he did much of his work during those morning hours.

According to legend, Descartes thought of his coordinate system while he was in bed watching a fly crawl around on the ceiling of the room. He noticed that the path of the fly could be described if he knew the relation connecting the fly's distances from the walls.

In 1649 Descartes was invited to tutor Queen Christina of Sweden. Since she was from a royal family, Descartes thought the job would be fun, and he accepted the invitation. However, when he arrived, he found that she wanted her lessons at daybreak. Since he was not used to getting up early, he soon became ill and died of pneumonia within a year.

It should be pointed out that 1,800 years earlier the Greek mathematician Apollonius (*c.* 262–190 B.C.) made a thorough investigation of the *conic section* introduced in this chapter. He considered himself a rival of Archimedes and was the first person to use the words *parabola, ellipse,* and *hyperbola.* His methods were expounded in an eight-volume work called *Conics,* and his work was so modern it is sometimes judged to be an analytic geometry preceding Descartes.

8

CONIC SECTIONS

Contents

8.1 Parabolas

8.2 Ellipses

8.3 Hyperbolas

8.4 General Conic Sections

8.5 Rotations

8.6 Conic Sections in Polar Form

8.7 Summary and Review

**Application for
Further Study**
    Planetary Orbits

Preview

By a **conic section,** we mean a curve that can be formed by intersecting a cone and a plane; these curves are called **parabolas, ellipses,** and **hyperbolas.** After introducing each of these curves and their equations, we will then translate each, and finally we use trigonometry to look at the rotated forms. In the last section of this chapter we look at the conic sections in polar form.

Perspective

Although many calculus books include coverage of analytic geometry, it is becoming more and more necessary for students to have an earlier knowledge of the conic sections. For that reason, a thorough introduction to this topic is presented in this chapter. As you will see in this chapter, examples of conic sections are common. The accompanying article shows the relationship between moiré patterns and the conic sections. When two geometric patterns are superimposed, and if additional geometric patterns become visible, then those additional patterns are known as *moiré patterns.*

Physical Applications

A ripple tank creates moiré fringes from the crests of two wave trains. A train of horizontal waves may be generated by dipping a ruler into the surface of the water with a certain frequency, while circular waves are easily formed by dipping a pencil into the tank. If wave trains are generated simultaneously from two such sources, the moiré fringes correspond to positions where *double crests* are formed. These appear as bright regions when the wave pattern is projected onto the screen of the ripple tank. Excellent ripple tank photos are given in many physics texts.

 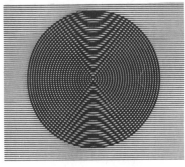

From ''Moiré Fringes and the Conic Sections,'' by M. R. Cullen, *The College Mathematics Journal*, Vol. 21, No. 5, pp. 370–378.

8.1 Parabolas

Conic Sections

In elementary algebra you learned about graphing the general **first-degree** (or **linear**) **equation**

$$Ax + By + C = 0$$

In this chapter, we consider graphing the general **second-degree** (or **quadratic**) **equation**

$$Ax^2 + Bxy + Cy^2 + Dx + Ey + F = 0$$

Historically, second-degree equations in two variables were first considered in a geometric context and were called **conic sections** because the curves they represent can be described as the intersections of a double-napped right circular cone and a plane. There are three general ways a plane can intersect a cone, as shown in Figure 8.1. (Several special cases are discussed later.)

Definition and Terminology

You have probably considered parabolas in your previous algebra courses, but here we will give a geometric definition.

PARABOLA

> A **parabola** is a set of all points in the plane equidistant from a given point (called the **focus**) and a given line (called the **directrix**).

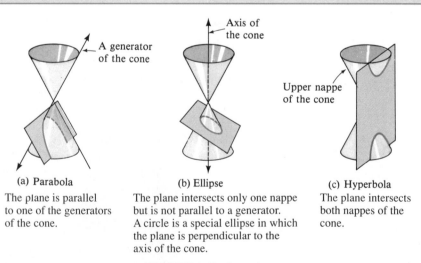

(a) Parabola
The plane is parallel to one of the generators of the cone.

(b) Ellipse
The plane intersects only one nappe but is not parallel to a generator. A circle is a special ellipse in which the plane is perpendicular to the axis of the cone.

(c) Hyperbola
The plane intersects both nappes of the cone.

FIGURE 8.1 Conic sections

To obtain the graph of a parabola from this definition, you can use the special type of graph paper shown in Figure 8.2a, where F is the focus and L is the directrix.

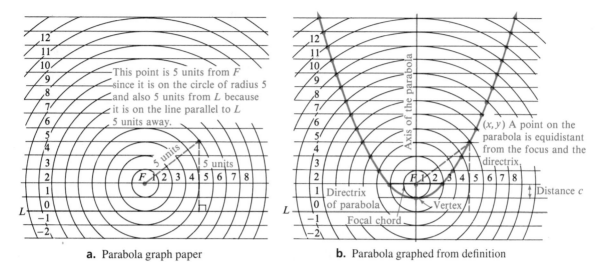

a. Parabola graph paper

b. Parabola graphed from definition

FIGURE 8.2

To sketch a parabola using the definition, let F be any point and let L be any line, as shown in Figure 8.2a. Plot points in the plane equidistant from the focus and the directrix. Draw a line through the focus and perpendicular to the directrix. This line is called the **axis** of the parabola. Let V be the point of this line halfway between the focus and the directrix. This is the point of the parabola nearest to both the focus and the directrix. It is called the **vertex** of the parabola. Plot other points equidistant from F and L, as shown in Figure 8.2b.

In Figure 8.2b, let c be the distance from the vertex to the focus. Notice that the distance from the vertex to the directrix is also c. Consider the segment that passes through the focus perpendicular to the axis and with endpoints on the parabola. This segment has length $4c$ and is called the **focal chord.**

Standard-Form Parabolas

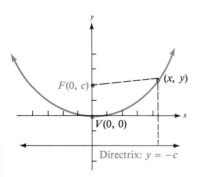

FIGURE 8.3 Graph of the parabola $x^2 = 4cy$

To obtain the equation of a parabola, first consider a special case—a parabola with focus $F(0, c)$ and directrix $y = -c$, where c is any positive number. This parabola must have its vertex at the origin (remember that the vertex is halfway between the focus and the directrix) and must open upward, as shown in Figure 8.3.

Let (x, y) be any point on the parabola. Then, from the definition of a parabola,

Distance from (x, y) to $(0, c)$ = distance from (x, y) to directrix

or

$$\sqrt{(x - 0)^2 + (y - c)^2} = y + c$$

Square both sides:

$$x^2 + (y - c)^2 = (y + c)^2$$
$$x^2 + y^2 - 2cy + c^2 = y^2 + 2cy + c^2$$
$$x^2 = 4cy$$

This is the equation of the parabola with vertex $(0, 0)$ and directrix $y = -c$.

You can repeat this argument for parabolas that have vertex at the origin and open downward, to the left, and to the right to obtain the results summarized below. A positive number c, the distance from the focus to the vertex, is assumed given. These are called the **standard-form parabolic equations** with vertex $(0, 0)$.

STANDARD-FORM EQUATIONS FOR PARABOLAS WITH VERTEX (0, 0)

PARABOLA	FOCUS	DIRECTRIX	VERTEX	EQUATION
Opens *upward*	$(0, c)$	$y = -c$	$(0, 0)$	$x^2 = 4cy$
Opens *downward*	$(0, -c)$	$y = c$	$(0, 0)$	$x^2 = -4cy$
Opens *right*	$(c, 0)$	$x = -c$	$(0, 0)$	$y^2 = 4cx$
Opens *left*	$(-c, 0)$	$x = c$	$(0, 0)$	$y^2 = -4cx$

Given Equation, Find Graph

EXAMPLE 1   **Graphing a parabola that opens up**

Graph $x^2 = 8y$.

SOLUTION   This equation represents a parabola that opens upward. The vertex is $(0, 0)$, and notice by inspection that

$$4c = 8$$
$$c = 2$$

Thus the focus is $(0, 2)$. After plotting the vertex $V(0, 0)$ and the focus $F(0, 2)$, the only question is the width of the parabola. We determine the width of the parabola by using the focal chord. Remember that **the length of the focal chord is 4c,** so that in this case it is 8. Do you see that in each case $4c$ is the absolute value of the coefficient of the first-degree term in the standard-form equations? Since a parabola is symmetric with respect to its axis, draw a segment of length 8 with the midpoint at F. Using these three points (the vertex and the endpoints of the focal chord), sketch the parabola, as shown in Figure 8.4.

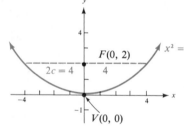

FIGURE 8.4 Graph of the parabola $x^2 = 8y$

EXAMPLE 2   **Graphing a parabola that opens to the left**

Graph $y^2 = -12x$.

SOLUTION   This equation represents a parabola that opens left. The vertex is $(0, 0)$ and

$$4c = 12$$
$$c = 3$$

(recall that c is positive), so the focus is $(-3, 0)$. The length of the focal chord is 12, and the parabola is drawn as in Figure 8.5.

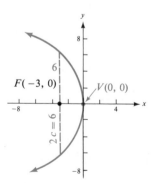

FIGURE 8.5 Graph of the parabola $y^2 = -12x$

EXAMPLE 3   **Graphing a parabola that opens to the right**

Graph $2y^2 - 5x = 0$.

SOLUTION   You must first put the equation into standard form by solving for the second-degree term:

$$y^2 = \frac{5}{2}x$$

The vertex is $(0, 0)$, and

$$4c = \frac{5}{2}$$

$$c = \frac{5}{8}$$

Thus the parabola opens to the right, the focus is $\left(\frac{5}{8}, 0\right)$, and the length of the focal chord is $\frac{5}{2}$, as shown in Figure 8.6.

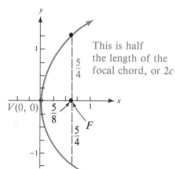

FIGURE 8.6 Graph of the parabola $2y^2 - 5x = 0$

CALCULATOR COMMENT

If you have a graphing calculator, you have been graphing functions like that shown in Example 1 since Chapter 2. However, the curve defined in Example 2 is not a function. To graph this curve using your graphing calculator, solve for y and define two functions:

$$Y1 = \sqrt{(-12X)}$$
$$Y2 = -\sqrt{(-12X)}$$

Given Graph, Find Equation

There are two basic types of problems you will need to solve concerning each curve in analytic geometry:

1.   Given the equation, draw the graph; this is what you did in Examples 1–3.
2.   Given the graph (or information about the graph), write the equation. Example 4 is a problem of this type.

EXAMPLE 4   **Finding the equation of a parabola**

Find the equation of the parabola with directrix $y = 4$ and focus at $F(0, -4)$.

SOLUTION   This curve is a parabola that opens downward with vertex at the origin, as shown in Figure 8.7. The value for c is found by inspection: $c = 4$. Thus $4c = 16$. Since the equation is of the form $x^2 = -4cy$, the desired equation is (by substitution)

$$x^2 = -16y$$

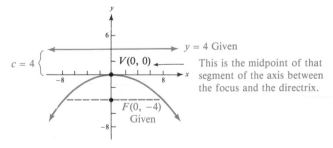

FIGURE 8.7 Graph of the parabola with focus at $(0, -4)$ and directrix $y = 4$

Translated Parabolas

The types of parabolas we have been considering are quite limited, since we have assumed that the vertex is at the origin and the directrix is parallel to one of the coordinate axes. Suppose, however, that you are given a parabola with vertex at (h, k) and a directrix parallel to one of the coordinate axes. In Section 2.2 we showed that you can translate the axes to (h, k) by a substitution:

$$x' = x - h$$
$$y' = y - k$$

Therefore the **standard-form parabola equations** with vertex (h, k) can be summarized by the following box.

STANDARD-FORM EQUATIONS FOR TRANSLATED PARABOLAS

PARABOLA	EQUATION
Opens *upward*	$(x - h)^2 = 4c(y - k)$
Opens *downward*	$(x - h)^2 = -4c(y - k)$
Opens *right*	$(y - k)^2 = 4c(x - h)$
Opens *left*	$(y - k)^2 = -4c(x - h)$

EXAMPLE 5   Graphing a translated parabola that opens downward

Sketch $(x + 2)^2 = -8(y - 4)$.

SOLUTION The procedure is identical to that used with the standard parabolas, except you count from the vertex point $(h, k) = (-2, 4)$ rather than from the origin. By inspection, $4c = 8$, $c = 2$, and the parabola opens downward from the vertex, as shown by Figure 8.8.

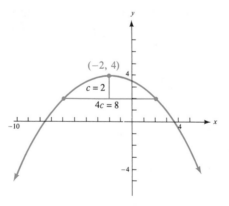

FIGURE 8.8 Graph of $(x + 2)^2 = -8(y - 4)$

EXAMPLE 6   **Finding the equation of a translated parabola**

Find the equation of the parabola with focus at $(4, -3)$ and directrix the line $x + 2 = 0$.

SOLUTION   Sketch the given information, as shown in Figure 8.9. The vertex is $(1, -3)$, since it must be equidistant from F and the directrix. Note that $c = 3$. Thus, substitute into the equation

$$(y - k)^2 = 4c(x - h)$$

since the parabola opens to the right. The desired equation is

$$(y + 3)^2 = 12(x - 1)$$

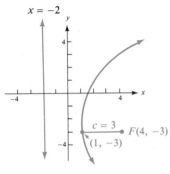

FIGURE 8.9

By sketching the focus and directrix, it is easy to find c and the vertex by inspection. The sketch of the curve is not necessary to find this equation.

8.1   Problem Set

A *Using the definition of a parabola, in Problems 1–6 sketch a parabola similar to the one in Figure 8.2a, such that the distance between the focus and the directrix is the given number.*

1. 4   **2.** 6   **3.** 8   **4.** 10   **5.** 3   **6.** 7

Sketch the curves given by the equations in Problems 7–24. Label the focus F and the vertex V.

7. $y^2 = 8x$ **8.** $y^2 = -12x$
9. $y^2 = -20x$ **10.** $x^2 = 4y$
11. $x^2 = -8y$ **12.** $x^2 = -10y$
13. $4x^2 = 10y$ **14.** $3x^2 = -12y$
15. $3y^2 = 18x$ **16.** $2x^2 + 5y = 0$
17. $5y^2 + 15x = 0$ **18.** $3y^2 - 15x = 0$
19. $(y - 1)^2 = 2(x + 2)$ **20.** $(y + 3)^2 = 3(x - 1)$
21. $(x + 3)^2 = -3(y - 1)$ **22.** $(x + 2)^2 = 2(y - 1)$
23. $(x - 1)^2 = 3(y + 3)$ **24.** $(x - 1)^2 = -2(y + 2)$

B *Find the equation of each curve in Problems 25–28.*

25. Directrix $x = -5$; focus at $(5, 0)$
26. Directrix $y = 4$; focus at $(0, -4)$

27. Vertex at $(-3, 2)$ and focus at $(-3, -1)$

28. Vertex at $(4, 2)$ and focus at $(-3, 2)$

29. ENGINEERING A radar antenna is constructed so that a cross section along its axis is a parabola with the receiver at the focus. Find the focus if the antenna is 12 m across and its depth is 4 m.

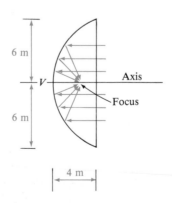

Dimensions for radar antenna in Problem 29

Courtesy of Wide World Photos, Inc.

FIGURE 8.10 Parabolic reflectors: A three-dimensional model of a parabola is a **parabolic reflector** (or **parabolic mirror**). If a source of light is placed at the focus of the mirror, the light rays will reflect from the mirror as rays parallel to the axis (as in an automobile headlamp). The radar antenna works in reverse—parallel incoming sound rays are focused at a single location.

30. ENGINEERING If the diameter of a parabolic reflector in Problem 29 is 16 cm and the depth is 8 cm, find the focus.

C

31. Derive the equation of a parabola with $F(0, -c)$, where c is a positive number and the directrix is the line $y = c$.

32. Derive the equation of a parabola with $F(c, 0)$, where c is a positive number and the directrix is the line $x = -c$.

33. Derive the equation of a parabola with $F(-c, 0)$, where c is a positive number and the directrix is the line $x = c$.

34. Show that the length of the focal chord for the parabola $y^2 = 4cx$ is $4c$ $(c > 0)$.

8.2 Ellipses

The second conic section considered in this chapter is called an *ellipse*.

Definition and Terminology

ELLIPSE

> An **ellipse** is the set of all points in the plane such that, for each point on the ellipse, the sum of its distances from two fixed points is a constant.

The fixed points are called the **foci** (plural for **focus**). To see what an ellipse looks like, we will use the special type of graph paper shown in Figure 8.11a, where F_1 and F_2 are the foci.

Let the given constant be 12. Plot all the points in the plane so that the sum of their distances from the foci is 12. If a point is 8 units from F_1, for example, then it is 4 units from F_2 and you can plot the points P_1 and P_2. The completed graph of this ellipse is shown in Figure 8.11b.

The line passing through F_1 and F_2 is called the **major axis.** The **center** is the midpoint of the segment $\overline{F_1F_2}$. The line passing through the center perpendicular to the major axis is called the **minor axis.** The ellipse is symmetric with respect to both the major and minor axes.

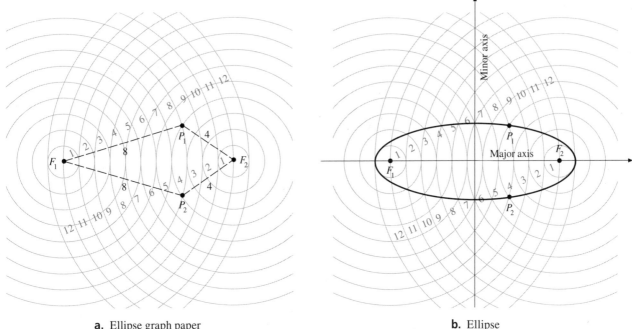

a. Ellipse graph paper

b. Ellipse

FIGURE 8.11

Standard-Form Ellipses

To find the equation of an ellipse, first consider a special case where the center is at the origin. Let the distance from the center to a focus be the positive number c; that is, let $F_1(-c, 0)$ and $F_2(c, 0)$ be the foci, and let the constant distance be $2a$, as shown in Figure 8.12.

Notice that the center of the ellipse in Figure 8.12 is $(0, 0)$. Let $P(x, y)$ be any point on the ellipse, and use the distance formula and the definition of an ellipse to derive the equation of this ellipse:

FIGURE 8.12 Developing the equation of an ellipse by using the definition

$$d_1 + d_2 = 2a$$

or

$$\sqrt{(x + c)^2 + (y - 0)^2} + \sqrt{(x - c)^2 + (y - 0)^2} = 2a$$

Simplifying,

$$\sqrt{(x + c)^2 + y^2} = 2a - \sqrt{(x - c)^2 + y^2} \qquad \text{Isolate one radical.}$$

$$(x + c)^2 + y^2 = 4a^2 - 4a\sqrt{(x - c)^2 + y^2} \qquad \text{Square both sides.}$$
$$+ (x - c)^2 + y^2$$

$$x^2 + 2cx + c^2 + y^2 = 4a^2 - 4a\sqrt{(x - c)^2 + y^2} + x^2 - 2cx + c^2 + y^2$$

$$4a\sqrt{(x - c)^2 + y^2} = 4a^2 - 4cx$$

$$\sqrt{(x-c)^2 + y^2} = a - \frac{c}{a}x$$

Since $a \neq 0$, divide by $4a$.

$$(x-c)^2 + y^2 = \left(a - \frac{c}{a}x\right)^2$$

Square both sides again.

$$x^2 - 2cx + c^2 + y^2 = a^2 - 2cx + \frac{c^2}{a^2}x^2$$

$$x^2 + y^2 = a^2 - c^2 + \frac{c^2}{a^2}x^2$$

$$x^2 - \frac{c^2}{a^2}x^2 + y^2 = a^2 - c^2$$

$$\left(1 - \frac{c^2}{a^2}\right)x^2 + y^2 = a^2 - c^2$$

$$\frac{a^2 - c^2}{a^2}x^2 + y^2 = a^2 - c^2$$

Divide both sides by $a^2 - c^2$.

$$\frac{x^2}{a^2} + \frac{y^2}{a^2 - c^2} = 1$$

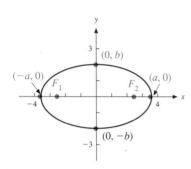

Let $b^2 = a^2 - c^2$; then $\dfrac{x^2}{a^2} + \dfrac{y^2}{b^2} = 1$. If $x = 0$, the y-intercepts are obtained:

$$\frac{y^2}{b^2} = 1$$

$$y = \pm b$$

If $y = 0$, the x-intercepts are obtained: $x = \pm a$. The intercepts on the major axis are called the **vertices** of the ellipse, and those intercepts on the minor axis are called the **covertices.**

The equation of the ellipse with major axis vertical, $F_1(0, c)$, $F_2(0, -c)$, and constant distance $2a$ is found in a similar fashion. Simplifying the equation as before,

$$\frac{y^2}{a^2} + \frac{x^2}{b^2} = 1$$

where $b^2 = a^2 - c^2$.

Notice that in both cases a^2 must be larger than both c^2 and b^2. If it were not, a square number would be equal to a negative number, which is a contradiction in the set of real numbers.

	ELLIPSE	FOCI	CONSTANT DISTANCE	CENTER	EQUATION
STANDARD-FORM EQUATIONS FOR ELLIPSES WITH CENTER (0, 0)	Horizontal	$(-c, 0), (c, 0)$	$2a$	$(0, 0)$	$\dfrac{x^2}{a^2} + \dfrac{y^2}{b^2} = 1$
	Vertical	$(0, c), (0, -c)$	$2a$	$(0, 0)$	$\dfrac{y^2}{a^2} + \dfrac{x^2}{b^2} = 1$
	where $b^2 = a^2 - c^2$ or $c^2 = a^2 - b^2$				

Given Equation, Find Graph

EXAMPLE 1 Graphing a horizontal ellipse

Sketch $\dfrac{x^2}{9} + \dfrac{y^2}{4} = 1$.

SOLUTION The center of the ellipse is $(0, 0)$. The x-intercepts are ± 3 (these are the vertices), and the y-intercepts are ± 2. Sketch the ellipse as shown in Figure 8.13. The foci can also be found, since

$$c^2 = a^2 - b^2$$
$$c^2 = 9 - 4$$
$$c = \sqrt{5}$$

FIGURE 8.13 Graph of $\dfrac{x^2}{9} + \dfrac{y^2}{4} = 1$

EXAMPLE 2 Graphing a vertical ellipse

Sketch $\dfrac{x^2}{4} + \dfrac{y^2}{9} = 1$.

⌐ION Here $a^2 = 9$ and $b^2 = 4$, which is an ellipse with major axis vertical. The x-intercepts are ± 2 and the y-intercepts are ± 3 (these are the vertices). The sketch is shown in Figure 8.14.

The foci are found by

$$c^2 = a^2 - b^2$$
$$c^2 = 9 - 4$$
$$c = \pm\sqrt{5}$$

Remember: The foci are always on the major axis, so plot $(0, \sqrt{5})$ and $(0, -\sqrt{5})$ for the foci.

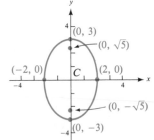

FIGURE 8.14 Graph of $\dfrac{x^2}{4} + \dfrac{y^2}{9} = 1$

Translated Ellipses

If the center of the ellipse is at (h, k), the equations can be written in terms of a translation, as shown by the following box.

STANDARD-FORM EQUATIONS FOR TRANSLATED ELLIPSES

ELLIPSE	EQUATION
Horizontal	$\dfrac{(x - h)^2}{a^2} + \dfrac{(y - k)^2}{b^2} = 1$
Vertical	$\dfrac{(y - k)^2}{a^2} + \dfrac{(x - h)^2}{b^2} = 1$

The segment from the center to a vertex on the major axis is called a **semimajor axis** and has length a; the segment from the center to an intercept on the minor axis is called a **semiminor axis** and has length b.

EXAMPLE 3   **Graphing a translated ellipse**

Graph $\dfrac{(x - 3)^2}{25} + \dfrac{(y - 1)^2}{16} = 1$.

SOLUTION   STEP 1:   Plot the center (h, k). By inspection, the center of this ellipse is $(3, 1)$. This becomes the center of a new translated coordinate system. The vertices and foci are now measured with reference to the new origin at $(3, 1)$.

STEP 2:   Plot the x'- and y'-intercepts. These are ± 5 and ± 4, respectively. Remember to measure these distances from $(3, 1)$, as shown in Figure 8.15.

The foci are found by

$$
\begin{aligned}
c^2 &= a^2 - b^2 \\
&= 25 - 16 \\
&= 9
\end{aligned}
$$

The distance from the center to either focus is 3, so the coordinates of the foci are $(6, 1)$ and $(0, 1)$.

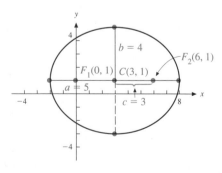

FIGURE 8.15 Graph of $\dfrac{(x - 3)^2}{25} + \dfrac{(y - 1)^2}{16} = 1$

Circles

We have seen that some ellipses are more circular and some more flat than others. A measure of the amount of flatness of an ellipse is called its **eccentricity,** which is defined as*

$$\epsilon = \frac{c}{a}$$

Notice that

$$\epsilon = \frac{c}{a} = \frac{\sqrt{a^2 - b^2}}{a} = \sqrt{\frac{a^2 - b^2}{a^2}} = \sqrt{1 - \left(\frac{b}{a}\right)^2}$$

Since $c < a$, ϵ is between 0 and 1. If $a = b$, then $\epsilon = 0$ and the conic section is a **circle.** Recall that a circle is the set of all points a given distance from a given point. If the ratio b/a is small, then the ellipse is very flat. Thus, for an ellipse,

$$0 \le \epsilon < 1$$

and ϵ measures the amount of roundness of the ellipse.

Consider a circle; that is, suppose $a = b$. In this case, let $r = a = b$ and call this distance the **radius.** You can see that a circle is a special case of an ellipse:

$$\frac{(x - h)^2}{r^2} + \frac{(y - k)^2}{r^2} = 1$$

STANDARD-FORM
EQUATION OF
A CIRCLE

> The equation of a circle with center at (h, k) and radius r is
>
> $$(x - h)^2 + (y - k)^2 = r^2$$

EXAMPLE 4   Graphing a circle

Graph $(x - 1)^2 + (y - 1)^2 = 25$.

SOLUTION   This is a circle with center at $(1, 1)$ and radius 5, as shown in Figure 8.16.

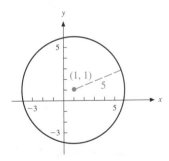

FIGURE 8.16

---

*Many books use e for eccentricity, but we reserve e to represent an irrational number approximately equal to 2.7182818.

Given Graph, Find Equation

EXAMPLE 5   Finding the equation of an ellipse given the vertices and foci

Find the equation of the ellipse with vertices at $(3, 2)$ and $(3, -4)$ and foci at $(3, \sqrt{5} - 1)$ and $(3, -\sqrt{5} - 1)$.

SOLUTION   By inspection, the ellipse is vertical, and it is centered at $(3, -1)$ where $a = 3$ and $c = \sqrt{5}$ (see Figure 8.17). Thus,

$$c^2 = a^2 - b^2$$
$$5 = 9 - b^2$$
$$b^2 = 4$$

The equation is

$$\frac{(y + 1)^2}{9} + \frac{(x - 3)^2}{4} = 1$$

FIGURE 8.17

EXAMPLE 6   Finding the equation of an ellipse using the definition

Find the equation of the set of all points with the sum of distances from $(-3, 2)$ and $(5, 2)$ equal to 16.

SOLUTION   By inspection, the ellipse is horizontal and is centered at $(1, 2)$. You are given

$$2a = 16 \qquad \text{This is the sum of the distances.}$$
$$a = 8$$

and $c = 4$—the distance from the center $(1, 2)$ to a focus $(5, 2)$. Thus,

$$c^2 = a^2 - b^2$$
$$16 = 64 - b^2$$
$$b^2 = 48$$

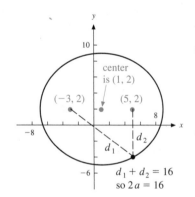

The equation is

$$\frac{(x - 1)^2}{64} + \frac{(y - 2)^2}{48} = 1$$

EXAMPLE 7   Finding the equation of an ellipse given the foci and eccentricity

Find the equation of the ellipse with foci at $(-3, 6)$ and $(-3, 2)$ with $\epsilon = \frac{1}{5}$.

SOLUTION   By inspection, the ellipse is vertical and is centered at $(h, k) = (-3, 4)$ with $c = 2$. Since

$$\epsilon = \frac{c}{a} = \frac{1}{5}$$

and $c = 2$, then

$$\frac{2}{a} = \frac{1}{5}$$

which implies $a = 10$. Just because

$$\frac{c}{a} = \frac{1}{5}$$

you cannot assume that $c = 1$ and $a = 5$; all you
know is that the reduced *ratio* of c to a is $\frac{1}{5}$. Also,

$$c^2 = a^2 - b^2$$
$$4 = 100 - b^2$$
$$b^2 = 96$$

Thus, the equation is

$$\frac{(y-4)^2}{100} + \frac{(x+3)^2}{96} = 1$$

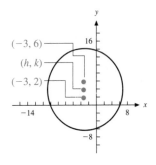

8.2   Problem Set

A *Sketch the curves in Problems 1–15.*

1. $\dfrac{x^2}{4} + \dfrac{y^2}{9} = 1$

2. $\dfrac{x^2}{25} + \dfrac{y^2}{36} = 1$

3. $x^2 + \dfrac{y^2}{9} = 1$

4. $4x^2 + 9y^2 = 36$

5. $25x^2 + 16y^2 = 400$

6. $36x^2 + 25y^2 = 900$

7. $3x^2 + 2y^2 = 6$

8. $4x^2 + 3y^2 = 12$

9. $5x^2 + 10y^2 = 7$

10. $(x-2)^2 + (y+3)^2 = 25$

11. $(x+4)^2 + (y-2)^2 = 49$

12. $(x-1)^2 + (y-1)^2 = \frac{1}{4}$

13. $\dfrac{(x+3)^2}{81} + \dfrac{(y-1)^2}{49} = 1$

14. $\dfrac{(x-3)^2}{16} + \dfrac{(y-2)^2}{9} = 1$

15. $\dfrac{(x+2)^2}{25} + \dfrac{(y+4)^2}{9} = 1$

*Use the definitions of an ellipse and a circle to sketch the curves
given in Problems 16–22.*

16. The set of points 6 units from the point $(4, 5)$

17. The set of points 3 units from the point $(-2, 3)$

18. The set of points 6 units from $(-1, -4)$

19. The set of points such that the sum of the distances from
$(-6, 0)$ and $(6, 0)$ is 20

20. The set of points such that the sum of the distances from
$(0, 4)$ and $(0, -4)$ is 10

21. The ellipse with vertices at $(0, 7)$ and $(0, -7)$ and foci at
$(0, 5)$ and $(0, -5)$

22. The ellipse with vertices at $(4, 3)$ and $(4, -5)$ and foci at
$(4, 2)$ and $(4, -4)$

B *Find the equations of the curves in Problems 23–30.*

23. The set of points 3 units from the point $(-2, 3)$

24. The set of points 5 units from $(-1, -4)$

25. The set of points such that the sum of the distances from
$(-6, 0)$ and $(6, 0)$ is 20

26. The set of points such that the sum of the distances from
$(0, 4)$ and $(0, -4)$ is 10

27. The ellipse with vertices at $(0, 7)$ and $(0, -7)$ and foci at
$(0, 5)$ and $(0, -5)$

28. The ellipse with vertices at $(4, 3)$ and $(4, -5)$ and foci at
$(4, 2)$ and $(4, -4)$

29. The ellipse with foci $(2, 0)$ and $(-2, 0)$ with $\epsilon = \frac{1}{5}$

30. The ellipse with foci $(0, 3)$ and $(0, -3)$ with $\epsilon = \frac{4}{5}$

The following information is needed for Problems 31–33.

The orbit of a planet (see Figure 8.18) can be described by

$$\frac{x^2}{a^2} + \frac{y^2}{b^2} = 1$$

with the sun at one focus. The orbit is commonly identified by its major axis, $2a$, and its eccentricity, ϵ:

$$\epsilon = \frac{c}{a} = \sqrt{1 - \frac{b^2}{a^2}} \qquad a \geq b > 0$$

where c is the distance between the center and a focus.

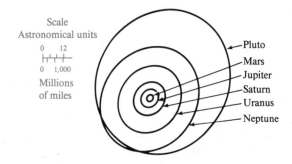

FIGURE 8.18 Planetary orbits: The orbits of planets or satellites serve as an example of ellipses.

31. ASTRONOMY The major axis of the earth's orbit is approximately 186,000,000 miles and its eccentricity is $\frac{1}{62}$. Find the

greatest distance, called the *aphelion*, and the smallest distance, called the *perihelion*, of the sun from the earth.

32. ASTRONOMY The *perihelion* (the greatest distance) of the planet Mercury with the sun is approximately 28 million miles, and the eccentricity of the orbit is $\frac{1}{5}$. Find the major axis of Mercury's orbit. (See Problem 31.)

33. ASTRONOMY The moon's orbit is elliptical with the earth at one focus. If the major axis of the orbit is approximately 378,000 miles and the *apogee* (the greatest distance) is approximately 199,000 miles, then what is the eccentricity of the moon's orbit? (See Problem 31.)

C

34. Derive the equation of the ellipse with foci at $(0, c)$ and $(0, -c)$ and constant distance $2a$. Let $b^2 = a^2 - c^2$. Show all your work.

35. If we are given an ellipse with foci at $(-c, 0)$ and $(c, 0)$ and vertices at $(-a, 0)$ and $(a, 0)$, we define the *directrices* of the ellipse as the lines $x = a/\epsilon$ and $x = -a/\epsilon$. Show that an ellipse is the set of all points with distances from $F(c, 0)$ equal to ϵ times their distances from the line $x = a/\epsilon$ ($a > 0, c > 0$).

36. A line segment through a focus parallel to a directrix (see Problem 35) and cut off by the ellipse is called the *focal chord*. Show that the length of the focal chord of the following ellipse is $2b^2/a$:

$$\frac{x^2}{a^2} + \frac{y^2}{b^2} = 1$$

8.3	**Hyperbolas**

The last of the conic sections to be considered has a definition similar to that of the ellipse.

Definition and Terminology

HYPERBOLA

> A **hyperbola** is the set of all points in the plane such that, for each point on the hyperbola, the difference of its distances from two fixed points is a constant.

---

*You might also see the Application for Further Study on page 333.

Transverse
axis   Focus   Center   Focus

FIGURE 8.19   Graph of a
hyperbola from the definition

The fixed points are called the **foci.** A hyperbola with foci at F_1 and F_2, where the given constant is 8, is shown in Figure 8.19.

The line passing through the foci is called the **transverse axis.** The **center** is the midpoint of the segment connecting the foci. The line passing through the center perpendicular to the transverse axis is called the **conjugate axis.** The hyperbola is symmetric with respect to both the transverse and the conjugate axes.

Standard-Form Hyperbolas

If you use the definition, you can derive the equation for a hyperbola with foci at $(-c, 0)$ and $(c, 0)$ with constant distance $2a$. If (x, y) is any point on the curve, then

$$\left| \sqrt{(x + c)^2 + (y - 0)^2} - \sqrt{(x - c)^2 + (y - 0)^2} \right| = 2a$$

The procedure for simplifying this expression is the same as that shown for the ellipse, so the details are left as a problem (see Problems 29 and 30). After several steps, you should obtain

$$\frac{x^2}{a^2} - \frac{y^2}{c^2 - a^2} = 1$$

If $b^2 = c^2 - a^2$, then

$$\frac{x^2}{a^2} - \frac{y^2}{b^2} = 1$$

which is the standard-form equation. Notice that $c^2 = a^2 - b^2$ for the ellipse and that $c^2 = a^2 + b^2$ for the hyperbola. For the ellipse it is necessary that $a^2 > b^2$, but for the hyperbola there is no restriction on the relative sizes for a and b.

Repeat the argument for a hyperbola with foci $(0, c)$ and $(0, -c)$, and you will obtain the other standard-form equation for a hyperbola with a vertical transverse axis.

**STANDARD-FORM
EQUATIONS FOR THE
HYPERBOLA WITH
CENTER (0, 0)**

HYPERBOLA	FOCI	CONSTANT DISTANCE	CENTER	EQUATION
Horizontal	$(-c, 0), (c, 0)$	$2a$	$(0, 0)$	$\dfrac{x^2}{a^2} - \dfrac{y^2}{b^2} = 1$
Vertical	$(0, c), (0, -c)$	$2a$	$(0, 0)$	$\dfrac{y^2}{a^2} - \dfrac{x^2}{b^2} = 1$
where $b^2 = c^2 - a^2$ or $c^2 = a^2 + b^2$				

As with the other conic sections, we will sketch a hyperbola by determining some information about the curve directly from the equation by inspection. The points of intersection of the hyperbola with the transverse axis are called the **vertices.** For

$$\frac{x^2}{a^2} - \frac{y^2}{b^2} = 1 \quad \text{and} \quad \frac{y^2}{a^2} - \frac{x^2}{b^2} = 1$$

notice that the vertices occur at $(a, 0)$, $(-a, 0)$ and $(0, a)$, $(0, -a)$, respectively. The number $2a$ is the **length of the transverse axis.** The hyperbola does not intersect the conjugate axis, but if you plot the points $(0, b)$, $(0, -b)$ and $(-b, 0)$, $(b, 0)$, respectively, you determine a segment on the conjugate axis called the **length of the conjugate axis.**

Given Equation, Find Graph

EXAMPLE 1   Graphing a hyperbola

Sketch $\dfrac{x^2}{4} - \dfrac{y^2}{9} = 1$.

SOLUTION   The center of the hyperbola is $(0, 0)$, $a = 2$, and $b = 3$. Plot the vertices at ± 2, as shown in Figure 8.20. The transverse axis is along the x-axis, and the conjugate axis is along the y-axis. Plot the length of the conjugate axis at ± 3. We call the points the **pseudovertices,** since the curve does not actually pass through these points.

Next draw lines through the vertices and pseudovertices parallel to the axes of the hyperbola. These lines form what we will call the **central rectangle.** The diagonal lines passing through the corners of the central rectangle are **slant asymptotes** for the hyperbola, as shown in Figure 8.21; they aid in sketching the hyperbola.

FIGURE 8.20

FIGURE 8.21 Graph of $\dfrac{x^2}{4} - \dfrac{y^2}{9} = 1$

Translated Hyperbolas

If the center of the hyperbola is (h, k), the following equations for a hyperbola are obtained:

STANDARD-FORM EQUATIONS FOR TRANSLATED HYPERBOLAS

HYPERBOLA	EQUATIONS
Horizontal	$\dfrac{(x - h)^2}{a^2} - \dfrac{(y - k)^2}{b^2} = 1$
Vertical	$\dfrac{(y - k)^2}{a^2} - \dfrac{(x - h)^2}{b^2} = 1$

EXAMPLE 2   **Graphing a translated hyperbola**

Sketch $16(x - 4)^2 - 9(y + 1)^2 = 144$.

SOLUTION   First, put the equation into standard form:

$$\frac{(x - 4)^2}{9} - \frac{(y + 1)^2}{16} = 1 \qquad \text{Divide both sides by 144.}$$

The graph is shown in Figure 8.22. You can also find the foci, since

$$c^2 = a^2 + b^2$$

Thus,

$$c^2 = 9 + 16 \qquad c = \pm 5$$

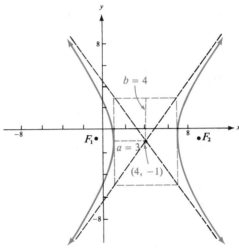

FIGURE 8.22 Sketch of $16(x - 4)^2 - 9(y + 1)^2 = 144$

Given Graph, Find Equation

EXAMPLE 3   **Finding the equation of a hyperbola given the vertices and foci**

Find the equation of the hyperbola with vertices at $(2, 4)$ and $(2, -2)$ and foci at $(2, 6)$ and $(2, -4)$.

SOLUTION   Plot the given points as shown in Figure 8.23. Notice that the center of the hyperbola is $(2, 1)$ since it is the midpoint of the segment connecting the foci. Also, $c = 5$ and $a = 3$. Since

$$c^2 = a^2 + b^2$$

you have

$$25 = 9 + b^2$$
$$b^2 = 16$$

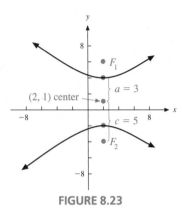

FIGURE 8.23

and the equation is

$$\frac{(y-1)^2}{9} - \frac{(x-2)^2}{16} = 1$$

∎

The eccentricity of the hyperbola and parabola is defined by the same equation that was used for the ellipse, namely

$$\epsilon = \frac{c}{a}$$

Remember that for the ellipse, $0 \le \epsilon < 1$; however, for the hyperbola $c > a$ so $\epsilon > 1$, and for the parabola $c = a$ so $\epsilon = 1$.

EXAMPLE 4　**Finding the equation of a hyperbola given the foci and eccentricity**

Find the equation of the hyperbola with foci at $(-3, 2)$ and $(5, 2)$ and with eccentricity $\frac{3}{2}$.

SOLUTION　This is a horizontal hyperbola with center at $(1, 2)$ and $c = 4$. Also, since

$$\epsilon = \frac{c}{a} = \frac{3}{2}$$

you have

$$\frac{4}{a} = \frac{3}{2}$$

$$a = \frac{8}{3}$$

And since $c^2 = a^2 + b^2$,

$$16 = \frac{64}{9} + b^2$$

$$b^2 = \frac{80}{9}$$

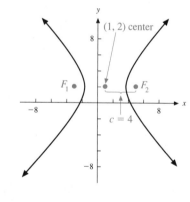

Thus the equation is

$$\frac{(x-1)^2}{\frac{64}{9}} - \frac{(y-2)^2}{\frac{80}{9}} = 1 \quad \text{or} \quad \frac{9(x-1)^2}{64} - \frac{9(y-2)^2}{80} = 1$$

∎

EXAMPLE 5　**Finding the equation of a hyperbola using the definition**

Find the set of points such that the difference of their distances from $(6, 2)$ and $(6, -5)$ is always 3.

SOLUTION This is a vertical hyperbola with center $\left(6, -\frac{3}{2}\right)$ and $c = \frac{7}{2}$. Also $2a = 3$, so $a = \frac{3}{2}$.

Since

$$c^2 = a^2 + b^2$$

you have

$$\frac{49}{4} = \frac{9}{4} + b^2$$

$$b^2 = 10$$

The equation is

$$\frac{\left(y + \frac{3}{2}\right)^2}{\frac{9}{4}} - \frac{(x-6)^2}{10} = 1$$

or $$\frac{4\left(y + \frac{3}{2}\right)^2}{9} - \frac{(x-6)^2}{10} = 1$$

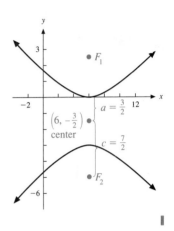

8.3 Problem Set

A *Sketch the curves in Problems 1–22.*

1. $x^2 - y^2 = 1$ **2.** $x^2 - y^2 = 4$

3. $y^2 - x^2 = 1$ **4.** $\dfrac{x^2}{4} - \dfrac{y^2}{9} = 1$

5. $\dfrac{x^2}{9} - \dfrac{y^2}{4} = 1$ **6.** $\dfrac{y^2}{9} - \dfrac{x^2}{4} = 1$

7. $\dfrac{x^2}{16} - \dfrac{y^2}{25} = 1$ **8.** $\dfrac{y^2}{16} - \dfrac{x^2}{25} = 1$

9. $\dfrac{x^2}{36} - \dfrac{y^2}{9} = 1$ **10.** $36y^2 - 25x^2 = 900$

11. $3x^2 - 4y^2 = 12$ **12.** $3y^2 = 4x^2 + 12$

13. $3x^2 - 4y^2 = 5$ **14.** $4y^2 - 4x^2 = 5$

15. $4y^2 - x^2 = 9$

16. $\dfrac{(x-2)^2}{4} - \dfrac{(y+3)^2}{16} = 1$

17. $\dfrac{(x+3)^2}{8} - \dfrac{(y-1)^2}{5} = 1$ **18.** $\dfrac{(y-1)^2}{6} - \dfrac{(x+2)^2}{8} = 1$

19. $\dfrac{(x-2)^2}{16} - \dfrac{(y+1)^2}{9} = 1$ **20.** $\dfrac{(y+2)^2}{25} - \dfrac{(x+1)^2}{16} = 1$

21. $5(x-2)^2 - 2(y+3)^2 = 10$

22. $4(x+4)^2 - 3(y+3)^2 = -12$

B *Find the equations of the curves in Problems 23–28.*

23. The hyperbola with vertices at $(0, 5)$ and $(0, -5)$ and foci at $(0, 7)$ and $(0, -7)$

24. The set of points such that the difference of their distances from $(-6, 0)$ and $(6, 0)$ is 10

25. The hyperbola with foci at $(5, 0)$ and $(-5, 0)$ and eccentricity 5

26. The hyperbola with vertices at $(4, 4)$ and $(4, 8)$ and foci at $(4, 3)$ and $(4, 9)$

27. The set of points such that the difference of their distances from $(4, -3)$ and $(-4, -3)$ is 6

28. The hyperbola with vertices at $(-2, 0)$ and $(6, 0)$ passing through $(10, 3)$

C

29. Derive the equation of the hyperbola with foci at $(-c, 0)$ and $(c, 0)$ and constant distance $2a$. Let $b^2 = c^2 - a^2$. Show all your work.

30. Derive the equation of the hyperbola with foci at $(0, c)$ and $(0, -c)$ and constant distance $2a$. Let $b^2 = c^2 - a^2$. Show all your work.

31. **PHYSICS** Consider a person A who fires a rifle at a distant gong B. Assuming that the ground is flat, where must you stand to hear the sound of the gun and the sound of the gong simultaneously? [*Hint:*   To answer this question, let x be the distance that sound travels in the length of time it takes the bullet to travel from the gun to the gong. Show that the person who hears the sounds simultaneously must stand on a branch of a hyperbola (the one nearest the target) so that the difference of the distances from A to B is x.]

8.4 General Conic Sections

Degenerate Conics

We have considered graphs of standard-form conic sections with centers at $(0, 0)$ or (h, k). We will now consider the general conics of the form

$$Ax^2 + Bxy + Cy^2 + Dx + Ey + F = 0$$

Geometrically they represent the intersection of a plane and a cone, usually resulting in a line, parabola, ellipse, or hyperbola. However, there are certain positions of the plane that result in what are called **degenerate conics.** To visualize some of these degenerate conics, consider Figure 8.24.

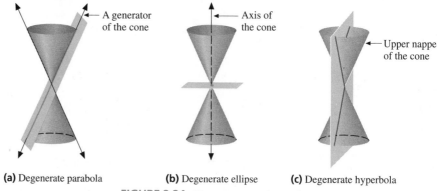

(a) Degenerate parabola   **(b)** Degenerate ellipse   **(c)** Degenerate hyperbola

FIGURE 8.24   Degenerate conic sections

For a **degenerate parabola,** visualize the plane (see Figure 8.24a) intersecting one of the generators of the cone; a line results. For a **degenerate ellipse,** visualize the plane intersecting at the vertex of the upper and lower nappes (see Figure 8.24b); a point results. And finally, for a **degenerate hyperbola,** visualize the plane situated so that the axis of the cone lies in the plane (see Figure 8.24c); a pair of intersecting lines results.

EXAMPLE 1   **Graphing a degenerate ellipse**

Sketch $\dfrac{(x - 2)^2}{4} + \dfrac{(y + 3)^2}{9} = 0$.

SOLUTION   There is only one point that satisfies this equation—namely, $(2, -3)$. This is an example of a *degenerate ellipse.* Notice that except for the zero the equation has the "form of an ellipse." See Figure 8.25.

FIGURE 8.25 Graph of
$\dfrac{(x - 2)^2}{4} + \dfrac{(y + 3)^2}{9} = 0$

EXAMPLE 2   Graphing a degenerate hyperbola

Sketch $\dfrac{x^2}{4} - \dfrac{y^2}{9} = 0$.

SOLUTION This equation has the "form of a hyperbola," but because of the zero it cannot be put into standard form. You can, however, treat this as a factored form:

$$\left(\frac{x}{2} - \frac{y}{3}\right)\left(\frac{x}{2} + \frac{y}{3}\right) = 0$$

$$\frac{x}{2} - \frac{y}{3} = 0 \quad \text{or} \quad \frac{x}{2} + \frac{y}{3} = 0$$

The graph (Figure 8.26) is a *degenerate hyperbola*.

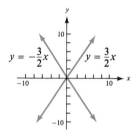

FIGURE 8.26

Graph of $\dfrac{x^2}{4} - \dfrac{y^2}{9} = 0$

Recognition of Conics by Inspection

It is important to be able to recognize the curve by inspection of the equation before you begin. The first thing to notice is whether there is an xy term.

If $B = 0$ (that is, no xy term), then:

TYPE OF CURVE	DEGREE OF EQUATION	DEGREE IN x	DEGREE IN y	RELATIONSHIP TO GENERAL EQUATION $Ax^2 + Bxy + Cy^2 + Dx + Ey + F = 0$
1. Line	First	First	First	$A = C = 0$
2. Parabola	Second	First	Second	$A = 0$ and $C \neq 0$
		Second	First	$A \neq 0$ and $C = 0$
3. Ellipse	Second	Second	Second	A and C have same sign ($A \neq 0$)
4. Circle	Second	Second	Second	$A = C$ ($A \neq 0$)
5. Hyperbola	Second	Second	Second	A and C have opposite signs ($A \neq 0$)

If $B \neq 0$ (that is, there is an xy term), then:

TYPE OF CURVE	RELATIONSHIP TO GENERAL EQUATION
1. Ellipse	$B^2 - 4AC < 0$
2. Parabola	$B^2 - 4AC = 0$
3. Hyperbola	$B^2 - 4AC > 0$

These tests do not distinguish degenerate cases. This means that the test may tell you the curve is an ellipse that may turn out to be a single point. Remember, too, that a circle is a special case of the ellipse. The expression $B^2 - 4AC$ is called the **discriminant**.

EXAMPLE 3   Identifying a conic by looking at its equation

Identify each curve.

SOLUTION   **a.** $x^2 + 4xy + 4y^2 = 9$

$B^2 - 4AC = 16 - 4(1)(4) = 0$; parabola

b. $2x^2 + 3xy + y^2 = 25$

$B^2 - 4AC = 9 - 4(2)(1) > 0$; hyperbola

c. $x^2 + xy + y^2 - 8x + 8y = 0$

$B^2 - 4AC = 1 - 4(1)(1) < 0$; ellipse

d. $xy = 5$

$B^2 - 4AC = 1 - 4(0)(0) > 0$; hyperbola

TABLE 8.1 Summary of Standard-Position Conic Sections

	PARABOLA	**ELLIPSE**	**HYPERBOLA**
Definition	All points equidistant from a given point and a given line	All points with the sum of distances from two fixed points constant	All points with the difference of distances from two fixed points constant
Equations	Up: $x^2 = 4cy$ Down: $x^2 = -4cy$ Right: $y^2 = 4cx$ Left: $y^2 = -4cx$	$c^2 = a^2 - b^2$ Horizontal axis: $$\frac{x^2}{a^2} + \frac{y^2}{b^2} = 1$$ Vertical axis: $$\frac{y^2}{a^2} + \frac{x^2}{b^2} = 1$$	$c^2 = a^2 + b^2$ Horizontal axis: $$\frac{x^2}{a^2} - \frac{y^2}{b^2} = 1$$ Vertical axis: $$\frac{y^2}{a^2} - \frac{x^2}{b^2} = 1$$
Recognition	Second-degree equation; linear in one variable, quadratic in the other variable	Second-degree equation; coefficients of x^2 and y^2 have same sign	Second-degree equation; coefficients of x^2 and y^2 have different signs
Eccentricity	$\epsilon = 1$	$0 \le \epsilon < 1$	$\epsilon > 1$
Directrix	Perpendicular to axis c units from the vertex (one directrix)	Perpendicular to major axis $\pm a/\epsilon$ units from center (two directrices)	Perpendicular to transverse axis $\pm a/\epsilon$ units from center (two directrices)

Translations to the point (h, k):

$$x' = x - h \quad \text{and} \quad y' = y - k$$

All conics can be reduced to a standard-form equation by means of a translation or (as we will see in Section 8.5) a rotation. It is important to remember these standard-position equations and some basic information about these curves. The important ideas are summarized in Table 8.1. Notice that the directrices of the ellipse and hyperbola are defined in this table to be consistent with the earlier definition we gave for the directrix of a parabola.

Translating Conics; Completing the Square

Now suppose $B = 0$ and that the rest of the equation is *not* in factored form. The procedure will be to rewrite the equation in factored form by carrying out a process that in algebra was called **completing the square.** This process allows you to determine (h, k) and is best understood by considering some examples.

EXAMPLE 4   **Graphing a translated parabola by completing the square**

Sketch $x^2 + 4y + 8x + 4 = 0$.

SOLUTION   First, notice this is a parabola (second-degree in x and first-degree in y). Now, complete the square in x.

STEP 1:   Associate together the terms involving the variable that is squared:

$$x^2 + 8x = -4y - 4$$

STEP 2:   Complete the square for the variable that is squared.

Coefficient is 1; if it is not, divide both sides by this coefficient.

$$\downarrow$$

$$x^2 + 8x + \left(\frac{1}{2} \cdot 8\right)^2 = -4y - 4 + \left(\frac{1}{2} \cdot 8\right)^2$$
$$\uparrow$$

Take one-half of this coefficient, square it, and add it to both sides.

$$x^2 + 8x + 16 = -4y - 4 + 16$$

$$(x + 4)^2 = -4y + 12$$

STEP 3:   Factor out the coefficient of the first-degree term:

$$(x + 4)^2 = -4(y - 3)$$

STEP 4:   Determine the vertex by inspection. Plot (h, k); in this example, the vertex is $(-4, 3)$. (See Figure 8.27.)

STEP 5:   Determine the focus. By inspection, $4c = 4$, $c = 1$, and the parabola opens downward from the vertex, as shown in Figure 8.27.

STEP 6:   Plot the endpoints of the focal chord; $4c = 4$. Draw the parabola as shown in Figure 8.27.   ▌

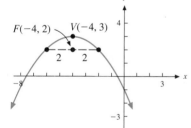

FIGURE 8.27 Graph of $x^2 + 4y + 8x + 4 = 0$

EXAMPLE 5   **Graphing a translated ellipse by completing the square**

Graph $3x^2 + 4y^2 + 24x - 16y + 52 = 0$.

SOLUTION First, recognize this as an ellipse (second-degree in x and y and coefficients are both positive). Next, complete the square for both squared terms.

STEP 1: Associate together the x and the y terms:

$$(3x^2 + 24x) + (4y^2 - 16y) = -52$$

STEP 2: Complete the squares in both x and y. This requires that the coefficients of the squared terms be 1. You can accomplish this by factoring:

$$3(x^2 + 8x \quad) + 4(y^2 - 4y \quad) = -52$$

Next complete the square for both x and y, being sure to add the same numbers to both sides:

Added 16 to both sides

$$3(x^2 + 8x + 16) + 4(y^2 - 4y + 4) = -52 + 48 + 16$$

Added 48 to both sides

STEP 3: Factor: $3(x + 4)^2 + 4(y - 2)^2 = 12$

STEP 4: Divide both sides by 12:

$$\frac{(x + 4)^2}{4} + \frac{(y - 2)^2}{3} = 1$$

STEP 5: Plot the center (h, k). By inspection, you can see the center is $(-4, 2)$. The vertices are at ± 2, and the length of the semiminor axis is $\sqrt{3}$, as shown in Figure 8.28.

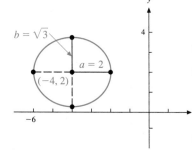

FIGURE 8.28 Graph of
$3x^2 + 4y^2 + 24x - 16y + 52 = 0$

8.4 Problem Set

A *In Problems 1–12 identify and sketch the curve.*

1. $2x - y - 8 = 0$

2. $2x + y - 10 = 0$

3. $4x^2 - 16y = 0$

4. $\dfrac{(x - 3)^2}{4} - \dfrac{(y + 2)}{6} = 1$

5. $\dfrac{(x - 3)^2}{9} - \dfrac{(y + 2)^2}{25} = 1$

6. $\dfrac{x - 3}{9} + \dfrac{y - 2}{25} = 1$

7. $(x + 3)^2 + (y - 2)^2 = 0$

8. $9(x + 3)^2 + 4(y - 2) = 0$

9. $9(x + 3)^2 - 4(y - 2)^2 = 0$

10. $x^2 + 8(y - 12)^2 = 16$

11. $x^2 + 64(y + 4)^2 = 16$

12. $x^2 + y^2 - 3y = 0$

B *In Problems 13–30 identify and sketch the curve.*

13. $4x^2 - 3y^2 - 24y - 112 = 0$

14. $y^2 - 4x + 2y + 21 = 0$

15. $9x^2 + 2y^2 - 48y + 270 = 0$

16. $x^2 + 4x + 12y + 64 = 0$

17. $y^2 - 6y - 4x + 5 = 0$

18. $100x^2 - 7y^2 + 98y - 368 = 0$

19. $x^2 + y^2 + 2x - 4y - 20 = 0$

20. $4x^2 + 12x + 4y^2 + 4y + 1 = 0$

21. $x^2 - 4y^2 - 6x - 8y - 11 = 0$

22. $9x^2 + 25y^2 - 54x - 200y + 256 = 0$

23. $2y^2 + 8y - 20x + 148 = 0$

24. $y^2 - 4x + 10y + 13 = 0$

25. $9x^2 + 6x + 18y - 23 = 0$

26. $x^2 + 9y - 6x + 18 = 0$

27. $x^2 + y^2 - 4x + 10y + 15 = 0$

28. $y^2 + 4x^2 + 2y - 8x + 1 = 0$

29. $x^2 + 9y^2 - 4x - 18y - 14 = 0$

30. $4x^2 + y^2 + 24x + 4y + 16 = 0$

31. ENGINEERING A stone tunnel is to be constructed such that the opening is a semielliptic arch as shown in Figure 8.29. It is necessary to know the height at 4-ft intervals from the

FIGURE 8.29 Semielliptic arch

center. That is, how high is the tunnel at 4, 8, 12, 16, and 20 ft from the center? (Answer to the nearest tenth foot.)

32. ENGINEERING A parabolic archway has the dimensions shown in Figure 8.30. Find the equation of the parabolic portion.

FIGURE 8.30 A parabolic archway

8.5 Rotations

Rotation of Axes

In Section 7.2 we rotated polar-form curves. In this section we consider the idea of **rotation** of the conic sections by writing the equation in terms of a rotated coordinate system. Suppose the coordinate axes are rotated through an angle θ ($0 < \theta < 90°$). The relationship between the old coordinates (x, y) and the new coordinates (x', y') can be found by considering Figure 8.31.

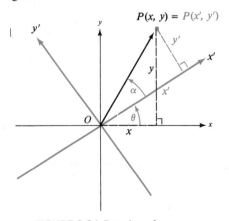

FIGURE 8.31 Rotation of axes

Let O be the origin and P be a point with coordinates (x, y) relative to the old coordinate system and (x', y') relative to the new rotated coordinate system. Let θ be the amount of rotation and let α be the angle between the x'-axis and $|OP|$. Then, using the definition of sine and cosine,

$$x = |OP| \cos(\theta + \alpha) \qquad x' = |OP| \cos \alpha$$
$$y = |OP| \sin(\theta + \alpha) \qquad y' = |OP| \sin \alpha$$

Now,

$$\begin{aligned} x = |OP| \cos(\theta + \alpha) &= |OP| [\cos \theta \cos \alpha - \sin \theta \sin \alpha] \\ &= |OP| \cos \theta \cos \alpha - |OP| \sin \theta \sin \alpha \\ &= \left(|OP| \cos \alpha\right) \cos \theta - \left(|OP| \sin \alpha\right) \sin \theta \\ &= x' \cos \theta - y' \sin \theta \end{aligned}$$

Also,

$$\begin{aligned} y = |OP| \sin(\theta + \alpha) &= |OP| \sin \theta \cos \alpha + |OP| \cos \theta \sin \alpha \\ &= x' \sin \theta + y' \cos \theta \end{aligned}$$

Therefore,

ROTATION OF AXES FORMULAS

$$x = x' \cos \theta - y' \sin \theta \qquad y = x' \sin \theta + y' \cos \theta$$

All the curves considered in this chapter can be characterized by the general second-degree equation

$$Ax^2 + Bxy + Cy^2 + Dx + Ey + F = 0$$

The presence of the xy-term indicates that the conic has been rotated. Thus, in this section we assume that $B \neq 0$ and now need to determine the amount this conic has been rotated from standard position. That is, the new axes should be rotated the same amount as the given conic so that it will be in standard position after the rotation. To find out how much to rotate the axes, substitute

$$x = x' \cos \theta - y' \sin \theta \qquad y = x' \sin \theta + y' \cos \theta$$

into

$$Ax^2 + Bxy + Cy^2 + Dx + Ey + F = 0 \qquad (B \neq 0)$$

After a lot of simplifying you will obtain

$$\begin{aligned} &(A \cos^2\theta + B \cos \theta \sin \theta + C \sin^2\theta)\, x'^2 \\ &\quad + [B(\cos^2\theta - \sin^2\theta) + 2(C - A) \sin \theta \cos \theta]\, x'y' \\ &\quad + (A \sin^2\theta - B \sin \theta \cos \theta + C \cos^2\theta)y'^2 + (D \cos \theta + E \sin \theta)x' \\ &\quad + (-D \sin \theta + E \cos \theta)y' + F = 0 \end{aligned}$$

This looks terrible, but you want to choose θ so that the coefficient of $x'y'$ is zero. This will give you a standard position relative to the new coordinate axes. That is,

$$B(\cos^2\theta - \sin^2\theta) + 2(C - A)\sin\theta\cos\theta = 0 \qquad \text{Using double-angle identities,}$$
$$B \neq 0, \theta \neq 0$$

$$B\cos 2\theta + (C - A)\sin 2\theta = 0$$
$$B\cos 2\theta = (A - C)\sin 2\theta$$
$$\frac{\cos 2\theta}{\sin 2\theta} = \frac{A - C}{B}$$

Simplifying, you obtain the following result:

AMOUNT OF ROTATION FORMULA

$$\cot\theta = \frac{A - C}{B}$$

Notice that we required $0 < \theta < 90°$, so 2θ is in Quadrant I or Quadrant II. This means that if $\cot 2\theta$ is positive, then 2θ must be in Quadrant I; if $\cot 2\theta$ is negative, then 2θ is in Quadrant II.

In Examples 1–4, find the appropriate rotation so that the given curve will be in standard position relative to the rotated axes. Also find the x and y values in the new coordinate system.

EXAMPLE 1   Finding the rotation formulas; rotation of 45°

$$xy = 6$$

SOLUTION

$$\cot 2\theta = \frac{A - C}{B} \qquad x = x'\cos\theta - y'\sin\theta \qquad y = x'\sin\theta + y'\cos\theta$$

$$= \frac{0 - 0}{1} \qquad\qquad = x'\cos 45° - y'\sin 45° \qquad = x'\sin 45° + y'\cos 45°$$

$$= 0 \qquad\qquad = x'\left(\frac{1}{\sqrt{2}}\right) - y'\left(\frac{1}{\sqrt{2}}\right) \qquad = x'\left(\frac{1}{\sqrt{2}}\right) + y'\left(\frac{1}{\sqrt{2}}\right)$$

Thus $2\theta = 90°$
and $\theta = 45°$.
$$= \frac{1}{\sqrt{2}}(x' - y') \qquad\qquad = \frac{1}{\sqrt{2}}(x' + y')$$

EXAMPLE 2   Finding the rotation formulas; rotation of 30°

$$7x^2 - 6\sqrt{3}\,xy + 13y^2 - 16 = 0$$

SOLUTION

$$\cot 2\theta = \frac{A - C}{B} \qquad x = x'\cos\theta - y'\sin\theta \qquad y = x'\sin\theta + y'\cos\theta$$

$$= \frac{7 - 13}{-6\sqrt{3}} \qquad = x'\left(\frac{\sqrt{3}}{2}\right) - y'\left(\frac{1}{2}\right) \qquad = \frac{1}{2}(x' + \sqrt{3}y')$$

$$= \frac{1}{\sqrt{3}} \qquad\qquad = \frac{1}{2}(\sqrt{3}x' - y')$$

Thus $2\theta = 60°$ and $\theta = 30°$.

EXAMPLE 3   **Finding the rotation formulas; cot 2θ is positive and is not exact**

$$x^2 - 4xy + 4y^2 + 5\sqrt{5}y - 10 = 0$$

SOLUTION

$$\cot 2\theta = \frac{1 - 4}{-4} = \frac{3}{4}$$

Since this is not an exact value for θ (as it was in Examples 1 and 2), you will need to use some trigonometric identities to find $\cos \theta$ and $\sin \theta$. If $\cot 2\theta = \frac{3}{4}$, then $\tan 2\theta = \frac{4}{3}$ and $\sec 2\theta = \sqrt{1 + (\frac{4}{3})^2} = \frac{5}{3}$. Then $\cos 2\theta = \frac{3}{5}$ and you can now apply the half-angle identities:

$$\cos \theta = \pm \sqrt{\frac{1 + \cos 2\theta}{2}} \qquad \sin \theta = \pm \sqrt{\frac{1 - \cos 2\theta}{2}}$$

$$\cos \theta = \sqrt{\frac{1 + (\frac{3}{5})}{2}} \qquad \sin \theta = \sqrt{\frac{1 - (\frac{3}{5})}{2}} \qquad \text{Positive because } \theta \text{ is in Quadrant I}$$

$$= \frac{2}{\sqrt{5}} \qquad\qquad = \frac{1}{\sqrt{5}}$$

To find the amount of rotation, use a calculator and one of the preceding equations to find $\theta \approx 26.6°$. Finally, the rotation of axes formulas provide

$$x = x' \cos \theta - y' \sin \theta \qquad\qquad y = x' \sin \theta + y' \cos \theta$$

$$= x'\left(\frac{2}{\sqrt{5}}\right) - y'\left(\frac{1}{\sqrt{5}}\right) \qquad = x'\left(\frac{1}{\sqrt{5}}\right) + y'\left(\frac{2}{\sqrt{5}}\right)$$

$$= \frac{1}{\sqrt{5}}(2x' - y') \qquad\qquad = \frac{1}{\sqrt{5}}(x' + 2y')$$

EXAMPLE 4   **Finding the rotation formulas; cot 2θ is negative and is not exact**

$$10x^2 + 24xy + 17y^2 - 9 = 0$$

SOLUTION

$$\cot 2\theta = \frac{10 - 17}{24}$$

$$= -\frac{7}{24}$$

Since this is negative, 2θ must be in Quadrant II; this means that $\sec 2\theta$ is negative in the following sequence of identities: Since $\cot 2\theta = -\frac{7}{24}$, then $\tan 2\theta = -\frac{24}{7}$ and $\sec 2\theta = -\sqrt{1 + \left(-\frac{24}{7}\right)^2} = -\frac{25}{7}$. Thus $\cos 2\theta = -\frac{7}{25}$, which gives

$$\cos \theta = \sqrt{\frac{1 + \left(-\frac{7}{25}\right)}{2}} \qquad \sin \theta = \sqrt{\frac{1 - \left(-\frac{7}{25}\right)}{2}}$$

$$= \frac{3}{5} \qquad\qquad = \frac{4}{5}$$

Using either of these equations and a calculator, you find that the rotation is $\theta \approx 53.1°$. The rotation of axes formulas provide

$$x = x' \cos \theta - y' \sin \theta \qquad y = x' \sin \theta + y' \cos \theta$$

$$= x'\left(\tfrac{3}{5}\right) - y'\left(\tfrac{4}{5}\right) \qquad\qquad = x'\left(\tfrac{4}{5}\right) + y'\left(\tfrac{3}{5}\right)$$

$$= \tfrac{1}{5}\left(3x' - 4y'\right) \qquad\qquad = \tfrac{1}{5}\left(4x' + 3y'\right)$$

Sketching Rotated Conics

PROCEDURE FOR SKETCHING A ROTATED CONIC

1. Identify curve.
2. a. Find the angle of rotation.
    b. Find x and y in the new coordinate system.
    c. Substitute the values found in Step 2b into the given equation and simplify.
3. Translate axes, if necessary.
4. Sketch the resulting equation relative to the new x' and y' axes.

EXAMPLE 5   Sketching a rotated hyperbola

Sketch $xy = 6$.

SOLUTION   From Example 1, the rotation is $\theta = 45°$ and

$$x = \frac{1}{\sqrt{2}}(x' - y') \qquad y = \frac{1}{\sqrt{2}}(x' + y')$$

CALCULATOR COMMENT

A curve such as the one shown in Example 5 can be sketched much more easily as $y = 6/x$. However, to use a calculator to sketch a curve like in Example 6 would require that you solve for y:

Y1=(6X√‾3 + √‾(36*3X^2 −
4*13(7X^2 − 16)))/26
Y2=(6X√‾3 − √‾(36*3X^2 −
4*13(7X^2 − 16)))/26

Substitute these values into the original equation $xy = 6$:

$$\left[\frac{1}{\sqrt{2}}(x' - y')\right]\left[\frac{1}{\sqrt{2}}(x' + y')\right] = 6$$

Simplify (see Problem 1 for the details) to obtain

$$x'^2 - y'^2 = 12$$

$$\frac{x'^2}{12} - \frac{y'^2}{12} = 1$$

This curve is a hyperbola that has been rotated 45°. Draw the rotated axis, and sketch this equation relative to the rotated axes. The result is shown in Figure 8.32.

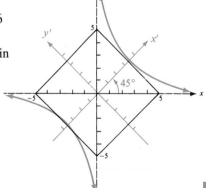

FIGURE 8.32 Graph of $xy = 6$

EXAMPLE 6   Sketching a rotated ellipse

Sketch $7x^2 - 6\sqrt{3}xy + 13y^2 - 16 = 0$.

SOLUTION   From Example 2, $\theta = 30°$ and

$$x = \tfrac{1}{2}(\sqrt{3}x' - y') \qquad y = \tfrac{1}{2}(x' + \sqrt{3}y')$$

Substitute into the original equation:

$$7\left(\tfrac{1}{2}\right)^2(\sqrt{3}x' - y')^2 - 6\sqrt{3}\left(\tfrac{1}{2}\right)(\sqrt{3}x' - y')\left(\tfrac{1}{2}\right)(x' + \sqrt{3}y')$$
$$+ 13\left(\tfrac{1}{2}\right)^2(x' + \sqrt{3}y')^2 - 16 = 0$$

Simplify (see Problem 2 for the details) to obtain

$$\frac{x'^2}{4} + \frac{y'^2}{1} = 1$$

This curve is a rotated ellipse with a 30° rotation. The sketch is shown in Figure 8.33.

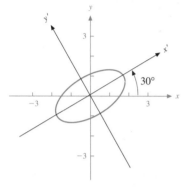

FIGURE 8.33 Graph of $7x^2 - 6\sqrt{3}xy + 13y^2 - 16 = 0$

EXAMPLE 7   Sketching a rotated parabola

Sketch $x^2 - 4xy + 4y^2 + 5\sqrt{5}y - 10 = 0$.

SOLUTION   From Example 3, the rotation is $\theta \approx 26.6°$ and

$$x = \frac{1}{\sqrt{5}}(2x' - y') \qquad y = \frac{1}{\sqrt{5}}(x' + 2y')$$

Substitute

$$\tfrac{1}{5}(2x' - y')^2 - 4\left(\tfrac{1}{5}\right)(2x' - y')(x' + 2y') + 4\left(\tfrac{1}{5}\right)(x' + 2y')^2$$
$$+ 5\sqrt{5}\left(\tfrac{1}{\sqrt{5}}\right)(x' + 2y') - 10 = 0$$

Simplify (see Problem 3 for the details) to obtain

$$y'^2 + 2y' = -x' + 2$$

This curve is a parabola with a rotation of about 26.6°. Next complete the square to obtain

$$(y' + 1)^2 = -(x' - 3)$$

This sketch is shown in Figure 8.34.

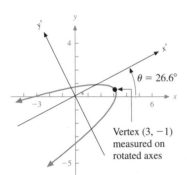

Vertex $(3, -1)$ measured on rotated axes

FIGURE 8.34 Graph of $x^2 - 4xy + 4y^2 + 5\sqrt{5}y - 10 = 0$

Conic Sections Graphing Summary

To graph the general second-degree equation

$$Ax^2 + Bxy + Cy^2 + Dx + Ey + F = 0$$

follow the steps shown in Figure 8.35.

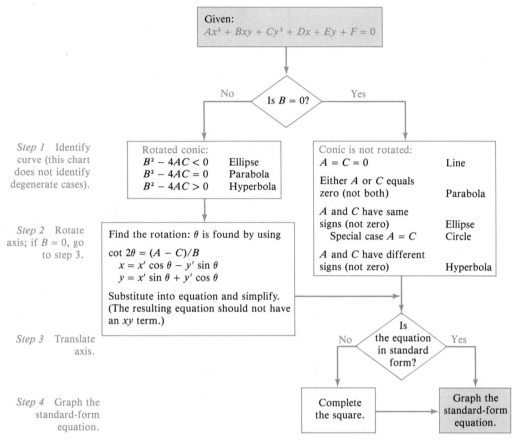

FIGURE 8.35 Procedure for graphing conics

8.5 Problem Set

A *Because of the amount of arithmetic and algebra involved, many algebraic steps were left out of the examples of this section. In Problems 1–3 fill in the details left out of the indicated example.*

1. Example 5   2. Example 6   3. Example 7

Identify the curves whose equations are given in Problems 4–17.

4. $xy = 10$ 5. $xy = -1$ 6. $xy = -4$
7. $13x^2 - 10xy + 13y^2 - 72 = 0$
8. $5x^2 - 26xy + 5y^2 + 72 = 0$
9. $x^2 + 4xy + 4y^2 + 10\sqrt{5}x = 9$
10. $5x^2 - 4xy + 8y^2 = 36$
11. $23x^2 + 26\sqrt{3}xy - 3y^2 - 144 = 0$

12. $3x^2 + 2\sqrt{3}xy + y^2 + 16x - 16\sqrt{3}y = 0$
13. $24x^2 + 16\sqrt{3}xy + 8y^2 - x + \sqrt{3}y - 8 = 0$
14. $3x^2 - 2\sqrt{3}xy + y^2 + 24x + 24\sqrt{3}y = 0$
15. $13x^2 - 6\sqrt{3}xy + 7y^2 + (16\sqrt{3} - 8)x$
    $+ (-16 - 8\sqrt{3})y + 16 = 0$
16. $3x^2 - 10xy + 3y^2 - 32 = 0$
17. $5x^2 - 3xy + y^2 + 65x - 25y + 203 = 0$

B *Find the appropriate rotation in Problems 18–33 so that the given curve will be in standard position relative to the rotated axes. Also find the x and y values in the new coordinate system by using the rotation of axes formulas. Finally, sketch the curve.*

18. $xy = 10$ 19. $xy = -1$ 20. $xy = -4$ 21. $xy = 8$

C

22. $8x^2 - 4xy + 5y^2 = 36$
23. $13x^2 - 10xy + 13y^2 - 72 = 0$
24. $5x^2 - 26xy + 5y^2 + 72 = 0$
25. $x^2 + 4xy + 4y^2 + 10\sqrt{5}x = 9$
26. $5x^2 - 4xy + 8y^2 = 36$
27. $23x^2 + 26\sqrt{3}xy - 3y^2 - 144 = 0$

28. $3x^2 + 2\sqrt{3}xy + y^2 + 16x - 16\sqrt{3}y = 0$
29. $24x^2 + 16\sqrt{3}xy + 8y^2 - x + \sqrt{3}y - 8 = 0$
30. $3x^2 - 2\sqrt{3}xy + y^2 + 24x + 24\sqrt{3}y = 0$
31. $3x^2 - 10xy + 3y^2 - 32 = 0$
32. $x^2 + 2xy + y^2 + 12\sqrt{2}x - 6 = 0$
33. $10x^2 + 24xy + 17y^2 - 9 = 0$

8.6 Conic Sections in Polar Form

In Table 8.1 of Section 8.4 it was noted that the conic sections could be distinguished by looking at the eccentricity, ϵ. Up to now in this chapter, we let the origin lie at the center of the conic, or else we translated the origin to point (h, k). However, there are many important applications of conics that put a focus at the origin (or pole). In this section, we give an alternate definition of the conic sections that puts a focus at the pole and also uses ϵ to distinguish one conic from another.

ALTERNATE DEFINITION OF A CONIC

The set of points in a plane such that the distance from a fixed point (called the **focus**) is in constant ratio to its distance from a fixed line (called the **directrix**) is called a **conic.** The constant ratio is called the **eccentricity** of the conic and is denoted by ϵ. Moreover, the conic is:

an **ellipse** if $0 < \epsilon < 1$;   a **parabola** if $\epsilon = 1$;   a **hyperbola** if $\epsilon > 1$.

(See Figure 8.36.) The single equation for all of these conic sections is

$$r = \frac{\epsilon p}{1 - \epsilon \cos \theta}$$

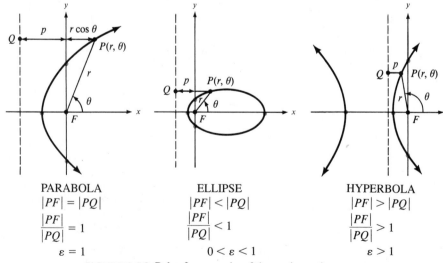

PARABOLA
$|PF| = |PQ|$
$\dfrac{|PF|}{|PQ|} = 1$
$\varepsilon = 1$

ELLIPSE
$|PF| < |PQ|$
$\dfrac{|PF|}{|PQ|} < 1$
$0 < \varepsilon < 1$

HYPERBOLA
$|PF| > |PQ|$
$\dfrac{|PF|}{|PQ|} > 1$
$\varepsilon > 1$

FIGURE 8.36 Polar-form graphs of the conic sections

We can show that this definition is equivalent for our previous definitions for the conics. For example, if $\epsilon = 1$, then

$$r = \frac{p}{1 - \cos \theta}$$

$$r(1 - \cos \theta) = p$$

$$r - r\cos \theta = p$$

To transform this equation into rectangular form, we need to remember the conversion formulas: $r = \sqrt{x^2 + y^2}$ and $x = r\cos \theta$:

$$\sqrt{x^2 + y^2} - x = p$$
$$\sqrt{x^2 + y^2} = x + p$$
$$x^2 + y^2 = (x + p)^2$$
$$x^2 + y^2 = x^2 + 2px + p^2$$
$$y^2 = 2px + p^2$$
$$y^2 = 2p\left(x + \frac{p}{2}\right)$$

This is the equation of a parabola with vertex (in rectangular coordinates) $(h, k) = \left(-\frac{p}{2}, 0\right)$ and focal chord length $2p$. The steps for verifying the polar-form equations for ellipses and hyperbolas are the same as illustrated for the parabola.

EXAMPLE 1   Sketching a parabola given a polar-form equation

Describe and sketch the graph of the equation $r = \dfrac{4}{3 - 3\cos \theta}$.

SOLUTION   The first thing to look at is the first term in the denominator; multiply the fraction by 1 written as the reciprocal of this first term divided by that same number. In this problem, the first term in the denominator is 3 so the fraction is to be multiplied by $\frac{1}{3}/\frac{1}{3}$.

$$r = \frac{4}{3 - 3\cos \theta} \cdot \frac{\frac{1}{3}}{\frac{1}{3}} = \frac{\frac{4}{3}}{1 - \cos \theta}$$

By inspection you can now see that $\epsilon = 1$ so this is a parabola. Furthermore, $\epsilon p = p = \frac{4}{3}$. Compare with Figure 8.36 to see that the directrix is $\frac{4}{3}$ to the left of the focus, and thus the vertex is $\frac{1}{2}\left(\frac{4}{3}\right) = \frac{2}{3}$ from the focus. Plot a few points; those shown in color in Figure 8.36 are generally the easiest to plot. If $\theta = \frac{\pi}{2}$, then

$$r = \frac{4}{3 - 3\cos \frac{\pi}{2}} = \frac{4}{3}$$

You can plot other points, say, $\theta = \pi$, or $\theta = \frac{3\pi}{2}$, but it is not necessary because you know where the vertex is located and you also know the point where the parabola crosses the negative y-axis by using symmetry. The graph is shown in Figure 8.37.

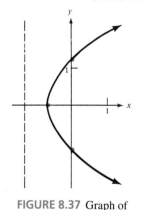

FIGURE 8.37 Graph of $r = \dfrac{4}{3 - 3\cos \theta}$

EXAMPLE 2 **Sketching an ellipse given a polar-form equation**

Describe and sketch the graph of the equation $r = \dfrac{4}{4 - 3 \cos \theta}$.

SOLUTION Notice the leading coefficient in the denominator:

$$r = \frac{4}{4 - 3 \cos \theta} \cdot \frac{\frac{1}{4}}{\frac{1}{4}} = \frac{1}{1 - \frac{3}{4} \cos \theta}$$

We see that $\epsilon = \frac{3}{4}$, which means this is an ellipse. Plot some values:

If $\theta = 0$, $r = \dfrac{4}{4 - 3(1)} = 4$; plot this point.

If $\theta = \frac{\pi}{2}$, $r = \dfrac{4}{4 - 3(0)} = 1$; plot this point and the one at $\theta = \frac{3\pi}{2}$ (by symmetry).

If $\theta = \pi$, $r = \dfrac{4}{4 - 3(-1)} = \frac{4}{7}$; plot this point.

It looks like we could use another point or two:

If $\theta = \frac{\pi}{4}$, $r = \dfrac{4}{4 - 3 \cos \frac{\pi}{4}} \approx 2.129$.

By symmetry, plot the point for $\theta = -\frac{\pi}{4}$ and connect the points as shown in Figure 8.38.

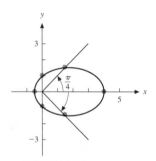

FIGURE 8.38 Graph of
$r = \dfrac{4}{4 - 3 \cos \theta}$

EXAMPLE 3 **Sketching a hyperbola given a polar-form equation**

Describe and sketch the graph of $r = \dfrac{4}{3 - 4 \cos \theta}$.

SOLUTION Look at the leading term in the denominator to decide the form of 1 for the multiplication step:

$$r = \frac{4}{3 - 4 \cos \theta} \cdot \frac{\frac{1}{3}}{\frac{1}{3}} = \frac{\frac{4}{3}}{1 - \frac{4}{3} \cos \theta}$$

This is called the standard form. You see that $\epsilon = \frac{4}{3}$ so this is a hyperbola. First find the intercepts:

If $\theta = 0$, then $r = \dfrac{4}{3 - 4(1)} = -4$.   This is $(-4, 0)$ or $(4, \pi)$.

If $\theta = \pi$, then $r = \dfrac{4}{3 - 4(-1)} = \frac{4}{7}$.   This is $\left(\frac{4}{7}, \pi\right)$.

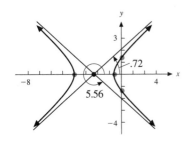

FIGURE 8.39 Graph of
$$r = \frac{4}{3 - 4\cos\theta}$$

The center of the hyperbola will be midway between these points: $\dfrac{4 + \frac{4}{7}}{2} = \dfrac{16}{7}$. The
y-intercepts are $\left(\frac{\pi}{2}, \frac{4}{3}\right)$ and $\left(\frac{3\pi}{2}, \frac{4}{3}\right)$. Even with these points you still do not have a clear
picture of the hyperbola. Hyperbolas are the most difficult conic to sketch because the
domain of θ determines on which branch of the hyperbola you are plotting points. It is
helpful to find the asymptotes. These are the values of θ that cause the denominator to
be 0. Solve this equation:

$$3 - 4\cos\theta = 0$$

$$\cos\theta = \tfrac{3}{4}$$

$$\theta = \cos^{-1} .75 \approx .72, 5.56$$

If you draw the lines through the center with these values of θ you will have the
asymptotes and can finish the sketch of the hyperbola as shown in Figure 8.39.   ∎

CALCULATOR COMMENT

Most software, such as CONVERGE, includes polar-form graphing. For Example 1, the graph
is shown:

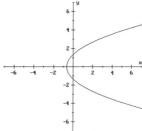

$$R = 4/[3 - 3\cos(\theta)]$$

Compare the graphs of the curves for Examples 2 and 3:

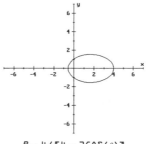

$$R = 4/[4 - 3\cos(\theta)]$$

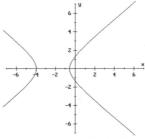

$$R = 4/[3 - 4\cos(\theta)]$$

Rotated Conics

It is easy (compared with what you needed to do in Section 8.5) to rotate the conic sections when the equation is given in polar form. Remember that if the argument of a trigonometric function is $(\theta - \alpha)$ then the curve is rotated through an angle α.

EXAMPLE 4   Sketching a rotated ellipse in polar form

Graph $r = \dfrac{16}{5 - 3\cos\left(\theta - \frac{\pi}{6}\right)}$.

SOLUTION   First determine the type of conic by writing the equation in standard form:

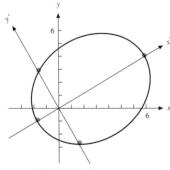

$$r = \frac{16}{5 - 3\cos\left(\theta - \frac{\pi}{6}\right)} \cdot \frac{\frac{1}{5}}{\frac{1}{5}} = \frac{\frac{16}{5}}{1 - \frac{3}{5}\cos\left(\theta - \frac{\pi}{6}\right)}$$

This is an ellipse $\left(\epsilon = \frac{3}{5}\right)$ that has been rotated $\frac{\pi}{6}$. First rotate the conic and then plot points on the rotated axis using the equation

$$r' = \frac{\frac{16}{5}}{1 - \frac{3}{5}\cos\theta'}$$

FIGURE 8.40 Graph of

$$r = \frac{16}{5 - 3\cos\left(\theta - \frac{\pi}{6}\right)}$$

If $\theta' = 0$, then $r' = 8$; if $\theta' = \frac{\pi}{2}$, then $r' = \frac{16}{5}$; if $\theta' = \pi$, then $r' = 2$. Plot these points as shown in Figure 8.40.

As we observed in the last chapter, a rotation of 90° changes the cosine function to a sine function; a rotation of 180° will change the sign of the cosine; and a rotation of 270° will change the sign and the function.

ALTERNATE EQUATIONS FOR POLAR-FORM CONICS

The conic sections can be written as one of the following polar-form equations:

$$r = \frac{\epsilon p}{1 \pm \epsilon \cos\theta} \quad \text{or} \quad r = \frac{\epsilon p}{1 \pm \epsilon \sin\theta}$$

EXAMPLE 5   Sketching a rotated parabola in polar form

Describe and graph $r = \dfrac{8}{4 + 4\sin\theta}$.

SOLUTION   Write in standard form:

$$r = \frac{8}{4 + 4\sin\theta} \cdot \frac{\frac{1}{4}}{\frac{1}{4}} = \frac{2}{1 + \sin\theta}$$

Recognize this as a parabola since $\epsilon = 1$; it is also rotated 270° so instead of opening to the right it opens down. The graph is shown in Figure 8.41.

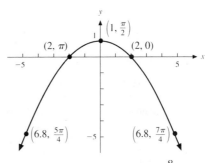

FIGURE 8.41 Graph of $r = \dfrac{8}{4 + 4 \sin \theta}$

<hr>

8.6   Problem Set

A *Identify the conic in each of Problems 1–16. In each case write the standard form of the equation and give* ϵ.

1. $r = \dfrac{3}{1 - \cos \theta}$

2. $r = \dfrac{5}{1 + \sin \theta}$

3. $r = \dfrac{5}{3 + 5 \cos \theta}$

4. $r = \dfrac{6}{2 - 5 \sin \theta}$

5. $r = \dfrac{5}{5 - 4 \cos \theta}$

6. $r = \dfrac{16}{8 + \cos \theta}$

7. $r(1 - 2 \cos \theta) = 4$

8. $r(1 + 3 \sin \theta) = 5$

9. $r(2 - 2 \cos \theta) = 8$

10. $r(4 - 4 \sin \theta) = 32$

11. $r = \dfrac{3}{1 - \cos\left(\theta - \frac{\pi}{4}\right)}$

12. $r = \dfrac{4}{2 - 2 \cos\left(\theta - \frac{\pi}{3}\right)}$

13. $r = \dfrac{5}{4 - 2 \cos\left(\theta - \frac{\pi}{6}\right)}$

14. $r = \dfrac{2}{2 - \cos(\theta + 1)}$

15. $r = \dfrac{\sqrt{2}}{1 + \cos(\theta + .5)}$

16. $r = \dfrac{\sqrt{3}}{1 + \sin(\theta - 1)}$

B *Graph each of the curves given in Problems 17–32. Notice that these are the equations you analyzed in Problems 1–16.*

17. $r = \dfrac{3}{1 - \cos \theta}$

18. $r = \dfrac{5}{1 + \sin \theta}$

19. $r = \dfrac{5}{3 + 5 \cos \theta}$

20. $r = \dfrac{6}{2 - 5 \sin \theta}$

21. $r = \dfrac{5}{5 - 4 \cos \theta}$

22. $r = \dfrac{16}{8 + \cos \theta}$

23. $r(1 - 2 \cos \theta) = 4$

24. $r(1 + 3 \sin \theta) = 5$

25. $r(2 - 2 \cos \theta) = 8$

26. $r(4 - 4 \sin \theta) = 32$

27. $r = \dfrac{3}{1 - \cos\left(\theta - \frac{\pi}{4}\right)}$

28. $r = \dfrac{4}{2 - 2 \cos\left(\theta - \frac{\pi}{3}\right)}$

29. $r = \dfrac{5}{4 - 2 \cos\left(\theta - \frac{\pi}{6}\right)}$

30. $r = \dfrac{2}{2 - \cos(\theta + 1)}$

31. $r = \dfrac{\sqrt{2}}{1 + \cos(\theta + .5)}$

32. $r = \dfrac{\sqrt{3}}{1 - \sin(\theta - 1)}$

C

33. Assume that $\epsilon = \frac{1}{2}$ in the definition. Convert the polar-form equation to rectangular form and show it is an ellipse.

34. Assume that $\epsilon = \frac{3}{2}$ in the definition. Convert the polar-form equation to rectangular form and show it is a hyperbola.

8.7   Summary and Review

Objectives

Pages / Examples / Problems

8.1   Parabolas

1. Graph standard-form parabolas centered at (0, 0).

pp. 293–295; Examples 1–3; Problems 7–18.

2. Graph standard-form parabolas centered at (h, k).

p. 296; Example 5; Problems 19–24.

3. Find the equations of parabolas given certain information about the graph.

pp. 295–297; Examples 4, 6; Problems 25–30.

8.2   Ellipses

4. Graph standard-form ellipses centered at (0, 0).

pp. 299–301; Examples 1–2; Problems 1–9.

5. Graph standard-form ellipses centered at (h, k).

p. 302; Example 3; Problems 13–15.

6. Graph standard-form circles with radius r centered at (h, k).

p. 303; Example 4; Problems 10–12.

7. Find the equations of ellipses and circles given certain information about the graph.

pp. 304–305; Examples 5–7; Problems 16–30.

8.3   Hyperbolas

8. Graph standard-form hyperbolas centered at (0, 0).

p. 308; Example 1; Problems 1–15.

9. Graph standard-form hyperbolas centered at (h, k).

p. 309; Example 2; Problems 16–22.

10. Find the equations of hyperbolas given certain information about the graph.

pp. 309–311; Examples 3–5; Problems 23–28.

8.4   General Conic Sections

11. Graph degenerate conics.

pp. 312–313; Examples 1–2; Problems 7–9.

12. Know the definitions and equations for the standard-position conics.

p. 314.

13. Identify the type of conic by looking at the equation.

p. 314; p. 323; Example 3; Problems 1–30; Section 8.5, Problems 4–17.

14. Graph general conic sections by completing the square.

pp. 315–316; Examples 4–5; Problems 13–30.

8.5   Rotations

15. Use the rotation of axes formulas and the amount of rotation formula.

pp. 318–321; Examples 1–4; Problems 1–3, 18–33.

16. Graph a conic section involving a rotation.

pp. 321–323; Examples 5–7; Problems 18–33.

8.6   Conic Sections in Polar Form

17. Identify and sketch a conic given a polar-form equation.

pp. 324–327; Examples 1–3; Problems 1–10, 17–26.

18. Identify and sketch a rotated conic given a polar-form equation

pp. 328–329; Examples 4–5; Problems 11–16, 27–32.

TERMS

Axis [8.1]
Center [8.2, 8.3]
Central rectangle [8.3]
Circle [8.2]
Completing the square [8.4]
Conic sections [8.1, 8.4, 8.6]
Conjugate axis [8.3]
Covertices [8.2]
Degenerate conic [8.4]
Directrix [8.1, 8.6]

Discriminant [8.4]
Eccentricity [8.2, 8.6]
Ellipse [8.1, 8.2, 8.4, 8.6]
First-degree equation [8.1]
Focal chord [8.1]
Focus [8.1, 8.2, 8.3, 8.6]
Hyperbola [8.1, 8.3, 8.4, 8.6]
Length of conjugate axis [8.3]
Length of focal chord [8.1]
Length of transverse axis [8.3]

Linear equation [8.1]
Major axis [8.2]
Minor axis [8.2]
Parabola [8.1, 8.4, 8.6]
Pseudovertices [8.3]
Quadratic equation [8.1]
Radius [8.2]
Rotation [8.5, 8.6]
Second-degree equation [8.1]

Semimajor axis [8.2]
Semiminor axis [8.2]
Slant asymptote [8.3]
Standard-form ellipse [8.2]
Standard-form hyperbola [8.3]
Standard-form parabola [8.1]
Translation [8.1, 8.2, 8.3, 8.4]
Transverse axis [8.3]
Vertex [8.1, 8.2, 8.3]

CHAPTER 8 REVIEW OF OBJECTIVES

*The problem numbers correspond to the objectives listed in
Section 8.7.*

1. Graph the given curves.
  a. $x^2 = y$ **b.** $2y^2 = x$
2. Graph the given curves.
  a. $(y - 1)^2 = 8(x + 2)$ **b.** $8(y - 2)^2 = x + 1$
3. Find the equations for the given parabolas.
  a. Vertex $(6, 3)$; directrix $x = 1$
  b. Directrix $y - 3 = 0$; focus $(-3, -2)$
  c. Vertex $(-3, 5)$; focus $(-3, -1)$
4. Graph the given curves.
  a. $4x^2 + 4y^2 = 9$ **b.** $(x - 3)^2 + (y + 1)^2 = 16$
5. Graph the given curves.
  a. $25x^2 + 16y^2 = 400$ **b.** $\dfrac{x^2}{25} + \dfrac{y^2}{4} = 1$
6. Graph the given curves.
  a. $9(x + 3)^2 + 5(y - 2)^2 = 45$
  b. $\dfrac{(x + 1)^2}{5} + \dfrac{(y - 2)^2}{9} = 1$
7. Find the equations of the given ellipses.
  a. The ellipse with center $(4, 1)$, a focus $(5, 1)$, and
  a semimajor axis 2
  b. The set of points with the sum of distances from
  $(-3, 4)$ and $(-7, 4)$ equal to 12
  c. The set of points 8 units from the point $(-1, -2)$
8. Graph the given curves.
  a. $y^2 - x^2 = 1$ **b.** $\dfrac{x^2}{4} - \dfrac{y^2}{6} = 1$

9. Graph the given curves.
  a. $5(x + 2)^2 - 3(y + 4)^2 = 60$
  b. $\dfrac{\left(x + \frac{1}{2}\right)^2}{3} - \dfrac{\left(y + \frac{1}{2}\right)^2}{3} = 1$
10. Find the equations of the given hyperbolas.
  a. The hyperbola with vertices at $(-3, 0)$ and $(3, 0)$ and
  foci at $(5, 0)$ and $(-5, 0)$
  b. Foci at $(0, -5)$ and $(0, 5)$; eccentricity $\frac{5}{3}$
  c. Vertices at $(-3, 1)$ and $(-5, 1)$; foci at $\left(-4 - \sqrt{6}, 1\right)$
  and $\left(-4 + \sqrt{6}, 1\right)$
11. Graph:
  a. $(x - 2)^2 + (y + 3)^2 = 0$ **b.** $\dfrac{x^2}{4} - \dfrac{y^2}{9} = 0$
12. State the appropriate standard-form equation.
  a. Horizontal ellipse   **b.** Vertical hyperbola
  c. Parabola opening right   **d.** Circle
13. Name the curve (by inspection).
  a. $3x - 2y^2 - 4y + 7 = 0$ **b.** $\dfrac{x}{16} + \dfrac{y}{4} = 1$
  c. $25x^2 + 9y^2 = 225$
  d. $25(x - 2)^2 + 25(y + 1)^2 = 400$
  e. $xy = 5$
  f. $5x^2 + 4xy + 5y^2 + 3x = 2$
  g. $xy + x^2 - 3x = 5$
  h. $x^2 + y^2 + 2xy + 3x - y = 3$
  i. $x^2 + 2xy + y^2 = 10$
  j. $(x - 1)(y + 1) = 7$

14. Graph the given curves.

   a. $y^2 + 4x + 4y = 0$

   b. $x^2 + y^2 = 4x + 2y - 3$

   c. $16x^2 + 9y^2 - 160x - 18y + 265 = 0$

   d. $3x^2 - 4y^2 + 6x - 8y - 16 = 0$

15. What is the rotation for each of the following conic sections?

   a. $5x^2 + 4xy + 5y^2 + 3x - 2y + 5 = 0$

   b. $4x^2 + 4xy + y^2 + 3x - 2y + 7 = 0$

16. Graph $10x^2 + 17y^2 = 9 - 24xy$.

17. Identify and graph $r = \dfrac{10}{5 - 5\cos\theta}$.

18. Identify and graph $r = \dfrac{6}{3 - 2\sin\theta}$.

DISCUSSION PROBLEMS

1. How do you recognize and identify the different conic sections?

2. Describe how you graph different conic sections.

3. What do we mean by degenerate conics?

4. Compare and contrast rotations in rectangular and polar coordinates.

5. Write a 500-word paper on the history of the conic sections.

MISCELLANEOUS PROBLEMS

Graph each of the curves in Problems 1–30.

1. $y = 10x^2$

2. $y = \frac{1}{5}x^2$

3. $x = 3y^2$

4. $x = .01y^2$

5. $y - 3 = (x + 1)^2$

6. $y = 2x^2 + 12x + 20$

7. $9x^2 + 25y^2 = 225$

8. $4x^2 + 9y^2 = 36$

9. $3x - 4y = 0$

10. $\dfrac{x}{16} + \dfrac{y}{4} = 1$

11. $\dfrac{x^2}{16} + \dfrac{y}{4} = 1$

12. $\dfrac{x^2}{16} + \dfrac{y^2}{4} = 1$

13. $\dfrac{x^2}{16} - \dfrac{y^2}{4} = 1$

14. $3y - 2x^2 - 4x + 7 = 0$

15. $x^2 + y^2 + 2x - 4y = 0$

16. $5x^2 + 4y^2 + 10x - 8y + 4 = 0$

17. $3x - 2y^2 - 4y + 7 = 0$

18. $25(x - 1)^2 + 25(y + 2)^2 = 400$

19. $3x^2 - y^2 + 6y - 4 = 0$

20. $2x^2 + 8x - 3y + 2 = 0$

21. $xy = 10$

22. $25x^2 + 14xy + 25y^2 + 4\sqrt{2}x - 20\sqrt{2}y + 2 = 2$

23. $r = \dfrac{\sqrt{3}}{2 - 2\cos\theta}$

24. $r = \dfrac{1}{4 - 5\cos\theta}$

25. $r = \dfrac{2}{6 - 5\cos\theta}$

26. $r = \dfrac{\sqrt{2}}{3 - 3\sin\theta}$

27. $r = \dfrac{3}{2 + 3\cos\theta}$

28. $r = \dfrac{4}{3 + 2\sin\theta}$

29. $r = \dfrac{5}{2 + 2\cos\left(\theta - \frac{\pi}{6}\right)}$

30. $r = \dfrac{3}{2 + \cos(\theta - .2)}$

▌ **Planetary Orbits**

A 'Planet X'
Way Out There?

LIVERMORE, Calif. (UPI)—A scientist at the Lawrence Livermore Laboratory suggested Friday the existence of a "Planet X" three times as massive as Saturn and nearly six billion miles from earth.

The planet, far beyond Pluto which is currently the outermost of the nine known planets of the solar system, was predicted on sophisticated mathematical computations of the movements of Halley's Comet.

Joseph L. Brady, a Lawrence mathematician and an authority on the comet, reported the calculations in the Journal of the Astronomical Society of the Pacific.

Brady said he and his colleagues, Edna M. Carpenter and Francis H. McMahon, used a computer to process mathematical observations of the strange deviations in Halley's Comet going back to before Christ.

Lawrence officials said the existence of a 10th planet has been predicted before, but Brady is the first to predict its orbit, mass and position.

Brady said the planet was about 65 times as far from the sun as earth, which is about 93 million miles from the sun. From earth "Planet X" would be located in the constellation Cassiopeia on the border of the Milky Way.

The size and location of "Planet X" were proposed to account for mysterious deviations in the orbit of Halley's Comet. But the calculations subsequently were found to account for deviations in the orbits of two other reappearing comets, Olbers and Pons-Brooks, Brady said.

No contradiction between the proposed planet and the known orbits of comets and other planets has been found.

The prediction of unseen planets is not new. The location of Neptune was predicted in 1846 on the basis of deviations in the orbit of Uranus. Deviations in Neptune's orbit led to a prediction of Pluto's location in 1915.

Although no such deviations of Pluto have been found, Brady pointed out that since its discovery in 1930, Pluto has been observed through less than one-fourth of its revolution around the sun and a complete picture of its orbit is not available.

Brady said "Planet X" may be as elusive as Pluto was a half century ago. It took 15 years to find it from the time of its prediction.

"The proposed planet is located in the densely populated Milky Way where even a tiny area encompasses thousands of stars, many of which are brighter than we expect this planet to be," he said. "If it exists, it will be extremely difficult to find."

"Planet X," in its huge orbit, takes 600 years to complete a revolution around the sun, he said.

After the planet Uranus was discovered in 1781, its motion revealed gravitational perturbations caused by an unknown planet. Independent mathematical calculations by Urbain Leverrier and John Couch Adams predicted the position of this unknown planet—the discovery of Neptune in 1846 is one of the greatest triumphs of celestial mechanics and mathematics in the history of astronomy. A similar search led to the discovery of Pluto in 1930. The article reproduced here is dated April 30, 1972, but at the time of this printing no planet has yet been found. Remember, though, that it took 15 years to find Pluto after the time of its prediction.

The orbits of the planets are elliptical in shape. If the sun is placed at one of the foci of a giant ellipse, the orbit of the earth is elliptical. The *perihelion* is the point where the planet comes closest to the sun; the *aphelion* is the farthest distance the planet travels from the sun. The eccentricity of a planet tells us the amount of roundness of that planet's orbit. The eccentricity of a circle is 0 and that of a parabola is 1. The eccentricity for each planet is our solar system is given here:

PLANET	ECCENTRICITY
Mercury	.194
Venus	.007
Earth	.017
Mars	.093
Jupiter	.048
Saturn	.056
Uranus	.047
Neptune	.009
Pluto	.249

EXAMPLE 1   Finding the perihelion and aphelion

The orbit of the earth around the sun is elliptical with the sun at one focus. If the semimajor axis of this orbit is 9.3×10^7 mi and the eccentricity is about .017, determine the greatest and least distance of the earth from the sun (correct to two significant digits).

SOLUTION   We are given $a = 9.3 \times 10^7$ and $\epsilon = .017$. Now

$$\epsilon = \frac{c}{a}$$

$$.017 = \frac{c}{9.3 \times 10^7}$$

$$c \approx 1.581 \times 10^6$$

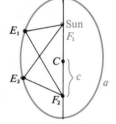

The greatest distance is

$$a + c \approx 9.5 \times 10^7$$

The least distance is

$$a - c \approx 9.1 \times 10^7$$

The orbit of a satellite can be calculated from Kepler's third law:

$$\frac{\text{Mass of (planet + satellite)}}{\text{Mass of (sun + planet + satellite)}} = \frac{(\text{semimajor axis of satellite orbit})^3}{(\text{semimajor axis of planet orbit})^3} \times \frac{(\text{period of planet})^2}{(\text{period of satellite})^2}$$

EXAMPLE 2   Calculating the orbit of a satellite

If the mass of the earth is 6.58×10^{21} tons, the sun 2.2×10^{27} tons, and the moon 8.1×10^{19} tons, calculate the orbit of the moon. Assume that the semimajor axis of the earth is 9.3×10^7 mi, the period of the earth 365.25 days, and the period of the moon 27.3 days.

SOLUTION   First solve the equation given as Kepler's third law for the unknown—the semimajor axis of the satellite orbit:

$(\text{Semimajor axis of satellite orbit})^3$

$$= \frac{\text{mass of (planet + satellite)}}{\text{mass of (sun + planet + satellite)}} \times \frac{(\text{period of satellite})^2}{(\text{period of planet})^2} \times (\text{semimajor axis of planet orbit})^3$$

Thus,

$(\text{Semimajor axis of satellite orbit})^3$

$$= \frac{(6.58 \cdot 10^{21}) + (8.1 \cdot 10^{19})}{(2.2 \cdot 10^{27}) + (6.58 \cdot 10^{21}) + (8.1 \cdot 10^{19})} \times \frac{(27.3)^2}{(365.25)^2} \times (9.3 \cdot 10^7)^3$$

$$\approx \frac{3.993131141 \cdot 10^{48}}{2.934975261 \cdot 10^{32}}$$

$$\approx 1.360533151 \cdot 10^{16}$$

Semimajor axis of satellite orbit $\approx 2.387278258 \cdot 10^5$

The moon's orbit has a semimajor axis of about 239,000 mi.

PROBLEMS FOR FURTHER STUDY: PLANETARY ORBITS

1. **Astronomy** The orbit of Mars about the sun is elliptical with the sun at one focus. If the semimajor axis of this orbit is 1.4×10^8 mi and the eccentricity is about .093, determine the greatest and least distance of Mars from the sun, correct to two significant digits.

2. **Astronomy** The orbit of Venus about the sun is elliptical with the sun at one focus. If the semimajor axis of this orbit is 6.7×10^7 mi and the eccentricity is about .007, determine the greatest and least distance of Venus from the sun, correct to two significant digits.

3. **Astronomy** The orbit of Neptune about the sun is elliptical with the sun at one focus. If the semimajor axis of this orbit is 3.66×10^9 mi and the eccentricity is about .009, determine the greatest and least distance of Neptune from the sun, correct to two significant digits.

4. **Astronomy** If a Planet X has an elliptical orbit about the sun with a semimajor axis of 6.89×10^{10} mi and eccentricity of about .35, what is the least distance between Planet X and the sun?

5. **Astronomy** If the mass of Mars is 7.05×10^{20} tons, its satellite Phobos 4.16×10^{12} tons, and the sun 2.2×10^{27}

tons, calculate the distance of the semimajor axis of Phobos. Assume that the semimajor axis of Mars is 1.4×10^8 mi, the period of Mars 693.5 days, and the period of Phobos .3 day.

6. **Astronomy** Use the information in Problem 5 to find the approximate mass of the Martian satellite Deimos if its elliptical orbit has a semimajor axis of 15,000 mi and a period of 1.26 days.

7. **Astronomy** If the perihelion distance of Mercury is 2.9×10^7 mi and the aphelion distance is 4.3×10^7 mi, write the equation for the orbit of Mercury.

8. **Astronomy** If the perihelion distance of Venus is 6.7234×10^7 mi and the aphelion distance is 6.8174×10^7 mi, write the equation for the orbit of Venus.

9. **Astronomy** If the perihelion distance of Earth is 9.225×10^7 mi and the aphelion distance is 9.542×10^7 mi, write the equation for the orbit of Earth.

10. **Astronomy** If the perihelion distance of Mars is 1.29×10^8 mi and the aphelion distance is 1.55×10^8 mi, write the equation for the orbit of Mars.

APPENDIXES

A   **Calculators**   338

B   **Accuracy and Rounding**   342

C   **Review of Geometry**   346

D   **Selected Answers**   350

CALCULATORS

Appendix A provides a brief introduction to calculators and calculator usage. In the last few years, pocket calculators have been one of the fastest growing items in the United States. There are probably two reasons for this increase in popularity. Most people (including mathematicians!) do not like to do arithmetic, and a good calculator can be purchased for under $20.

This book was written with the assumption that students have access to a calculator. Throughout the text, references concerning the use of calculators are made. Trigonometry is the first course most students take that *requires* the use of a calculator, and it is important that students know how to use a calculator. This appendix is included to help the student choose a calculator and understand the calculator comments in this book.

Calculators are classified by the types of problems they are equipped to handle, as well as by the type of logic for which they are programmed. The task of selecting a calculator is compounded by the multiplicity of brands from which to choose.

The different types of calculators are distinguished primarily by their price.

1. *Four-function calculators* (*under* $10).
   These calculators have a keyboard consisting of the numerals and the four arithmetic operations or functions: addition $\boxed{+}$, subtraction $\boxed{-}$, multiplication $\boxed{\times}$, and division $\boxed{\div}$.

2. *Four-function calculators with memory* ($10–$20).
   Usually no more expensive than four-function calculators, these offer a memory register: $\boxed{M}$, $\boxed{STO}$, or $\boxed{M^+}$. The more expensive models may have more than one memory register. Memory registers allow you to store partial calculations for later recall. Some models will even remember the total when they are turned off.

3. *Scientific calculators* ($20–$50).
   These calculators have additional mathematical functions, such as square root $\boxed{\sqrt{\ }}$, trigonometric $\boxed{\sin}$, $\boxed{\cos}$, and $\boxed{\tan}$, and logarithmic $\boxed{\log}$ and $\boxed{\exp}$. Depending on the particular brand, a scientific model may have other keys as well.

4. *Special-purpose calculators* ($40–$400).
   Special-use calculators for business, statistics, surveying, medicine, or even gambling and chess are available.

5. *Programmable calculators* ($50–$600).
   With these calculators you can enter a *sequence* of steps for the calculator to repeat on your command. Some of these calculators allow the insertion of different cards that "remember" the sequence of steps for complex calculations.

6. *Graphics calculators* ($60–$300).

These calculators (which include the Texas Instruments TI-81, TI-82, TI-85; the Casio fx-7000ga, fx-7700g, fx-8700g; Sharp EL-9200C, EL-9300C; and HP 48SX) not only can do arithmetic operations, but also can graph and do some algebraic manipulations.

For most nonscientific purposes, a four-function calculator with memory will be sufficient for everyday usage. **For this book you will need a scientific or graphing calculator.** Calculator logic continues to be improved as newer models are offered. However, for classroom use, not all students will have access to the latest technology. For this reason, it is important that you understand the logic of the calculator you are using. You might ask, "What is calculator logic, and why do I need to know anything about it?" The quick answer is that the logic used by your calculator determines the way you must input information. There are now four types of calculator logic in common use: arithmetic (older, inexpensive calculators), RPN, algebraic, and direct algebraic logic (newer calculators, including most graphing calculators). Here is a test problem, which you can use to test the type of logic used by your calculator:

$$\log(2 + 3 \times 4)$$

The correct answer is 1.146128036. To obtain the correct answer for this test problem requires different keystrokes with each type of logic:

Arithmetic logic: $\boxed{3}\ \boxed{\times}\ \boxed{4}\ \boxed{=}\ \boxed{+}\ \boxed{2}\ \boxed{=}\ \boxed{\log}$

RPN logic: $\boxed{2}\ \boxed{\text{ENTER}}\ \boxed{3}\ \boxed{\text{ENTER}}\ \boxed{4}\ \boxed{\times}\ \boxed{+}\ \boxed{\text{LOG}}$

Algebraic logic: $\boxed{2}\ \boxed{+}\ \boxed{3}\ \boxed{\times}\ \boxed{4}\ \boxed{=}\ \boxed{\text{LOG}}$

Direct algebraic (or graphing) logic: $\boxed{\text{LOG}}\ \boxed{(}\ \boxed{2}\ \boxed{+}\ \boxed{3}\ \boxed{\times}\ \boxed{4}\ \boxed{)}\ \boxed{\text{ENTER or} =}$

Notice that direct algebraic logic inputs information just as it is written in algebra. Also, note that many keys on today's calculators are reached by using a $\boxed{\text{2nd}}$ or "uppercase" key. In this book, we generally do not show particular keystrokes (because of the variety of logic types), but will expect that you understand how to input algebraic expressions for the calculator you use. We do, however, show typical calculator displays using the following numerals:

0 1 2 3 4 5 6 7 8 9

The number of digits shown (as well as the rounding for the last digit displayed) also varies according to the model of the calculator you use.

Even though most of you may not yet have a graphics calculator, I have included some Calculator Comment applications in the text so you can see what you can do with these calculators. These sections are set apart in a box and are optional. In writing this book, I have been using a Sharp EL-531A (I paid $9.95) and a Texas Instruments TI-81 for graphing applications (I paid $125, but have seen them advertised for under $60). You should always consult the owner's manual for your calculator. In the previous editions of this book, I showed keystrokes for both algebraic and RPN logic. However, with the new generation of calculators, I now see much more diversity in possible keystrokes for a particular process. For that reason, I have generally removed specific keystrokes and instead have included only general remarks about process or procedure. For example, to find $-3 + \cos 2$ you might press

Algebraic logic: [3] [+/−] [+] [2] [COS] [=]

RPN logic: [3] [CHS] [ENTER] [2] [COS] [+]

Direct algebraic (graphing) logic: [(-)] [3] [+] [COS] [2] [ENTER]

There is also a limit to the accuracy of your calculator. You may have a calculator with a 6-, 8-, or 10-digit display. You can test the accuracy with the following example.

EXAMPLE 1   Using a calculator for ordinary operations

Find $2 \div 3$.

SOLUTION   Graphing: [2] [÷] [3] [ENTER]

Algebraic: [2] [÷] [3] [=]

RPN logic: [2] [ENTER] [3] [÷]

DISPLAY: 0.6666666667

There may be some discrepancy between this answer and the one you obtain on your calculator. Some machines will not round the answer as shown here but will show the display

$$0.6666666666$$

Others will show a display such as

$$6.6666-01$$

This is a number in scientific notation and should be interpreted as

$$6.6666 \times 10^{-1} \quad \text{or} \quad .66666$$

Most calculators will also use scientific notation when the numbers become larger than allowed by their display register. Example 2 will show you how your calculator handles large numbers. Some calculators will simply show an overflow and will not accept large numbers. You will want a calculator that will accept and handle large and small numbers.

E X A M P L E 2   Using a calculator for exponents

Find 50^6.

SOLUTION Check your owner's manual to find out how to use an exponent key. Most calculators will work as shown below:

DISPLAY

Algebraic: 50 y^x 6 =   1.5625   10

RPN: 50 ENTER 6 y^x 1.562500   10

Graphing: 50 ^ 6 ENTER   1.5625E10

When the maximum size of the display has been reached, the calculator should automatically switch to scientific notation. The point at which a calculator will do this varies from one type or brand to another. The answer for this example is

$$1.5625 \times 10^{10} = 15,625,000,000$$

Answers on a calculator can sometimes be inverted so that a word is shown in the display.

E X A M P L E 3   Using a calculator with mixed operations

Find $5 \times .9547 + 2 \times 353$.

SOLUTION   Graphing: 5 × .9547 + 2 × 353 ENTER

Algebraic: 5 × .9547 + 2 × 353 =

RPN: 5 ENTER .9547 × 2 ENTER 353 × +

Some problems involving calculator computations are given in Appendix B.

ACCURACY AND ROUNDING

Applications involving measurements can never be exact. Appendix B discusses some agreements concerning the accuracy of our calculations. It is particularly important that you pay attention to the accuracy of your measurements when you have access to a calculator, because calculator processes can give you a false sense of security about the accuracy in a particular problem. For example, if you measure a triangle and find the sides are approximately 1.2 and 3.4 and then find the ratio of $1.2/3.4 \approx .35294117$, it appears that the result is more accurate than the original measurements! Some discussion about the accuracy of results follows in this section.

The digits known to be correct in a number obtained by a measurement are called **significant digits.** The digits 1, 2, 3, 4, 5, 6, 7, 8, and 9 are always significant, whereas the digit 0 may or may not be significant.

1. Zeros that come between two other digits are significant, as in 203 or 10.04.

2. If the zero's only function is to place the decimal point, it is not significant, as in

If it does more than fix the decimal point, it is significant, as in

This second rule can, of course, result in certain ambiguities, such as in 23,000 (measured to the *exact* unit). To avoid such confusion, we use scientific notation in this case.

$$2.3 \times 10^4 \text{ has two significant digits}$$
$$2.3000 \times 10^4 \text{ has five significant digits}$$

Numbers that come about by counting are considered to be exact and are correct to any number of significant digits. Since exponents are usually counting numbers, in this book we will not consider exponents in deciding on the proper number of significant digits used in a particular calculation.

EXAMPLE 1   Recognizing significant digits

a. Two significant digits: 46, .00083, 4.0×10^1, .050

b. Three significant digits: 523, 403, 4.00×10^2, .000800

c. Four significant digits: 600.1, 4.000×10^1, .0002345

When we are doing calculations with approximate numbers (particularly when using a calculator), it is often necessary to round off results. In this book we use the following rounding procedure.

ROUNDING PROCEDURE

To round off numbers:

1. Increase the last retained digit by 1 if the remainder is greater than or equal to 5; or

2. Retain the last digit unchanged if the remainder is less than 5.

3. In problems requiring rounding but involving several steps, round only once, at the end. That is, do not work with rounded results, since round-off errors can accumulate.

Note this agreement about rounding.

Elaborate rules for computation of approximate data can be developed (when it is necessary for some applications, such as in chemistry), but there are two simple rules that will work satisfactorily for the material in this text.

RULES FOR SIGNIFICANT DIGITS

Addition–Subtraction: Add or subtract in the usual fashion, and then round off the result so that the last digit retained is in the column farthest to the right in which both given numbers have significant digits.

Multiplication–Division: Multiply or divide in the usual fashion, and then round off the results to the smaller number of significant digits found in either of the given numbers.

Counting numbers: Numbers used to count or whole numbers used as exponents are considered to be correct to any number of significant digits.

Multiple operations: Apply the rounding rules previously given for significant digits only when stating your final answer, in order to avoid errors resulting from repeated rounding.

EXAMPLE 2   Writing significant digits with a calculation

Calculate $b = \dfrac{50}{\tan 35°}$

by division, by multiplication, and by calculator. Compare the results and then give the answer to the correct number of significant digits.

SOLUTION By division: tan 35° = .7002. (When we say tan 35° = .7002, we mean, of course, tan 35° ≈ .7002. We often write the former for convenience, but keep in mind that many calculator values are approximate.) Thus,

$$b \approx \frac{50}{.7002} = 71.408169090 \ldots$$

By multiplication:

$$b = \frac{50}{\tan 35°}$$

$$= 50 \cot 35°$$

$$\approx 50(1.428)$$

$$= 71.400$$

By calculator: DISPLAY: 71.40740034

Notice that all these answers differ. We now use the multiplication–division rule.

50:   This number has one or two significant digits; there is no ambiguity if we write 5×10^1 or 5.0×10^1. *In this book,* if the given data include a number whose degree of accuracy is doubtful, *we assume the maximum degree of accuracy.* Thus, 50 has two significant digits.

tan 35°:   On a calculator you may have 8, 10, or 12 significant digits (depending on the calculator).

The result of this division is correct to two significant digits—namely,

$$b = \frac{50}{\tan 35°} = 71$$

which agrees with all the previous methods of solution.

⊗ Note this agreement about accuracy. ⊗

In working with triangles in this text, we assume a certain relationship in the accuracy of the measurement between the sides and the angles.

ACCURACY IN SIDES	EQUIVALENT ACCURACY IN ANGLES
Two significant digits	Nearest degree
Three significant digits	Nearest tenth of a degree
Four significant digits	Nearest hundredth of a degree

This chart means that, if the data include one side given with two significant digits and another with three significant digits, the angle would be computed to the nearest degree. If one side is given to four significant digits and an angle to the nearest tenth of a degree, then the other sides would be given to three significant digits and the angles computed to the nearest tenth of a degree. In general, results computed from this table should not be more accurate than the least accurate item of the given data.

B   Problem Set

A *Use a calculator to evaluate each of the expressions in Problems 1–12. Be sure to round off each answer to the appropriate number of significant digits.*

1. $(14)(351)$

2. $(218)(263)$

3. $(2.00)^4(1,245)(277)$

4. $(3.00)^3(182)$

5. $\dfrac{(1,979)(1,356)}{452}$

6. $\dfrac{(515)(20,600)}{200}$

7. $(990)(1,117)(342) - 89$

8. $[.14 + (197)(25.08)](19)$

9. $\dfrac{1.00}{.005 + .020}$

10. $\dfrac{1.500 \times 10^4 + (7.000)(67.00)}{20,000}$

11. $3,478.06 + 2,256.028 + 1,979.919 + .00091$

12. $57,300 + .094 + 32.3 + 2.09 + .0074$

B *Use a calculator to evaluate each of the expressions in Problems 13–30. Be sure to round off each answer to the appropriate number of significant digits.*

13. $(6.28)^{1/2}(4.85)$

14. $(8.23)^{1/2}(6.14)$

15. $\dfrac{1.00}{\sqrt{4.83} + \sqrt{2.51}}$

16. $\dfrac{1.00}{\sqrt{8.48} + \sqrt{21.3}}$

17. $[(4.083)^2(4.283)^3]^{-2/3}$

18. $[(6.128)^4(3.412)^2]^{-1/2}$

19. $\dfrac{(2.51)^2 + (6.48)^2 - (2.51)(6.48)(.3462)}{(2.51)(6.48)}$

20. $\dfrac{241^2 + 568^2 - (241)(568)(.5213)}{(241)(568)}$

21. $\dfrac{23.7 \sin 36.2°}{45.1}$

22. $\dfrac{461 \sin 43.8°}{1,215}$

23. $\dfrac{62.8 \sin 81.5°}{\sin 42.3°}$

24. $\dfrac{4.381 \sin 49.86°}{\sin 71.32°}$

25. $\dfrac{16 \sin 22°}{25}$

26. $\dfrac{42 \sin 52°}{61}$

27. $\dfrac{16^2 + 25^2 - 9.0^2}{(2.0)(16)(25)}$

28. $\dfrac{216^2 + 418^2 - 315^2}{(2.00)(216)(418)}$

29. $\dfrac{4.82^2 + 6.14^2 - 9.13^2}{(2.00)(4.82)(6.14)}$

30. $\dfrac{18.361^2 + 15.215^2 - 13.815^2}{(2.0000)(18.361)(15.215)}$

REVIEW OF GEOMETRY

In this appendix we will review those ideas from plane geometry that are necessary for the study of trigonometry.

Mensuration Formulas

Let A = area, P = perimeter, C = circumference, S = surface area, and V = volume.

POLYGON	
TYPE	NUMBER OF SIDES
triangle	3
quadrilateral	4
pentagon	5
hexagon	6
heptagon	7
octagon	8
nonagon	9
decagon	10
undecagon	11
dodecagon	12

RECTANGLE

$A = lw$

$P = 2l + 2w$

SQUARE

$A = s^2$

$P = 4s$

TRIANGLE

$A = \frac{1}{2}bh$

$P = a + b + c$

CIRCLE

$A = \pi r^2$

$C = 2\pi r = \pi D$

SPHERE

$S = 4\pi r^2$

$V = \frac{4}{3}\pi r^3$

RIGHT-CIRCULAR CYLINDER

$S = 2\pi rh + 2\pi r^2$

$V = \pi r^2 h$

RIGHT-CIRCULAR CONE

$S = \pi r^2 + \pi rs$

$V = \frac{1}{3}\pi r^2 h$

$s = \sqrt{r^2 + h^2}$

Classification of Angles

Complementary angles are two angles whose measures add up to $\frac{\pi}{2}$ or 90°.

Supplementary angles are two angles whose measures add up to π or 180°.

Vertical angles are two nonadjacent angles formed by two intersecting lines.

Vertical angles are equal. The measures of angles 1 and 3 are equal and the measures of angles 2 and 4 are equal. (See Figure C.1.)

FIGURE C.1 Vertical angles. $\angle 1$ and $\angle 3$ are vertical, and $\angle 2$ and $\angle 4$ are also vertical.

Adjacent angles are two angles that have the same vertex and a common side between them. In Figure C.1, ∠1 and ∠2 are adjacent, and ∠2 and ∠3 are adjacent. Can you name other adjacent pairs?

Parallel Lines

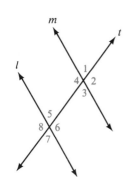

FIGURE C.2 Parallel lines *l* and *m* cut by a transversal *t*

Definition: **Parallel lines** are straight lines in the same plane that do not intersect. A **transversal** of two or more lines is a line that cuts across these lines (see Figure C.2).
Properties:

1. **Interior angles** formed by two lines cut by a transversal are the angles between the two lines.
    a. ∠3, ∠4, ∠5, and ∠6 in Figure C.2 are interior angles.
    b. Adjacent interior angles are supplementary.
    c. If the lines are parallel, then each pair of interior angles on the same side of the transversal is supplementary.

2. **Exterior angles** formed by two lines cut by a transversal are the angles outside the two lines.
    a. ∠1, ∠2, ∠7, and ∠8 in Figure C.2 are exterior angles.
    b. Adjacent exterior angles are supplementary.
    c. If the lines are parallel, then each pair of exterior angles on the same side of the transversal is supplementary.

3. **Corresponding angles** formed by two lines cut by a transversal are the angles on the same side of the transversal and on the same side of the lines.
    a. ∠1 and ∠5, ∠2 and ∠6, ∠3 and ∠7, and ∠4 and ∠8 in Figure C.2 are corresponding angles.
    b. If two lines are parallel, then each pair of corresponding angles is equal.

4. **Alternate interior angles** of two lines cut by a transversal are nonadjacent angles between the two lines and on opposite sides of the transversal.
    a. ∠4 and ∠6, and ∠3 and ∠5 in Figure C.2 are alternate interior angles.
    b. If two lines are parallel, then each pair of alternate interior angles is equal.

Triangles

Definition: A **triangle** is a figure formed by three line segments (called **sides**) joining three noncollinear points (called **vertices**—plural of **vertex**). The **height** or **altitude** of a triangle is the length of a segment drawn from any vertex perpendicular to the opposite side, called the **base.**
    Classification of triangles:

BY SIDES	BY ANGLES
scalene—no equal sides	**acute**—three acute angles
isosceles—at least two equal sides	**right**—one right angle
equilateral—three equal sides	**obtuse**—one obtuse angle

Congruent triangles: **Congruent triangles** are triangles that have the same size and shape (symbolized by ≅). (Remember, in the text it was agreed that when we say "angles are equal," we mean "the measures of the angles are equal," and when we say "line segments are equal," we mean "the measures of the line segments are equal.")

1. Corresponding parts of congruent triangles are equal.
2. (SAS)   If two sides and the included angle of one triangle are equal to the corresponding parts of another, then the triangles are congruent.
3. (ASA)   If two angles and the included side of one triangle are equal to the corresponding parts of another, then the triangles are congruent.
4. (SSS)   If three sides of a triangle are equal to three sides of another, then the triangles are congruent.

Isosceles triangles:

1. If two sides of a triangle are equal, then the angles opposite these sides are equal.
2. If two angles of a triangle are equal, then the sides opposite these angles are equal.
3. An altitude of an isosceles triangle bisects its base.

Equilateral triangles:

1. If a triangle is equilateral, then it is equiangular.
2. If a triangle is equiangular, then it is equilateral.

Angles of a triangle:

1. The sum of the angles of a triangle is π or 180°.
2. If two angles of one triangle are equal to two angles of another triangle, respectively, then the remaining angles are equal.
3. The acute angles of a right triangle are complementary.
4. A triangle can have no more than one right angle.
5. A triangle can have no more than one obtuse angle.

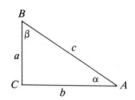

FIGURE C.3 Pythagorean theorem. If C is a right angle in $\triangle ABC$, then $a^2 + b^2 = c^2$.

Right triangles:

1. *Pythagorean theorem:*   In a right triangle, the square of the hypotenuse equals the sum of the squares of the legs. (See Figure C.3.)
2. 30°–60°–90° *triangle:*   The leg opposite the 30° angle equals one-half the hypotenuse, and the leg opposite the 60° angle equals one-half the hypotenuse times the square root of 3. In Figure C.3, if $\alpha = 30°$, then $\beta = 60°$ and $a = \frac{1}{2}c$ and $b = a\sqrt{3}$.
3. 45°–45°–90° *triangle:*   The leg opposite a 45° angle equals one-half the hypotenuse times the square root of 2. In Figure C.3, if $\alpha = 45°$, then $\beta = 45°$ and $a = \frac{1}{2}c\sqrt{2}$ and $b = \frac{1}{2}c\sqrt{2}$.

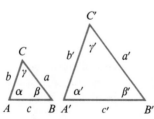

FIGURE C.4 Similar triangles. If $\triangle ABC \sim \triangle A'B'C'$, then $\alpha = \alpha'$, $\beta = \beta'$, $\gamma = \gamma'$, and
$$\frac{a}{a'} = \frac{b}{b'} = \frac{c}{c'} = \cdots.$$

Similar triangles: Triangles whose corresponding angles are congruent and whose corresponding sides are in proportion (symbolized by ∼) are similar triangles.

1. The **ratio** of two quantities is the first divided by the second.
2. A **proportion** is a statement of equality of two ratios. (See Figure C.4.)
3. Two triangles are similar if two angles of one triangle are equal, respectively, to two angles of the other.
4. If two triangles are similar, then their corresponding sides are proportional:

$$\frac{a}{a'} = \frac{b}{b'}; \frac{a}{a'} = \frac{c}{c'}; \frac{b}{b'} = \frac{c}{c'}; \frac{a}{b} = \frac{a'}{b'}; \cdots$$

Parallelograms

Definition: A **parallelogram** is a quadrilateral whose opposite sides are parallel.
Properties:

1. The opposite sides of a parallelogram are parallel.
2. A diagonal of a parallelogram divides it into two congruent triangles.
3. The opposite sides of a parallelogram are equal.
4. The opposite angles of a parallelogram are equal.
5. The consecutive angles of a parallelogram are supplementary.

Circles

Definition: A **circle** is the set of all points in a plane a given distance (called the **radius**) from a given point (called the **center**). (See Figure C.5.)
Terminology:

1. **Circumference:**   distance around the circle, $2\pi r$
2. **Chord:**   a line joining two points of the circle
3. **Diameter:**   a chord through the center; AB in Figure C.5
4. **Arc:**   part of a circle: $\overset{\frown}{BC}$, $\overset{\frown}{AC}$, or $\overset{\frown}{ACB}$ in Figure C.5
5. To **intercept an arc** is to cut off the arc; in Figure C.5, $\angle COB$ intercepts $\overset{\frown}{BC}$.
6. **Tangent** of a circle is a line that intersects the circle at one and only one point.
7. **Secant** of a circle is a line that intersects the circle at two points.
8. **Inscribed polygon** is a polygon all of whose sides are chords of a circle.
9. **Inscribed circle** is a circle to which all the sides of a polygon are tangents.
10. **Circumscribed polygon** is a polygon all of whose sides are tangents to a circle.
11. **Circumscribed circle** is a circle passing through each vertex of a polygon.

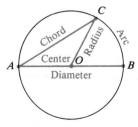

FIGURE C.5 Circle terminology

D

SELECTED ANSWERS

1.1 Problem Set, page 8

1. a. theta **b.** alpha **c.** phi **d.** omega **e.** mu
3. a. δ **b.** ϕ **c.** θ **d.** ω **e.** μ

5. a. 60° **b.** 180° **c.** 45°

d. 360° **e.** 300°

7. a. 120° **b.** 210° **c.** 135°

d. 330° **e.** 240°

9. a. −60° **b.** 315° **c.** −330°

d. −45° **e.** −90°

11. a. 40° **b.** 180° **c.** 80° **d.** 30° **e.** 120°
13. a. 240° **b.** 10° **c.** 220° **d.** 280° **e.** 40°
15. a. 305° **b.** 40° **c.** 90° **d.** 320° **e.** 190°
17. 65.67° **19.** 85.33° **21.** 240.50° **23.** 16.70°
25. 143.38° **27.** 128.178° **29.** 48.469° **31.** 210.671°
33. .645° **35.** .102° **37.** Answers vary.

1.2 Problem Set, page 15

1. a. $\frac{\pi}{6}$ **b.** $\frac{\pi}{2}$ **c.** $\frac{3\pi}{2}$ **d.** $\frac{\pi}{4}$ **e.** 2π **f.** $\frac{\pi}{3}$ **g.** π

3. a. $\frac{\pi}{2}$ **b.** $\frac{\pi}{6}$

c. 2.5 **d.** 6

5. a. $-\frac{\pi}{4}$ **b.** $\frac{7\pi}{6}$

c. −2.76 **d.** $\sqrt{17}$

7. a. π **b.** $\frac{\pi}{6}$ **c.** π **d.** $7 - 2\pi \approx .7168$
9. a. $9 - 2\pi \approx 2.7168$ **b.** $2\pi - 5 \approx 1.2832$
    c. $\sqrt{50} - 2\pi \approx .7879$ **d.** $2\pi - 6 \approx .2832$
11. a. $\frac{2\pi}{9}$ **b.** $\frac{\pi}{9}$ **13. a.** $\frac{127\pi}{90}$ **b.** $\frac{17\pi}{36}$
15. a. −1.10 **b.** 6.11 **17. a.** 40.00° **b.** 18.00°
19. a. −171.89° **b.** −14.32° **21.** 14.04 cm **23.** 70.69 cm
25. 12.57 dm **27.** 87 cm **29.** 2,900 mi **31.** 10
33. about 6,366 km **35.** about 890 km

1.3 Problem Set, pages 25–27

1. Let θ be an angle in standard position with the point (a, b) the intersection of the terminal side of θ and the unit circle. Then the six trigonometric functions are defined as follows: $\cos \theta = a$; $\sin \theta = b$; $\tan \theta = \frac{b}{a}, a \neq 0$; $\sec \theta = \frac{1}{a}, a \neq 0$; $\csc \theta = \frac{1}{b}, b \neq 0$; and $\cot \theta = \frac{a}{b}, b \neq 0$.

3. $\cos \theta = \pm\sqrt{1 - \sin^2\theta}$ **5.** $\cot \theta = \pm\sqrt{\csc^2\theta - 1}$
7. a. −.3 **b.** −.3 **c.** −.4 **d.** −7.0
9. a. .5592 **b.** 1.7883 **c.** .2679 **d.** .8788
11. a. .8776 **b.** 4.0420 **c.** 2.1804 **d.** 1.3154
13. a. $\sec 3$ **b.** $\cos 1$ **c.** $\tan 2$ **d.** $\frac{1}{\csc 19°}$ **e.** $\frac{1}{\cos 181°}$
    f. $\frac{1}{\tan 210°}$

15. a. I, IV **b.** III, IV **c.** I, II **d.** II, III
17. a. II **b.** II **c.** III **d.** II
19. a. $\cot(A + B)$ **b.** $\cos 75°$ **c.** $-\sec 210°$
21. a. 1 **b.** 1 **c.** -1
23. $\cos \theta = \cos \theta$; $\sec \theta = \dfrac{1}{\cos \theta}$; $\sin \theta = \pm\sqrt{1 - \cos^2\theta}$;

$\csc \theta = \dfrac{\pm 1}{\sqrt{1 - \cos^2\theta}}$; $\tan \theta = \dfrac{\pm\sqrt{1 - \cos^2\theta}}{\cos \theta}$;

$\cot \theta = \dfrac{\pm\cos \theta}{\sqrt{1 - \cos^2\theta}}$

25. $\cot \theta = \cot \theta$; $\tan \theta = \dfrac{1}{\cot \theta}$; $\csc \theta = \pm\sqrt{1 + \cot^2\theta}$;

$\sin \theta = \dfrac{\pm 1}{\sqrt{1 - \cot^2\theta}}$; $\cos \theta = \dfrac{\pm\cot \theta}{\sqrt{1 + \cot^2\theta}}$;

$\sec \theta = \dfrac{\pm\sqrt{1 + \cot^2\theta}}{\cot \theta}$

27. $\csc \theta = \csc \theta$; $\sin \theta = \dfrac{1}{\csc \theta}$; $\cot \theta = \pm\sqrt{\csc^2\theta - 1}$;

$\tan \theta = \dfrac{\pm 1}{\sqrt{\csc^2\theta - 1}}$; $\sec \theta = \dfrac{\pm\csc \theta}{\sqrt{\csc^2\theta - 1}}$;

$\cos \theta = \dfrac{\pm\sqrt{\csc^2\theta - 1}}{\csc \theta}$

29. $\cos \theta \approx .92$; $\sin \theta \approx .38$; $\tan \theta \approx .42$; $\sec \theta \approx 1.08$;
$\csc \theta = 2.60$; $\cot \theta = 2.40$
31. $\cos \theta \approx .78$; $\sin \theta \approx .62$; $\tan \theta \approx .80$; $\sec \theta \approx 1.28$;
$\csc \theta = 1.60$; $\cot \theta = 1.25$
33. $\cot \theta = \dfrac{a}{b}$ Definition

$\quad = \dfrac{\cos \theta}{\sin \theta}$ Definition and substitution

35. $a^2 + b^2 = 1$

$\dfrac{a^2}{b^2} + \dfrac{b^2}{b^2} = \dfrac{1}{b^2}$ Divide both sides by b^2.

$\left(\dfrac{a}{b}\right)^2 + 1 = \left(\dfrac{1}{b}\right)^2$ Properties of exponents

$(\cot \theta)^2 + 1 = (\csc \theta)^2$ Definition and substitution

$\cot^2\theta + 1 = \csc^2\theta$

37. Answers vary.

1.4   Problem Set, pages 31–32

1. a. 30° **b.** 60° **c.** 60° **d.** 60° **e.** 50°
3. a. 40° **b.** 70° **c.** 20° **d.** 50° **e.** 30°
5. a. 30° **b.** 60° **c.** 90° **d.** 60° **e.** 40°
7. a. .642788 **b.** .342020 **c.** 5.671282 **d.** .529919
e. .087156
9. a. 1.220775 **b.** 1.345633 **c.** 4.346886 **d.** 3.765682
e. 3.077684
11. a. $-.363970$ **b.** $-.642788$ **c.** .642788 **d.** 1.051462
e. -1.279942
13. a. $-\cos 70°$ **b.** $-\sin 40°$ **c.** $\tan 80°$ **d.** $\sin 32°$
e. $-\cos 5°$

15. a. $-\cos 80°$ **b.** $-\tan 88°$ **c.** $-\sin 84°$ **d.** $-\cos 42°$
e. $\tan 29°$
17. a. $\cot 85°$ **b.** $\sec 5°$ **c.** $\csc 28°$ **d.** $-\cot 15°$
e. $-\sec 13°$
19. $x = 3, y = 4, r = 5$

$\cos \theta = \dfrac{3}{5}$ $\sec \theta = \dfrac{5}{3}$

$\sin \theta = \dfrac{4}{5}$ $\csc \theta = \dfrac{5}{4}$

$\tan \theta = \dfrac{4}{3}$ $\cot \theta = \dfrac{3}{4}$

21. $x = -5, y = -12, r = 13$

$\cos \theta = \dfrac{-5}{13}$ $\sec \theta = \dfrac{-13}{5}$

$\sin \theta = \dfrac{-12}{13}$ $\csc \theta = \dfrac{-13}{12}$

$\tan \theta = \dfrac{12}{5}$ $\cot \theta = \dfrac{5}{12}$

23. $x = -6, y = 1, r = \sqrt{37}$

$\cos \theta = \dfrac{-6}{\sqrt{37}} = \dfrac{-6}{37}\sqrt{37}$ $\sec \theta = -\dfrac{\sqrt{37}}{6}$

$\sin \theta = \dfrac{1}{\sqrt{37}} = \dfrac{1}{37}\sqrt{37}$ $\csc \theta = \sqrt{37}$

$\tan \theta = -\dfrac{1}{6}$ $\cot \theta = -6$

25. a. .5000 **b.** 1.7321 **c.** $-.2500$ **d.** .5000
27. a. 1.0000 **b.** 1.0000 **c.** 1.0000
29. a. .5000 **b.** .2969 **c.** .5000
31. $\theta^g = \dfrac{10}{9}\theta°$ or $\theta° = .9\theta^g$
33. a. .2181 **b.** .2639 **c.** 1.4229

1.5   Problem Set, pages 36–37

1. a. .8415 **b.** .9801 **c.** .3093 **d.** 1.1395 **e.** 1.3940
3. a. .9004 **b.** 4.4552 **c.** .8820 **d.** 1.2658 **e.** 1.7465
5. a. 2.1850 **b.** $-.9975$ **c.** .6600 **d.** -1.6310
e. -4.0420
7. a. 4.8097 **b.** .8855 **c.** $-.5440$ **d.** 1.5882
e. -3.2361
9. a. 1 **b.** 1 **c.** $\dfrac{\sqrt{3}}{2}$ **d.** 2 **e.** undefined
11. a. 1 **b.** 1 **c.** 0 **d.** $\dfrac{2}{\sqrt{3}} = \dfrac{2}{3}\sqrt{3}$ **e.** $\dfrac{\sqrt{2}}{2}$
13. a. 1 **b.** -1 **c.** -1 **d.** 0 **e.** undefined
15. a. $\dfrac{\sqrt{3}}{3}$ **b.** 0 **c.** 0 **d.** 2 **e.** undefined
17. a. -1 **b.** $-\dfrac{\sqrt{2}}{2}$ **c.** $-\dfrac{\sqrt{2}}{2}$ **d.** 1 **e.** -1
19. a. -1 **b.** $\dfrac{1}{2}$ **c.** $\dfrac{\sqrt{2}}{2}$ **d.** 1 **e.** $\dfrac{\sqrt{2}}{2}$
21. a. $\dfrac{\sqrt{3}}{2}$ **b.** $\dfrac{1}{2}$ **c.** $-\sqrt{3}$ **d.** 2 **e.** 2
23. a. 1 **b.** $\dfrac{1}{2}$ **25. a.** $\sqrt{3}$ **b.** $\dfrac{2\sqrt{3}}{3}$ **27. a.** $\dfrac{1}{2}$ **b.** $\dfrac{1}{2}$

29. a. $\frac{\sqrt{3}}{2}$ **b.** $\frac{\sqrt{3}}{2}$ **31. a.** $\frac{\sqrt{3}}{2}$ **b.** $\frac{\sqrt{3}}{2}$

33. $\cos = 1 - 1^2/2! + 1^4/4! - 1^6/6! + \cdots$
$\approx 1 - .5 + .041667 - .001389 + \cdots$
$\approx .5403$

Chapter 1 Summary and Review

Review of Objectives, pages 39–40

1. a. alpha **b.** theta **c.** gamma **d.** lambda **e.** beta

2. a.

b. 45° **c.** 60°

d. 180° **e.** 300° **3.** 50.60°

4. a. 40° **b.** 210° **c.** $7 - 2\pi \approx .7168$ **d.** $\frac{7\pi}{6}$ **e.** π

5. a. 1 **b.** 3 **c.** $\frac{3\pi}{4}$

d. $\frac{2\pi}{3}$ **e.** $\frac{5\pi}{6}$

6. 8.73 **7.** $-218°$ **8.** 50π m; 157 m
9. $(\cos \alpha, \sin \alpha)$ **10.** See inside back cover.
11.

$a^2 + b^2 = 1$ Pythagorean theorem

$\dfrac{a^2}{a^2} + \dfrac{b^2}{a^2} = \dfrac{1}{a^2}$ Divide both sides by a^2.

$1 + \left(\dfrac{b}{a}\right)^2 = \left(\dfrac{1}{a}\right)^2$ Property of exponents

$1 + (\tan \theta)^2 = (\sec \theta)^2$ Definition of tangent and secant
$1 + \tan^2\theta = \sec^2\theta$

12. -1
13. $\cos 340° \approx .9$; $\sin 340° \approx -.3$; $\tan 340° \approx -.3$; $\sec 340° \approx 1.1$;
$\csc 340° \approx -2.9$; $\cot 340° \approx -2.7$
14. a. all **b.** none **c.** sine and cosecant **d.** cosine, secant,
tangent, and cotangent **e.** tangent and cotangent **f.** cosine,
secant, sine, and cosecant **g.** cosine and secant **h.** sine,
cosecant, tangent, and cotangent
15. $\cos 42.5° \approx .7373$; $\sin 42.5° \approx .6756$; $\tan 42.5° \approx .9163$;
$\sec 42.5° \approx 1.3563$; $\csc 42.5° \approx 1.4802$; $\cot 42.5° \approx 1.0913$
16. Let θ be any angle in standard position with a point $P(x, y)$
on the terminal side a distance of r from the origin $(r \neq 0)$.
Then, $\cos \theta = \dfrac{x}{r}$; $\sin \theta = \dfrac{y}{r}$; $\tan \theta = \dfrac{y}{x}$, $x \neq 0$; $\sec \theta = \dfrac{r}{x}$, $x \neq 0$;

$\csc \theta = \dfrac{r}{y}$, $y \neq 0$; $\cot \theta = \dfrac{x}{y}$, $y \neq 0$.

17. $\cos \delta = \dfrac{-1}{3}$; $\sin \delta = \dfrac{2\sqrt{2}}{3}$; $\tan \delta = -2\sqrt{2}$; $\sec \delta = -3$;
$\csc \delta = \dfrac{3\sqrt{2}}{4}$; $\cot \delta = \dfrac{-\sqrt{2}}{4}$

18. a. 40° **b.** 60° **c.** $\frac{\pi}{3}$ **d.** $\frac{\pi}{6}$ **e.** 82°
19. a. $-\cos 20°$ **b.** $-\sin 20°$ **c.** $-\tan 12°$ **d.** $-\cot 20°$
e. $-\csc 60°$
20. a. $-.4161$ **b.** $-.7568$ **c.** 1.1383 **d.** -1.0025
e. -6.8729

21. a. $\frac{1}{2}$ **b.** 0 **c.** $-\frac{\sqrt{3}}{2}$ **d.** $\frac{1}{2}$ **e.** $-\frac{\sqrt{3}}{2}$ **f.** $\frac{1}{2}$ **g.** -1

h. $-\frac{\sqrt{2}}{2}$ **i.** $\frac{\sqrt{3}}{2}$ **j.** 1 **k.** $\frac{1}{2}$ **l.** $-\frac{\sqrt{3}}{2}$ **m.** $\frac{1}{2}$ **n.** $\frac{\sqrt{3}}{2}$

o. 0 **p.** $-\frac{\sqrt{2}}{2}$ **q.** $\sqrt{3}$ **r.** undefined **s.** $-\frac{\sqrt{3}}{3}$

t. $-\sqrt{3}$ **u.** $-\frac{\sqrt{3}}{3}$ **v.** $\sqrt{3}$ **w.** 0 **x.** 1

Miscellaneous Problems, pages 40–41

1. $(2 \cos \alpha, 2 \sin \alpha)$ **3.** $(\cos 23°, \sin 23°)$ **5.** 2.62
7. -1.75 **9.** 120° **11.** $-143°$ **13.** 30 in. **15.** 7 ft
17. I; 15° **19.** II; 95° **21.** .9965 **23.** $-.4161$
25. $-.9365$ **27.** -3.4364 **29.** $-.6873$ **31.** .4794
33. $\cos \theta = \dfrac{4}{\sqrt{41}}$; $\sin \theta = \dfrac{5}{\sqrt{41}}$; $\tan \theta = \dfrac{5}{4}$; $\sec \theta = \dfrac{\sqrt{41}}{4}$; $\csc \theta = \dfrac{\sqrt{41}}{5}$;

$\cot \theta = \dfrac{4}{5}$

35. $\cos \theta = \dfrac{5}{6}$; $\sin \theta = \dfrac{\sqrt{11}}{6}$; $\tan \theta = \dfrac{\sqrt{11}}{5}$; $\sec \theta = \dfrac{6}{5}$; $\csc \theta = \dfrac{6}{\sqrt{11}}$;

$\cot \theta = \dfrac{5}{\sqrt{11}}$

37. $\frac{\sqrt{3}}{2}$ **39.** $\sqrt{3}$ **41.** $\frac{-\sqrt{2}}{2}$ **43.** -1 **45.** 1 **47.** 0

49. $-\frac{2}{\sqrt{3}}$ **51.** $\sqrt{3}$

Application for Further Study:

Angular and Linear Velocity, page 47

1. 8 radians per second

3. The angular velocity is $3,600\pi$ radians per minute or 678,584 radians per hour, and the linear velocity is 1,357 kph.

5. It is at the top of the stroke.

7. $\omega \approx 4.73$ radians per hour

9. The angular velocity is $4,800\pi$ radians per hour, and the linear velocity is about 357 mph.

11. 6.8 cm from the top of the stroke.

2.1 **Problem Set, pages 57–58**

1. a. undefined **b.** 0 **c.** 0 **d.** $\frac{\sqrt{3}}{2}$ **e.** undefined **f.** 0

3. a. undefined **b.** -1 **c.** 0 **d.** 1 **e.** 1 **f.** $\frac{1}{2}$

5. a. 1 **b.** $\sqrt{2}$ **c.** $\sqrt{3}$ **d.** 1 **e.** 0 **f.** -1

7. See Figure 2.4 in text.

9.

11.

13.

15.

17.

19.

21. $(-1, .54)$, $(0, 1)$, $(1, .54)$, $(2, -.42)$, $(3, -.99)$, $(4, -.65)$,
$(5, .28)$, $(6, .96)$, $(7, .75)$

23. $(-1, -1.56)$, $(0, 0)$, $(1, 1.56)$, $(2, -2.19)$, $(3, -.14)$, $(4, 1.16)$,
$(5, -3.38)$, $(6, -.29)$, $(7, .87)$

25. $(-1, 1.85)$, $(0, 1)$, $(1, 1.85)$, $(2, -2.40)$, $(3, -1.01)$, $(4, -1.53)$,
$(5, 3.53)$, $(6, 1.04)$, $(7, 1.33)$

27. $(-1, -1.19)$, undefined at $x = 0$; $(1, 1.19)$, $(2, 1.10)$, $(3, 7.09)$,
$(4, -1.32)$, $(5, -1.04)$, $(6, -3.58)$, $(7, 1.52)$

29. $(-1, -.64)$, undefined at $x = 0$, $(1, .64)$, $(2, -.46)$, $(3, -7.02)$,
$(4, .86)$, $(5, -.30)$, $(6, -3.44)$, $(7, 1.15)$

31. Answers vary.

33.

35. a. Answers vary. **b.** Cut A will produce an edge that is a line;
cut B will produce an edge that is a sine or a cosine curve.

37. $\cos \theta = \dfrac{x'}{r'}$ By definition

$\quad = \dfrac{|CQ|}{|PC|}$

$\quad = |CQ|$ Since $|PC| = 1$

$\quad = |PR|$ Since $|CQ| = |PR|$

39. See Figure 2.2 in text. **41.** See Figure 2.5 in text.

2.2 **Problem Set, pages 67–68**

1.

3.

5.

7.

9.

11.

13.

15.

17.

19.

21.

23.

25.

27.

15.

17.

29.

31.

19.

21.

33.

35.

23.

25.

37. $y = 60 \sin \frac{\pi}{15}(x - 7.5)$

39. a.

27.

29.

b. 3,000 miles   **c.** 2 hours

2.4 **Problem Set, pages 77–79**

2.3 **Problem Set, pages 73–74**

1. a. 1   **b.** undefined   **c.** 0   **d.** undefined   **e.** −1   **f.** 0

3. a. $\frac{\sqrt{3}}{3}$ **b.** 2   **c.** $\frac{\sqrt{2}}{2}$ **d.** $\frac{\sqrt{2}}{2}$ **e.** $\sqrt{3}$ **f.** undefined

5. a. $\sqrt{3}$ **b.** 0   **c.** $\frac{2\sqrt{3}}{3}$ **d.** $\sqrt{2}$ **e.** −1   **f.** $\frac{\sqrt{3}}{3}$

7. See inside back cover.   **9.** See inside back cover.

11.

13.

1.

3.

5.

7.

9.

11.

35.

37.

13.

15.

39.

41.

17.

19.

Chapter 2 Summary and Review

Review of Objectives, pages 80–81

1. a.

b.

21.

23.

c.

2. a.

25.

27.

$y = \cos \left(x + \dfrac{\pi}{2} \right)$ $y = \sin x$

$y = \sin x + \cos \left(x + \dfrac{\pi}{2} \right)$

b.

c.

29. a. Sun curve: $y = \cos \dfrac{\pi}{6} x$ **b.** Moon curve: $y = 4 \cos \dfrac{\pi}{6}(x - 6)$

  c. Combined curve: $y = \cos \dfrac{\pi}{6}(x - 6) + 4 \cos \dfrac{\pi}{6}(x - 6) =$

  $5 \cos \dfrac{\pi}{6}(x - 6)$

31.

33.

d.

3. a.

$y = \sin x + \cos x$

b.

7. a.

b.

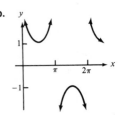

4 a. (h, k) **b.** $2|a|$ **c.** $\frac{2\pi}{b}$ **d.** $\frac{2\pi}{b}$ **e.** $\frac{\pi}{b}$ **f.** amplitude
g. period

c.

8. a.

5. a.

$p = 2\pi$
$a = 1$
$(\frac{\pi}{6}, 2)$

b.

$p = 2\pi$
$a = 1$
$(-\frac{\pi}{4}, 0)$

b.

c.

c.

$p = 2\pi$
$(0, 0)$
$a = \frac{1}{2}$

d.

$p = \pi$
$(\frac{2\pi}{3}, -2)$
$a = 1$

b.

c.

e.

$p = 3\pi$
$(0, 0)$
$a = 2$

6. a.

d.

9. a.

b.

c.

b.

d.

e.

10. a.

b.

Miscellaneous Problems, page 81

1.

3.

5.

7.

9.

11.

13.

15.

17.

19.

21.

23.

25.

27.

29.

Application for Further Study:

Computer Graphing of Trigonometric Curves, pages 87–88

1.

3.

5.

7.

9.

11.

13.

15.

17.

6. a.   **b.**

19.

21.

c.

23.

7.   **8.**

25.

9.   **10.**

Cumulative Review for Chapters 1 and 2

Objective Test, page 89

1. See inside front cover.   **2. a.** sine   **b.** secant
  c. cotangent   **d.** cotangent   **e.** cosecant   **f.** cosine
  g. secant   **h.** sine   **i.** cosine   **j.** cosecant

3. $\cos \alpha = \dfrac{\sqrt{7}}{4}$; $\sin \alpha = \dfrac{3}{4}$; $\tan \alpha = \dfrac{3\sqrt{7}}{7}$; $\sec \alpha = \dfrac{4\sqrt{7}}{7}$; $\csc \alpha = \dfrac{4}{3}$;

  $\cot \alpha = \dfrac{\sqrt{7}}{3}$;   **4.** 2,800 mi   **5. a.** 180°   **b.** −240°

  c. $-\dfrac{\pi}{4}$ **d.** $\dfrac{7\pi}{6}$ **e.** $\dfrac{\pi}{2}$ **f.** $-\dfrac{\sqrt{2}}{2}$ **g.** 0   **h.** $\dfrac{\sqrt{3}}{2}$ **i.** $-\dfrac{1}{2}$

  j. 1   **k.** $\dfrac{\sqrt{2}}{2}$ **l.** −1   **m.** $-\dfrac{1}{2}$ **n.** $-\dfrac{\sqrt{3}}{2}$ **o.** 0

  p. −1   **q.** 0   **r.** $-\sqrt{3}$ **s.** $\dfrac{\sqrt{3}}{3}$ **t.** undefined

Multiple-Choice Test , 90–91

1. $\dfrac{y}{r}; \dfrac{r}{x}, x \neq 0; \dfrac{x}{y}, y \neq 0$ **2.** II   **3.** negative

4. cosine (or secant); tangent (or cotangent)
5. 2π **6.** C   **7.** A   **8.** D   **9.** D   **10.** C
11. A   **12.** C   **13.** C   **14.** E   **15.** E   **16.** A
17. B   **18.** C   **19.** E   **20.** D

3.1 Problem Set, pages 98–99

1. $\dfrac{1 - \cos^2\theta}{\sin\theta} = \dfrac{\sin^2\theta}{\sin\theta}$
$$= \sin\theta$$

3. $\dfrac{\dfrac{\sin\theta}{\cos\theta} + \dfrac{\cos\theta}{\sin\theta}}{\dfrac{1}{\sin\theta\cos\theta}} = \dfrac{\sin^2\theta + \cos^2\theta}{\cos\theta\sin\theta} \cdot \dfrac{\sin\theta\cos\theta}{1}$

5. $\dfrac{\cos\theta + \dfrac{\sin^2\theta}{\cos\theta}}{\sin\theta} = \dfrac{\cos^2\theta + \sin^2\theta}{\cos\theta\sin\theta}$

$\qquad\qquad = \dfrac{1}{\cos\theta\sin\theta}$

7. $\sin\theta + \cot\theta = \sin\theta + \dfrac{\cos\theta}{\sin\theta}$

$\qquad\qquad\qquad = \dfrac{\sin^2\theta + \cos\theta}{\sin\theta}$

9. $\dfrac{\tan\theta + \cot\theta}{\sec\theta\csc\theta} = \dfrac{\dfrac{\sin\theta}{\cos\theta} + \dfrac{\cos\theta}{\sin\theta}}{\dfrac{1}{\cos\theta}\cdot\dfrac{1}{\sin\theta}}$

$\qquad\qquad\qquad = \dfrac{\sin^2\theta + \cos^2\theta}{\cos\theta\sin\theta}\cdot\dfrac{\cos\theta\sin\theta}{1}$

$\qquad\qquad\qquad = 1$

11. $\sec^2\theta + \tan^2\theta = \dfrac{1}{\cos^2\theta} + \dfrac{\sin^2\theta}{\cos^2\theta}$

$\qquad\qquad\qquad = \dfrac{1 + \sin^2\theta}{\cos^2\theta}$

13–41. Proofs vary.

3.2 Problem Set, page 101

Proofs vary.

3.3 Problem Set, page 104

1–10. Answers vary.   **11–31.** Proofs vary.   **11.** identity
13. identity   **15.** identity   **17.** not an identity
19. not an identity   **21.** identity   **23.** identity   **25.** identity
27. identity   **29.** identity   **31.** identity

3.4 Problem Set, pages 111–112

1 $\sqrt{89}$ **3.** $2\sqrt{53}$ **5.** $\sqrt{2 - 2\sin\alpha}$ **7.** $\sin 75°$
9. $\tan 49°$ **11.** $-\sin\dfrac{\pi}{3}$ **13.** $\cos 18°$ **15.** $-\sin 41°$
17. $\cos 39°$
19. a. $\cos\left(\theta - \dfrac{\pi}{3}\right)$ **b.** $-\sin\left(\theta - \dfrac{\pi}{3}\right)$ **c.** $-\tan\left(\theta - \dfrac{\pi}{3}\right)$
21. a. $\cos(\alpha - \beta)$ **b.** $-\sin(\alpha - \beta)$ **c.** $-\tan(\alpha - \beta)$

23.

25.

27.

29.

31. Proof varies.

3.5 Problem Set, pages 114–115

1. a. $\dfrac{\sqrt{3}\cos\theta - \sin\theta}{2}$ **b.** $\dfrac{\sqrt{2}(\sin\theta - \cos\theta)}{2}$

3. a. $\dfrac{1}{2}(\cos\theta + \sqrt{3}\sin\theta)$ **b.** $\dfrac{\sqrt{2}}{2}(\cos\theta + \sin\theta)$

5. a. $\cos^2\theta - \sin^2\theta$ **b.** $2\sin\theta\cos\theta$

7. .9135   **9.** .2079   **11.** .6428   **13.** 1.1918

15. .6009   **17.** $\cos(-15°) = \dfrac{\sqrt{6} + \sqrt{2}}{4}$; $\sin(-15°) = \dfrac{\sqrt{2} - \sqrt{6}}{4}$;

$\tan(-15°) = -2 + \sqrt{3}$ **19.** $\cos 75° = \dfrac{\sqrt{6} - \sqrt{2}}{4}$;

$\sin 75° = \dfrac{\sqrt{6} + \sqrt{2}}{4}$; $\tan 75° = 2 + \sqrt{3}$

21. $\cos 345° = \dfrac{\sqrt{6} + \sqrt{2}}{4}$; $\sin 345° = \dfrac{\sqrt{2} - \sqrt{6}}{4}$;

$\tan 345° = -2 + \sqrt{3}$

23. $\sin(\alpha + \beta) = \sin\alpha\cos\beta + \cos\alpha\sin\beta$ $\qquad$ Identity 17

$\sin[\alpha + (-\beta)] = \sin\alpha\cos(-\beta) + \cos\alpha\sin(-\beta)$ $\quad$ Replace β by $-\beta$.

$\sin(\alpha - \beta) = \sin\alpha\cos\beta + \cos\alpha(-\sin\beta)$ $\qquad$ Identities 12 and 13

$\therefore \sin(\alpha - \beta) = \sin\alpha\cos\beta - \cos\alpha\sin\beta$

25. $\cot(\alpha + \beta) = \dfrac{\cos(\alpha + \beta)}{\sin(\alpha + \beta)}$

$\qquad\qquad = \dfrac{\cos\alpha\cos\beta - \sin\alpha\sin\beta}{\sin\alpha\cos\beta + \cos\alpha\sin\beta}\cdot\dfrac{\dfrac{1}{\sin\alpha\sin\beta}}{\dfrac{1}{\sin\alpha\sin\beta}}$

$\qquad\qquad = \dfrac{\cot\alpha\cot\beta - 1}{\cot\beta\cot\alpha}$

27. $\dfrac{\cos 5\theta}{\sin\theta} - \dfrac{\sin 5\theta}{\cos\theta} = \dfrac{\cos 5\theta\cos\theta - \sin 5\theta\sin\theta}{\sin\theta\cos\theta}$

$\qquad\qquad\qquad = \dfrac{\cos(5\theta + \theta)}{\sin\theta\cos\theta}$

$\qquad\qquad\qquad = \dfrac{\cos 6\theta}{\sin\theta\cos\theta}$

29. $\sin(\alpha + \beta)\cos\beta - \cos(\alpha + \beta)\sin\beta$
$= (\sin\alpha\cos\beta + \cos\alpha\sin\beta)\cos\beta - (\cos\alpha\cos\beta - \sin\alpha\sin\beta)\sin\beta$
$= \sin\alpha\cos^2\beta + \cos\alpha\cos\beta\sin\beta - \cos\alpha\cos\beta\sin\beta + \sin\alpha\sin^2\beta$
$= \sin\alpha(\cos^2\beta + \sin^2\beta)$
$= \sin\alpha$

31–49. Proofs vary.

3.6 Problem Set, pages 121–122

1. $\frac{\sqrt{2}}{2}$ **3.** $\frac{1}{2}$ **5.** -1 **7.** $\frac{1}{2}\sqrt{2-\sqrt{2}}$ **9.** $\sqrt{2}-1$

	$\cos 2\theta$	$\sin 2\theta$	$\tan 2\theta$
11.	$\frac{119}{169}$	$\frac{-120}{169}$	$\frac{-120}{119}$
13.	$\frac{7}{25}$	$\frac{-24}{25}$	$\frac{-24}{7}$
15.	$\frac{-119}{169}$	$\frac{120}{169}$	$\frac{-120}{119}$

	$\cos\frac{1}{2}\theta$	$\sin\frac{1}{2}\theta$	$\tan\frac{1}{2}\theta$
17.	$\frac{\sqrt{26}}{26}$	$\frac{5\sqrt{26}}{26}$	5
19.	$\frac{\sqrt{10}}{10}$	$\frac{3\sqrt{10}}{10}$	3
21.	$\frac{-2\sqrt{13}}{13}$	$\frac{3\sqrt{13}}{13}$	$\frac{-3}{2}$

	$\cos\theta$	$\sin\theta$	$\tan\theta$
23.	$\frac{\sqrt{2}}{2}$	$\frac{\sqrt{2}}{2}$	1
25.	$\frac{1}{2}$	$\frac{\sqrt{3}}{2}$	$\sqrt{3}$
27.	$\frac{3\sqrt{10}}{10}$	$\frac{\sqrt{10}}{10}$	$\frac{1}{3}$

29. a. $r \approx 2.6$ **b.** $r = 2\sqrt{2\sqrt{2}+4}$. This is $\dfrac{2}{\sqrt{2-\sqrt{2}}}$ after is has been rationalized.

31.

33.

35.

37. $\cos 2\theta = \cos^2\theta - \sin^2\theta$
Let $\theta = 2\theta$; then $\cos 4\theta = \cos^2 2\theta - \sin^2 2\theta$.

39. $\tan 2\theta = \dfrac{2\tan\theta}{1-\tan^2\theta}$
Let $\theta = \frac{3\beta}{4}$; then $2\theta = \frac{3\beta}{2}$ and the result is proved.

41. $\tan\frac{1}{2}\theta = \dfrac{1-\cos\theta}{\sin\theta}$ (Problem 40)

$$= \frac{(1-\cos\theta)(1+\cos\theta)}{\sin\theta(1+\cos\theta)}$$

$$= \frac{1-\cos^2\theta}{\sin\theta(1+\cos\theta)}$$

$$= \frac{\sin\theta}{1+\cos\theta}$$

43–51. Proofs vary.

3.7 Problem Set, pages 126–127

1. $\cos 40° + \cos 110°$ **3.** $\cos 11° - \cos 59°$

5. $\frac{1}{2}\cos 18° - \frac{1}{2}\cos 158°$ **7.** $\frac{1}{2}\sin 60° + \frac{1}{2}\sin 22°$

9. $\frac{1}{2}\cos 75° - \frac{1}{2}\cos 165°$ **11.** $\frac{1}{2}\cos 2\theta + \frac{1}{2}\cos 4\theta$

13. $2\sin 53°\cos 10°$ **15.** $-2\sin 80°\sin 1°$

17. $2\sin 257.5°\cos 42.5°$ **19.** $-2\sin\frac{x}{2}\cos\frac{3x}{2}$

21. $2\sin\frac{3x}{2}\cos\frac{x}{2}$ **23.** $2\cos 7y\cos 2y$

25. $2\sin\alpha\sin\beta = \cos(\alpha-\beta) - \cos(\alpha+\beta)$ Product formula

Let $x = \alpha - \beta$ and $y = \alpha + \beta$; then $\alpha = \dfrac{x+y}{2}$ and $\beta = \dfrac{y-x}{2}$.

$2\sin\left(\dfrac{x+y}{2}\right)\sin\left(\dfrac{y-x}{2}\right) = \cos x - \cos y$ Substitution

$-2\sin\left(\dfrac{x+y}{2}\right)\sin\left(\dfrac{x-y}{2}\right) = \cos x - \cos y$ Since $\sin(-\theta) = -\sin$

27. $2\cos\alpha\sin\beta = \sin(\alpha+\beta) - \sin(\alpha-\beta)$ Product formula

Let $x = \alpha + \beta$ and $y = \alpha - \beta$; then $\alpha = \dfrac{x+y}{2}$ and $\beta = \dfrac{x-y}{2}$.

$2\cos\left(\dfrac{x+y}{2}\right)\sin\left(\dfrac{x-y}{2}\right) = \sin x - \sin y$ Substitution

29–31. Answers vary.

Chapter 3 Summary and Review

Review of Objectives, pages 128–129

1. $\dfrac{\sec^2 x + \tan^2 x + 1}{\sec x} = \dfrac{\sec^2 x + (\sec^2 x - 1) + 1}{\sec x}$

$$= \frac{2\sec^2 x}{\sec x}$$

$$= 2\sec x$$

2. $\dfrac{\csc^2\alpha}{1 + \cot^2\alpha} = \dfrac{\csc^2\alpha}{\csc^2\alpha}$

$$= 1$$

3. $\dfrac{1}{\sin\theta + \cos\theta} + \dfrac{1}{\sin\theta - \cos\theta} = \dfrac{\sin\theta - \cos\theta + \sin\theta + \cos\theta}{\sin^2\theta - \cos^2\theta}$

$= \dfrac{2\sin\theta}{\sin^2\theta - \cos^2\theta} \cdot \dfrac{\sin^2\theta + \cos^2\theta}{\sin^2\theta + \cos^2\theta}$

$= \dfrac{2\sin\theta(\sin^2\theta + \cos^2\theta)}{\sin^4\theta - \cos^4\theta}$

$= \dfrac{2\sin\theta}{\sin^4\theta - \cos^4\theta}$

4. $\dfrac{\cos\theta}{\sec\theta} - \dfrac{\sin\theta}{\cot\theta} = \cos^2\theta - \dfrac{\sin^2\theta}{\cos\theta}$

$= \dfrac{\cos^3\theta - \sin^2\theta}{\cos\theta}$ Also,

$\dfrac{\cos\theta\cot\theta - \tan\theta}{\csc\theta} = \dfrac{\cos\theta \cdot \dfrac{\cos\theta}{\sin\theta} - \dfrac{\sin\theta}{\cos\theta}}{\dfrac{1}{\sin\theta}}$

$= \left(\dfrac{\cos^2\theta}{\sin\theta} - \dfrac{\sin\theta}{\cos\theta}\right)\sin\theta$

$= \dfrac{(\cos^3\theta - \sin^2\theta)\sin\theta}{\sin\theta\cos\theta}$

$= \dfrac{\cos^3\theta - \sin^2\theta}{\cos\theta}$

5. $\dfrac{1 + \csc\beta}{\cos\beta\csc\beta} = \dfrac{1}{\cos\beta\csc\beta} + \dfrac{\csc\beta}{\cos\beta\csc\beta}$

$= \dfrac{1}{\dfrac{\cos\beta}{\sin\beta}} + \dfrac{1}{\cos\beta}$

$= \dfrac{\sin\beta}{\cos\beta} + \sec\beta$

$= \tan\beta + \sec\beta$

6. $\dfrac{1 - \cos x}{1 + \cos x} = \dfrac{1 - \cos x}{1 + \cos x} \cdot \dfrac{1 - \cos x}{1 - \cos x}$

$= \dfrac{1 - 2\cos x + \cos^2 x}{1 - \cos^2 x}$

$= \dfrac{1 - 2\cos x + (1 - \sin^2 x)}{\sin^2 x}$

$= \dfrac{2 - 2\cos x - \sin^2 x}{\sin^2 x}$

7. $\dfrac{\sin^2\theta - \cos^2\theta}{\sin\theta + \cos\theta} = \dfrac{(\sin\theta + \cos\theta)(\sin\theta - \cos\theta)}{\sin\theta + \cos\theta}$

$= \sin\theta - \cos\theta$

8. $\dfrac{1 + \tan^2\theta}{\csc\theta} = \dfrac{\sec^2\theta}{\csc\theta}$

$= \sec\theta \cdot \dfrac{\sec\theta}{\csc\theta}$

$= \sec\theta \cdot \dfrac{\sin\theta}{\cos\theta}$

$= \sec\theta\tan\theta$

9. Answers vary.

10. a. False; let $t = 1$. Then $(\sin t + \cos t)^2 \approx 1.9$.
  b. False; let $t = 1$. Then $\cos t\tan t\csc t\sec t \approx 1.85$.

11. $|P_\alpha P_\beta| = \sqrt{(\cos\beta - \cos\alpha)^2 + (\sin\beta - \sin\alpha)^2}$
$= \sqrt{2 - 2(\cos\alpha\cos\beta + \sin\alpha\sin\beta)}$

12. $|AP_\theta| = \sqrt{2 - 2\cos\theta}$

13. a. $\cos 52°$ **b.** $\cot\dfrac{3\pi}{8}$ **c.** $\sin\left(\dfrac{\pi}{2} - 2\right)$ or $-\sin(.4292)$

14. a. $\cos\left(\theta - \dfrac{\pi}{6}\right)$ **b.** $-\sin\left(\theta - \dfrac{\pi}{6}\right)$ **c.** $-\tan\left(\theta - \dfrac{\pi}{6}\right)$

15.

16. $\dfrac{\sqrt{3}}{2}\cos\theta + \dfrac{1}{2}\sin\theta$ **17.** $\dfrac{\sqrt{6} + \sqrt{2}}{4}$

18. Let P_α and P_β be defined as in Problem 11. Then,
$$|P_\alpha P_\beta| = \sqrt{2 - 2(\cos\alpha\cos\beta + \sin\alpha\sin\beta)}$$
The angle between rays through P_α and P_β is $\alpha - \beta$, so
$$|P_\alpha P_\beta| = \sqrt{2 - 2\cos(\alpha - \beta)} \text{ from Problem 12.}$$
Then, $\sqrt{2 - 2\cos(\alpha - \beta)} = \sqrt{2 - 2(\cos\alpha\cos\beta + \sin\alpha\sin\beta)}$
$$2 - 2\cos(\alpha - \beta) = 2 - 2(\cos\alpha\cos\beta + \sin\alpha\sin\beta)$$
$$\cos(\alpha - \beta) = \cos\alpha\cos\beta + \sin\alpha\sin\beta$$

19. $\dfrac{2\tan\dfrac{\pi}{6}}{1 - \tan^2\dfrac{\pi}{6}} = \tan\left(2 \cdot \dfrac{\pi}{6}\right)$

$= \tan\dfrac{\pi}{6}$

$= \sqrt{3}$

20. $\cos 2\theta = \dfrac{7}{25}$; $\sin 2\theta = -\dfrac{24}{25}$; $\tan 2\theta = -\dfrac{24}{7}$

21. $-\sqrt{\dfrac{1 + \cos 240°}{2}} = \cos 120°$

$= -\dfrac{1}{2}$

22. $\cos\theta = \dfrac{1}{10}\sqrt{10}$; $\sin\theta = \dfrac{3}{10}\sqrt{10}$; $\tan\theta = 3$

23. $y = \cos^2 x = \frac{1}{2}\cos 2x + \frac{1}{2}$; $y - \frac{1}{2} = \frac{1}{2}\cos 2x$

24. $\cos 2\alpha = 2\cos^2\alpha - 1$

$2\cos^2\alpha = \cos 2\alpha + 1$

$\cos^2\alpha = \dfrac{\cos 2\alpha + 1}{2}$

$\cos\alpha = \pm\sqrt{\dfrac{1 + \cos 2\alpha}{2}}$

Let $\alpha = \frac{1}{2}\theta$; then $\cos\frac{1}{2}\theta = \pm\sqrt{\dfrac{1 + \cos\theta}{2}}$

25. $\frac{1}{2}\sin 4\theta + \frac{1}{2}\sin 2\theta$ **26.** $2\sin\left(\dfrac{h}{2}\right)\cos\left(\dfrac{2x + h}{2}\right)$

27. $\dfrac{\sin 5\theta + \sin 3\theta}{\cos 5\theta - \cos 3\theta} = \dfrac{2\sin 4\theta\cos\theta}{-2\sin 4\theta\sin\theta}$

$\qquad\qquad\qquad = -\cot\theta$

Miscellaneous Problems, page 130

1 –37. Proofs vary.   **39.** $\dfrac{\sqrt{6} + \sqrt{2}}{4}$ **41.** $\dfrac{1}{2}$ **43.** 1

45. $\sqrt{3}$ **47.** $\dfrac{\sqrt{2}}{2}$ **49.** $\dfrac{-7}{25}$

Application for Further Study:
Sunrise, Sunset, and Linear Interpolation, page 133

1. 16 hr 50 min or 4:50 P.M.   **3.** 19 hr 1 min or 7:01 P.M.
5. 6 hr 18 min or 6:18 A.M.
7.   **11.** Answers vary.

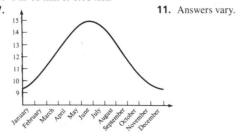

4.1 **Problem Set, pages 145–146**

1. a. $30° + 360°k, 150° + 360°k$ **b.** $60° + 360°k, 300° + 360°k$
3. a. $23° + 360°k, 157° + 360°k$ **b.** $73° + 360°k, 287° + 360°k$

5. a. $\frac{7\pi}{6} + 2\pi k, \frac{11\pi}{6} + 2\pi k$ **b.** $\frac{2\pi}{3} + 2\pi k, \frac{4\pi}{3} + 2\pi k$

7. a. $\frac{4\pi}{3} + 2\pi k, \frac{5\pi}{3} + 2\pi k$ **b.** $\frac{3\pi}{4} + 2\pi k + 2\pi k$

9. a. $\frac{\pi}{6}$ **b.** $30°$ **11. a.** $\frac{\pi}{6}, \frac{5\pi}{6}$ **b.** no solution

13. $0°, 180°$ **15.** $\frac{\pi}{6}, \frac{5\pi}{6}, \frac{7\pi}{6}, \frac{11\pi}{6}$ **17.** $\frac{\pi}{6}, \frac{5\pi}{6}, \frac{\pi}{3}, \frac{5\pi}{3}$

19. $0, \pi, \frac{\pi}{3}, \frac{4\pi}{3}$ **21.** $\frac{\pi}{4}, \frac{3\pi}{4}, \frac{5\pi}{4}, \frac{7\pi}{4}$ **23.** $0, \pi, \frac{\pi}{3}, \frac{5\pi}{3}$
25. 2.2370, 4.0461   **27.** .3649, 1.2059, 3.5065, 4.3475
29. 12.1°   **31.** 3.8079, 5.6169   **33.** .8213, 2.3203

35. $\frac{\pi}{2}$ **37.** $\frac{\pi}{3}, \frac{2\pi}{3}, \frac{4\pi}{3}, \frac{5\pi}{3}$ **39.** $\frac{\pi}{2}, \frac{3\pi}{2}$ **41.** 65

4.2 **Problem Set, pages 150–152**

1. $\frac{\pi}{6}, \frac{5\pi}{6}, \frac{7\pi}{6}, \frac{11\pi}{6}$ **3.** $\frac{\pi}{3}, \frac{2\pi}{3}, \frac{4\pi}{3}, \frac{5\pi}{3}$ **5.** $\frac{2\pi}{3}, \frac{5\pi}{6}, \frac{5\pi}{3}, \frac{11\pi}{6}$

7. $\frac{\pi}{12}, \frac{5\pi}{12}, \frac{3\pi}{4}, \frac{13\pi}{12}, \frac{17\pi}{12}, \frac{7\pi}{4}$ **9.** $\frac{5\pi}{12}, \frac{7\pi}{12}, \frac{17\pi}{12}, \frac{19\pi}{12}$

11. 0, 1.57, 2.09, 3.14, 4.19, 4.71   **13.** .39, 1.96, 3.53, 5.11
15. 1.57   **17.** no solution   **19.** .81, 2.90, 3.95, 6.04
21. .02, 1.59, 3.17, 4.74   **23.** .52, 2.62, 4.71
25. 0.00, .52, 2.62, 3.14
27. 0.00, 1.05, 1.22, 1.92, 2.09, 3.14, 3.32, 4.01, 4.19, 5.24, 5.41, 6.11
29. .52, 1.05, 3.67, 4.19   **31.** .0002
33. 32.5 min and 77.5 min

4.3 **Problem Set, pages 160–161**

1. a. 0   **b.** $\frac{\pi}{6}$ **c.** $\frac{\pi}{2}$ **d.** 0

3. a. $-\frac{\pi}{2}$ **b.** $\frac{3\pi}{4}$ **c.** $-\frac{\pi}{4}$ **d.** $\frac{2\pi}{3}$

5. a. $\frac{\pi}{3}$ **b.** $\frac{\pi}{4}$ **c.** $\frac{\pi}{4}$ **d.** $\frac{3\pi}{4}$ **7. a.** 15°   **b.** 67°   **c.** 20°
9. a. 164°   **b.** −43°   **c.** 111°
11. a. 21°   **b.** 15°   **c.** 135°   **13. a.** .21   **b.** .58   **c.** .33
15. a. 1.94   **b.** 2.50   **c.** −.40

17. a. .17   **b.** .34   **c.** 2.82   **19. a.** $\frac{\pi}{6}$ **b.** $\frac{1}{3}$ **c.** $\frac{2\pi}{15}$

21. $-2\sqrt{2}$ **23.** $\frac{4}{5}$ **25.** $\frac{4}{5}$ **27.** 0   **29.** $\dfrac{2\sqrt{2} + \sqrt{15}}{12}$

31. Let $\theta = \text{Arcsec } x$ where $x \geq 1$ or $x \leq -1$. Then,

$\sec\theta = x$ $0 \leq \theta \leq \pi$ Definition of Arcsec x, $\theta \neq \frac{\pi}{2}$

$\dfrac{1}{\cos\theta} = x$ $0 \leq \theta \leq \pi, \theta \neq \frac{\pi}{2}$

$\cos\theta = \dfrac{1}{x}$ $0 \leq \theta \leq \pi, \theta \neq \frac{\pi}{2}$

$\theta = \text{Arccos}\dfrac{1}{x}$ $0 \leq \theta \leq \pi$ Definition of Arccos x

33. Let $\theta = $ Arctan x where $x \geq 0$. Then,

$\tan \theta = x$ $\qquad 0 \leq \theta < \frac{\pi}{2}$ $\quad$ Definition of Arctan x

$\sec^2\theta = \tan^2\theta + 1$ $\qquad$ Fundamental identity

$\sec \theta = \sqrt{\tan^2\theta + 1}$ $\qquad$ Positive value since $0 \leq \theta < \frac{\pi}{2}$

By substitution, $\sec(\text{Arctan } x) = \sqrt{x^2 + 1}$.

35. Let $\theta = $ Arctan x where $x \geq 0$. Then,

$\tan \theta = x$ $\qquad 0 \leq \theta < \frac{\pi}{2}$ $\quad$ Definition of Arctan x

$\sec \theta = \sqrt{x^2 + 1}$ $\qquad$ From Problem 33

$\cos \theta = \dfrac{1}{\sqrt{x^2 + 1}}$ $\qquad$ Since $\sec \theta = \dfrac{1}{\cos \theta}$

By substitution, $\cos(\text{Arctan } x) = \dfrac{1}{\sqrt{1 + x^2}}$.

37. Answers vary.

4.4 Problem Set, page 164

1. $\sqrt{2} \sin\left(\theta + \frac{\pi}{4}\right)$ $\quad$ **3.** $\sqrt{2} \sin\left(\frac{\theta}{2} + \frac{3\pi}{4}\right)$ $\quad$ **5.** $\sin\left(\theta + \frac{\pi}{3}\right)$

7. $\sin\left(\theta + \frac{2\pi}{3}\right)$ $\quad$ **9.** $\sqrt{2} \sin\left(\theta + \frac{5\pi}{4}\right)$ $\quad$ **11.** $2 \sin\left(\pi\theta + \frac{2\pi}{3}\right)$

13. $\sqrt{13} \sin(\theta + 1.0)$ $\quad$ **15.** $\sqrt{13} \sin(\theta + 2.6)$

17. $5 \sin(\theta + 2.2)$ $\quad$ **19.** $17 \sin(\theta + 4.2)$

21. $\sqrt{29} \sin(\theta + 2.0)$

23.

25.

27.

29.

31.

33.

Chapter 4 Summary and Review

Review of Objectives, pages 165–166

1. $1, 1 + \pi$ $\quad$ **2.** $\frac{\pi}{3}, \frac{2\pi}{3}, \frac{4\pi}{3}, \frac{5\pi}{3}$ $\quad$ **3.** $1.25, 2.36, 4.39, 5.50$

4. $.31, 2.83$ $\quad$ **5.** $.48, 1.09, 2.05, 2.66, 3.62, 4.23, 5.19, 5.81$

6. $\frac{\pi}{6}, \frac{\pi}{3}, \frac{2\pi}{3}, \frac{5\pi}{6}, \frac{7\pi}{6}, \frac{4\pi}{3}, \frac{5\pi}{3}, \frac{11\pi}{6}$

7. a. $30°$ $\quad$ **b.** $45°$ $\quad$ **c.** $60°$ $\quad$ **d.** $45°$ $\quad$ **e.** $\frac{3\pi}{4}$ $\quad$ **f.** $\frac{5\pi}{6}$

$\quad$ **g.** $\frac{\pi}{6}$ $\quad$ **h.** $\frac{\pi}{6}$ $\quad$ **i.** 1.16 $\quad$ **j.** 1.28

8. a. I $\quad$ **b.** θ between -1.57 and 0 $\quad$ **c.** I

$\quad$ **d.** θ between 0 and 1.57 $\quad$ **e.** II $\quad$ **f.** θ between 0 and 1.57

$\quad$ **g.** IV $\quad$ **h.** θ between -1.57 and 0 $\quad$ **i.** II

$\quad$ **j.** θ between 1.57 and 3.14

9. $2\sqrt{2}$ $\quad$ **10. a.** I $\quad$ **b.** III $\quad$ **c.** IV $\quad$ **d.** II

$\quad$ **e.** $13 \sin(\theta + 5.9)$

11.

Miscellaneous Problems, page 166

1. $-1 \leq x \leq 1; 0 \leq y \leq \pi$ $\quad$ **3.** all reals; $-\frac{\pi}{2} < y < \frac{\pi}{2}$

5. $x \geq 1$ or $x \leq -1$; $-\frac{\pi}{2} \leq y \leq \frac{\pi}{2}, y \neq 0$

7. $\frac{4\pi}{3} + 2\pi k, \frac{5\pi}{3} + 2\pi k$ $\quad$ **9.** $.95 + 6.28k, 2.19 + 6.28k$

11. $1.94 + 6.28k, 4.34 + 6.28k$ $\quad$ **13.** $\frac{3\pi}{4}$ $\quad$ **15.** 2.83

17. 1.26 $\quad$ **19.** $-\frac{\pi}{4}$ $\quad$ **21.** 1.80 $\quad$ **23.** 2.11 $\quad$ **25.** $-.86$

27. $-.28$ $\quad$ **29.** $.28$ $\quad$ **31.** $\frac{\pi}{6}, \frac{5\pi}{6}$ $\quad$ **33.** 0 $\quad$ **35.** $1.32, 4.97$

37. $1.23, 5.05, \frac{2\pi}{3}, \frac{4\pi}{3}$ $\quad$ **39.** $\frac{\pi}{2}, \frac{3\pi}{2}$ $\quad$ **41.** $\frac{\pi}{3}, \frac{5\pi}{3}, .84, 5.44$

43. no solution $\quad$ **45.** $.73, 2.41$ $\quad$ **47.** $.98, 4.12, .46, 3.61$

49. a. $\frac{\pi}{3}, \frac{5\pi}{3}$ $\quad$ **b.** $\frac{\pi}{2} + 1$ (or 2.57), $\frac{3\pi}{2} + 1$ (or 5.71)

Application for Further Study:

Waves, Harmonic Motion, and Resonance, pages 173–174

1. $y = 3 \sin .03(x - 30.54t)$

3. 24 ft from trough to crest; wavelength $\lambda \approx \frac{2\pi}{1.32} \approx 4.76$ ft; phase velocity is about 4.75 ft/sec ≈ 3.24 mph.

5. $t = .21x - .16 \sin^{-1}\left(\frac{y}{12}\right)$

7. Since $\omega = 10\pi, n = 5$, and $T = \frac{1}{5}$, we know that $y = -\sqrt{3}$ when $t = 0$. Thus, the period is $\frac{1}{5}$, the frequency is 5 cycles per unit of time, and the maximum distance from the origin is $\sqrt{3}$.

9. a.

b.

c.

11. Maximum value is 60, minimum is -60.

13. Resonance

15. Harmonic motion

17. Harmonic motion

4. $(\cos \beta - \cos \alpha)^2 + (\sin \beta - \sin \alpha)^2$
$= \cos^2\beta - 2 \cos \alpha \cos \beta + \cos^2\alpha + \sin^2\beta$
$\qquad - 2 \sin \alpha \sin \beta + \sin^2\alpha$
$= (\cos^2\beta + \sin^2\beta) + (\cos^2\alpha + \sin^2\alpha)$
$\qquad - 2(\cos \alpha \cos \beta + \sin \alpha \sin \beta)$
$= 1 + 1 - 2(\cos \alpha \cos \beta + \sin \alpha \sin \beta)$
$= 2 - 2(\cos \alpha \cos \beta + \sin \alpha \sin \beta)$

5. $\dfrac{\sin 3\theta + \cos 3\theta + 1}{\cos 3\theta} = \dfrac{\sin 3\theta}{\cos 3\theta} + \dfrac{\cos 3\theta}{\cos 3\theta} + \dfrac{1}{\cos 3\theta}$

$\qquad = \tan 3\theta + 1 + \sec 3\theta$

6. $\sec A - \tan^2 A = \dfrac{1}{\cos A} - \dfrac{\sin^2 A}{\cos^2 A}$

$\qquad = \dfrac{\cos A - \sin^2 A}{\cos^2 A}$

$\qquad = \dfrac{\cos A - (1 - \cos^2 A)}{\cos^2 A}$

$\qquad = \dfrac{1}{\cos^2 A}(\cos^2 A + \cos A - 1)$

$\qquad = \sec^2 A(\cos^2 A + \cos A - 1)$

7. $\tan 2\theta \cos 2\theta = \dfrac{\sin 2\theta}{\cos 2\theta} \cdot \cos 2\theta$

$\qquad = \sin 2\theta$

$\qquad = 2 \sin \theta \cos \theta$

8. .308, 2.834

9.

10. $\cos \theta = \dfrac{4}{5}$; $\sin \theta = -\dfrac{3}{5}$; $\tan \theta = -\dfrac{3}{4}$; $\sec \theta = \dfrac{5}{4}$; $\csc \theta = -\dfrac{5}{3}$; $\cot \theta = -\dfrac{4}{3}$

Cumulative Review for Chapters 3 and 4

Objective Test, page 175

1. See endpaper facing inside front cover.

2. $x^2 + y^2 = r^2$ — Pythagorean theorem

$\dfrac{x^2}{y^2} + \dfrac{y^2}{y^2} = \dfrac{r^2}{y^2}$ — Divide both sides by y^2, $y \neq 0$.

$\left(\dfrac{x}{y}\right)^2 + 1 = \left(\dfrac{r}{y}\right)^2$ — Properties of exponents

$\cot^2\theta + 1 = \csc^2\theta$ — Definition of trigonometric functions

3. a. $A(1, 0)$, $P_\theta(\cos \theta, \sin \theta)$ **b.** $\sqrt{2 - 2 \cos \theta}$

Multiple-Choice Test, pages 176–177

1. $\cos \theta \sec \theta = 1$; $\sin \theta \csc \theta = 1$; $\tan \theta \cot \theta = 1$

2. $\tan \theta = \dfrac{\sin \theta}{\cos \theta}$; $\cot \theta = \dfrac{\cos \theta}{\sin \theta}$

3. $\cos^2\theta + \sin^2\theta = 1$; $1 + \tan^2\theta = \sec^2\theta$; $\cot^2\theta + 1 = \csc^2\theta$

4. $x^2 + y^2 = r^2$ by the Pythagorean theorem.

5. Multiply by the conjugate.   **6.** B   **7.** E   **8.** D

9. D   **10.** A   **11.** A   **12.** A   **13.** B   **14.** D

15. E   **16.** C   **17.** B   **18.** D   **19.** B   **20.** B

5.1 Problem Set, pages 186–189

1. $\alpha = 30°$; $\beta = 60°$; $\gamma = 90°$; $a = 80$; $b = 140$; $c = 160$
3. $\alpha = 76°$; $\beta = 14°$; $\gamma = 90°$; $a = 29$; $b = 7.2$; $c = 30$
5. $\alpha = 45°$; $\beta = 45°$; $\gamma = 90°$; $a = 49$; $b = 49$; $c = 69$
7. $\alpha = 69.2°$; $\beta = 20.8°$; $\gamma = 90.0°$; $a = 26.5$; $b = 10.0$; $c = 28.3$
9. $\alpha = 56.00°$; $\beta = 34.00°$; $\gamma = 90.00°$; $a = 3{,}484$; $b = 2{,}350$;
  $c = 4{,}202$
11. $\alpha = 42°$; $\beta = 48°$; $\gamma = 90°$; $a = 320$; $b = 350$; $c = 470$
13. $\alpha = 67.8°$; $\beta = 22.2°$; $\gamma = 90.0°$; $a = 26.6$; $b = 10.8$; $c = 28.7$
15. $\alpha = 28.95°$; $\beta = 61.05°$; $\gamma = 90.00°$; $a = 202.7$; $b = 366.4$;
  $c = 418.7$
17. $\alpha = 48.3°$; $\beta = 41.7°$; $\gamma = 90.0°$; $a = .355$; $b = .316$; $c = .475$
19. 45 m   **21.** 240 m   **23.** 23 m   **25.** 540 ft   **27.** 350 ft
29. 780 ft   **31.** The height of the tower is 222.0 ft.   **33.** 995 m
35. 6.79×10^7 mi   **37.** 921.1 ft
39. a. $m \approx .7$ **b.** Using exact values, $m = \sqrt{3}$ so the equation
  is $\sqrt{3}\, x - y - 2\sqrt{3} - 3 = 0$; using approximate values,
  $m = \tan 60° \approx 1.7$; $1.7x - y - 6.4 = 0$ or $17x - 10y - 64 = 0$.

  c. Using exact values, $m = \frac{\sqrt{3}}{3}$ so the equation is
  $\sqrt{3}x - 3y - 4\sqrt{3} - 3 = 0$; using approximate values,
  $m = \tan 30° \approx .6$; $.6x - y - 3.4 = 0$ or $3x - 5y - 17 = 0$.

41. Area $= \frac{1}{2}$(base)(height)

  $= \frac{1}{2}(b)(a)$

  But $\sin A = \frac{a}{c}$

  $a = c \sin A$

  Thus, Area $= \frac{1}{2}bc \sin A$

5.2 Problem Set, pages 196–197

1. 54°   **3.** 49°   **5.** 80°   **7.** 13   **9.** 10   **11.** 17
13. $a = 14.2$; $b = 16.3$; $c = 4.00$; $\alpha = 52.1°$; $\beta = 115.0°$; $\gamma = 12.9°$
15. $a = 5.0$; $b = 4.0$; $c = 1.5$; $\alpha = 125°$; $\beta = 41°$; $\gamma = 14°$
17. Two solutions: $a_1 = 98.2$; $a_2 = 41.6$; $b = 82.5$; $c = 52.2$;
  $\alpha_1 = 90.8°$; $\alpha_2 = 25.0°$; $\beta_1 = 57.1°$; $\beta_2 = 122.9°$; $\gamma = 32.1°$
19. Two solutions: $a = 10.2$; $b = 11.8$; $c_1 = 13.5$; $c_2 = 2.61$;
  $\alpha = 47.0°$; $\beta_1 = 57.8°$; $\beta_2 = 122.2°$; $\gamma_1 = 75.2°$; $\gamma_2 = 10.8°$
21. No solution since the sum of the two smaller sides must be larger
  than the third side.
23. $a = 4.3$; $b = 5.2$; $c = 3.4$; $\alpha = 55°$; $\beta = 85°$; $\gamma = 41°$
25. $a = 214$; $b = 320$; $c = 126$; $\alpha = 25.8°$; $\beta = 139.4°$; $\gamma = 14.8°$
27. $a = 140$; $b = 85.0$; $c = 105$; $\alpha = 94.3°$; $\beta = 57.3°$; $\gamma = 48.4°$
29. 249 mi   **31.** 42 ft (right-triangle solution)
33. The distance is 454 feet.   **35.** 26 in.   **37.** Answers vary.
39. Answers vary.   **41.** Answers vary.   **43.** Answers vary.
45. Answers vary.

47. Find one example. The following list was generated by using
a computer program.

PRIMITIVE QUADRUPLES FOR THE LAW OF COSINES			
(a, b, c LESS THAN 100; $A = 60, 120$)			
a	b	c	ANGLE
1	1	1	60°
7	3	5	120°
7	3	8	60°
7	5	8	60°
13	7	8	120°
13	7	15	60°
13	8	15	60°
19	5	16	120°
19	5	21	60°
19	16	21	60°
31	11	24	120°
31	11	35	60°
31	24	35	60°
37	7	33	120°
37	7	40	60°
37	33	40	60°
43	13	35	120°
43	13	48	60°
43	35	48	60°
49	16	39	120°
49	16	55	60°
49	39	55	60°
61	9	56	120°
61	9	65	60°
61	56	65	60°
67	32	45	120°
67	32	77	60°
67	45	77	60°
73	17	63	120°
73	17	80	60°
73	63	80	60°
79	40	51	120°
79	40	91	60°
79	51	91	60°
91	11	85	120°
91	11	96	60°
91	19	80	120°
91	19	99	60°
91	80	99	60°
91	85	96	60°
97	55	57	120°

49. Answers vary.

5.3 Problem Set, pages 200–204

1. $\alpha = 48°; \beta = 62°; \gamma = 70°; a = 10; b = 12; c = 13$
3. $\alpha = 30°; \beta = 50°; \gamma = 100°; a = 30; b = 46; c = 59$
5. $\alpha = 50°; \beta = 70°; \gamma = 60°; a = 33; b = 40; c = 37$
7. $\alpha = 82°; \beta = 19°; \gamma = 79°; a = 53; b = 18; c = 53$
9. $\alpha = 35.0°; \beta = 81.0°; \gamma = 64.0°; a = 73.4°; b = 126; c = 115$
11. $\alpha = 71°; \beta = 85°; \gamma = 24°; a = 210; b = 220; c = 91$
13. $\alpha = 45.2°; \beta = 21.5°; \gamma = 113.3°; a = 41.0; b = 21.2; c = 53.1$
15. Cannot be solved since the sum of α and γ is more than 180°.
17. $\alpha = 14°; \beta = 42°; \gamma = 123°; a = 26; b = 71; c = 88$
19. $\alpha = 74.0°; \beta = 81.0°; \gamma = 25.0°; a = 25.0; b = 25.7; c = 11.0$
21. $\alpha = 74.7°; \beta = 16.1°; \gamma = 89.2°; a = 80.6; b = 23.2; c = 83.6$
23. $\alpha = 35.60°; \beta = 42.10°; \gamma = 102.30°; a = 16.90; b = 19.46; c = 28.36$
25. The height of the building is about 260 ft.
27. Luke's craft is 970.4 m from the first observation point and 1,532 m from the second.
29. The row boat is about 16 ft long.
31. The length of the tower is 179 ft.
33. The jetty is about 5,030 ft long.
35. Drop a perpendicular from C and let h be the length. From the definition of the trigonometric functions,

$$\sin \alpha = \frac{h}{b} \quad \text{and} \quad \sin \beta = \frac{h}{a}$$

Solve for h and equate:

$$b \sin \alpha = a \sin \beta$$

Divide by ab:

$$\frac{\sin \alpha}{a} = \frac{\sin \beta}{b}$$

37. From Problem 35, $\dfrac{\sin \alpha}{a} = \dfrac{\sin \beta}{b}$

Multiply by a: $\sin \alpha = \dfrac{a \sin \beta}{b}$

Divide by $\sin \beta$: $\dfrac{\sin \alpha}{\sin \beta} = \dfrac{a}{b}$

39–45. Proofs vary.   47. About 39 mph

5.4 Problem Set, pages 208–209

Answers to 1–23 are in square units and are given to the correct number of significant digits.

1. 37   3. 83   5. 680   7. 560   9. 6.4
11. 35   13. 54.0   15. 58.2 and 11.3
17. 2,150 and 911   19. 180   21. no triangle formed
23. 20,100   25. 7.82 cm   27. 18,000 m^2
29. 21,000 ft^3 or 760 yd^3 31. 400 ft^3 33. Proofs vary.
35. Answers vary.

5.5 Problem Set, pages 214–216

1. A vector with a direction N37°W and a magnitude of 5.0
3. A vector with a direction N38°E and a magnitude of 23
5. The horizontal component has a magnitude of 22 and the vertical component a magnitude of 7.1.
7. The boat is traveling at 13 mph with a course of S22°W.
9. The horizontal component is 2,090 fps and the vertical component is 368 fps.
11. In one hour it has traveled 590 miles south.
13. The direction is S32.8°W with a speed of 22.6 knots.
15. The magnitude is 360 lb, and it makes an angle of 23° with the 220-lb force, and a 29° angle with the 180-lb force.
17. 56° with the 30-lb force, and 30° with the 50-lb force
19. The planes are 370 miles apart.
21. At 10 kph it would be there at about 12:25 P.M. and 2:00 P.M.
23. The boats are 67.2 miles apart.
25. A force of 4.71 tons is needed to keep it from sliding down the hill. The force against the hill is 51.8 tons.
27. The angle of inclination is 39.2°.
29. The heaviest piece of cargo is 12,200 lb.
31. The weight of the astronaut is resolved into two components: one parallel to the inclined plane with length y and the other perpendicular to it with length x. The weight of the astronaut is $|\mathbf{x}|$.

5.6 Problem Set, pages 222–223

Answers to Problems 1–5 should also include a vector diagram.

1. $\mathbf{v} = 6\mathbf{i} + 6\sqrt{3}\mathbf{j}$
3. $\mathbf{v} = 4 \cos 112°\mathbf{i} + 4 \sin 112°\mathbf{j} \approx -1.4984\mathbf{i} + 3.7087\mathbf{j}$
5. $\mathbf{v} = -2\mathbf{i} + 2\mathbf{j}$ 7. 5   9. $\sqrt{85} \approx 9.2195$
11. $2\sqrt{2} \approx 2.8284$ 13. orthogonal   15. not orthogonal

Answers to Problems 17–25 are given in the form $\mathbf{v} \cdot \mathbf{w}$; $|\mathbf{v}|$; $|\mathbf{w}|$; cos θ.

17. $63; 5; 13; \dfrac{63}{65}$ 19. $3\sqrt{5}; 3; 3\sqrt{6}; \dfrac{\sqrt{30}}{18}$

21. $3; \sqrt{13}; \sqrt{61}; \dfrac{3\sqrt{793}}{793}$ 23. $1; 1; 1; 1$ 25. $0; 1; 1; 0$

27. $75°$ 29. $45°$ 31. $90°$ 33. $-\dfrac{8}{5}$

35. Answers vary.

Chapter 5 Summary and Review

Review of Objectives, pages 224–225

1. a. $\dfrac{a}{c}$ b. $\dfrac{b}{c}$ c. $\dfrac{b}{a}$
2. a. $a = 59.8, b = 77.9, c = 112; \alpha = 30.7°, \beta = 41.7°, \gamma = 107.6°$ b. $a = 3.5, b = .89, c = 3.6; \alpha = 79°, \beta = 87°, \gamma = 14°$ c. $a = .60, b = 7.7, c = 8.0; \alpha = 4°, \beta = 58°, \gamma = 118°$
3. a. $a = 5.30, b = 12.2, c = 13.3, \alpha = 23.5°, \beta = 66.5°, \gamma = 90.0°$

4. a. The tower is 2,063 ft.   **b.** It is 278 ft across the bridge.

5. a. $b^2 + c^2 - 2bc \cos \alpha$ **b.** $\dfrac{a^2 + c^2 - b^2}{2ac}$

6. $a = 14, b = 27, c = 19, \alpha = 29°, \beta = 109°, \gamma = 42°$

7. $a = 19, b = 7.2, c = 15, \alpha = 113°, \beta = 20°, \gamma = 47°$

8. (ambiguous case) At 12:42 P.M. and at 4:40 P.M.

9. It is about 430 ft.   **10.** $\dfrac{\sin \alpha}{a} = \dfrac{\sin \beta}{b} = \dfrac{\sin \gamma}{c}$

11. $a = 92.6, b = 273, c = 224, \alpha = 18.3°, \beta = 112.4°, \gamma = 49.3°$

12. About 113 m of fence is needed.

13. a. $\cos \alpha = \dfrac{b^2 + c^2 - a^2}{2bc}$ **b.** $\alpha = \cos^{-1}\!\left(\dfrac{b^2 + c^2 - a^2}{2bc}\right)$

   c. $\dfrac{\sin \alpha}{a} = \dfrac{\sin \beta}{b}$ **d.** $\alpha = \sin^{-1}\!\left(\dfrac{a \sin \beta}{b}\right)$

   e. $\alpha + \beta + \gamma = 180°$ **f.** $\alpha = 180° - \beta - \gamma$

   g. $\dfrac{\sin \alpha}{a} = \dfrac{\sin \beta}{b}$ **h.** $b = \dfrac{a \sin \beta}{\sin \alpha}$

   i. $b^2 = a^2 + c^2 - 2ac \cos \beta$ **j.** $b = \sqrt{a^2 + c^2 - 2ac \cos \beta}$

14. a. 320   **b.** 51.7   **c.** 1,900   **15.** 130,900 ft^2

16. 1,800,000 ft^3 ≈ 66,800 yd^3 **17.** 7.1; N47°W

18. 44 fps ≈ 30 mph; S60°E

19. Horizontal component is 2.8 and the vertical component is 3.5.

20. .2 ton ≈ 420 lb

21. a. $\mathbf{v} = \mathbf{i} + \mathbf{j}$ **b.** $-4.4\mathbf{i} + 8.2\mathbf{j}$ **c.** $-6\mathbf{i} - 5\mathbf{j}$

22. $|\mathbf{v}| = \sqrt{13}; |\mathbf{w}| = \sqrt{2}$ **23.** 5   **24.** 11°

25. $a = -\dfrac{3}{2}$

Miscellaneous Problems, pages 226–227

1. $a = 24; b = 52; c = 61; \alpha = 23°; \beta = 58°; \gamma = 99°$

3. $a = 30; b = 46; c = 59; \alpha = 30°; \beta = 50°; \gamma = 100°$

5. $a = 33; b = 40; c = 37; \alpha = 50°; \beta = 70°; \gamma = 60°$

7. $a = 15; b = 8.0; c = 10; \alpha = 112°; \beta = 30°; \gamma = 38°$

9. $a = 17; b = 14; c = 12; \alpha = 82°; \beta = 54°; \gamma = 44°$

11. $a = 7.0; b = 8.0; c = 2.0; \alpha = 54°; \beta = 113°; \gamma = 13°$

13. no solution

15. ambiguous case: $a = 14.5; b = 17.2; c = 24.5$ and $c' = 3.47$; $\alpha = 35.5°; \beta = 43.5°$ and $\beta' = 136.5°; \gamma = 101.0°$ and $\gamma' = 8.0°$

17. $a = 121; b = 315; c = 255; \alpha = 21.3°; \beta = 108.7°; \gamma = 50.0°$

19. $a = 11; b = 9.0; c = 8.0; \alpha = 80°; \beta = 54°; \gamma = 46°$

21. $a = 12.4; b = 32.5; c = 35.1; \alpha = 20.7°; \beta = 67.8°; \gamma = 91.6°$

23. no solution

25. $a = 49.5; b = 23.5; c = 28.1; \alpha = 147.0°; \beta = 15.0°; \gamma = 18.0°$

27. ambiguous case: $a = 117$ and $a' = 34.9; b = 82.5; c = 52.2;$ $\alpha = 118.6°$ and $\alpha' = 15.2°; \beta = 38.3°$ and $\beta' = 141.7°; \gamma = 23.1°$

29. ambiguous case: $a = 10.2; b = 11.8; c = 13.5$ and $c' = 2.61$; $\alpha = 47.0°; \beta = 57.8°$ and $\beta' = 122.2°; \gamma = 75.2°$ and $\gamma' = 10.8°$

31. The tunnel is 979 ft.

33. The log should show 12:45 P.M. and 1:11 P.M.

35. It is 7,200 ft from the first tower and 6,220 ft from the second tower.

37. 10.03°

Application for Further Study: Parking-Lot Design, page 229

1.

θ	c	w
10°	57.6	13.5
20°	29.2	16.6
30°	20.0	19.2
40°	15.6	21.2
50°	13.1	22.5
60°	11.5	23.2
70°	10.6	23.2
80°	10.2	22.4
90°	10.0	21.0

3.

θ	USABLE LENGTH
30°	3.7
40°	9.1
50°	12.6
60°	15.2
70°	17.4
80°	19.2
90°	21.0

5. To minimize c, θ should be 90°.

6.1   Problem Set, pages 235–236

1. a. $8 + 7i$ **b.** $11 + i$ **c.** $1 - 6i$

3. a. -4 **b.** $8 - 10i$ **c.** $11 - i$

5. a. $7 + i$ **b.** $9 + 2i$ **c.** 29

7. a. -1 **b.** 1   **c.** -1 **9. a.** 1   **b.** -1 **c.** 1

11. a. -1 **b.** 1   **c.** $-i$ **13. a.** 1   **b.** i **c.** 1

15. $32 - 24i$ **17.** $-9 + 40i$ **19.** $-\dfrac{3}{2} + \dfrac{3}{2}i$

21. $-2i$ **23.** $\dfrac{-6}{29} + \dfrac{15}{29}i$ **25.** $1 + i$

27. $\dfrac{-35}{37} - \dfrac{12}{37}i$ **29.** $\dfrac{1}{2} - \dfrac{1}{2}i$ **31.** $.3131 + 2.2281i$

33. $(-1 - \sqrt{3}) + 2i$ **35.** $\left(\dfrac{5 + 2\sqrt{3}}{13}\right) + \left(\dfrac{1 + 3\sqrt{3}}{13}\right)i$

37. $(13 + 8\sqrt{3}) + (12 + 4\sqrt{3})i$ **39.** Answers vary.

6.2   Problem Set, pages 240–241

1. $\sqrt{10}; 5\sqrt{2}$ **3.** $\sqrt{13}; \sqrt{29}$ **5.** $2\sqrt{2}; \sqrt{2}$ **7.** 2; 4

9. 2; 5   **11.** $\sqrt{2}$ cis 315°   **13.** 2 cis 30°   **15.** cis 0°

17. cis 270°   **19.** 3 cis 55°   **21.** 4 cis 100°   **23.** $\dfrac{3}{2} + \dfrac{3\sqrt{3}}{2}i$

25. $-\dfrac{\sqrt{3}}{2} + \dfrac{1}{2}i$ **27.** $2\sqrt{3} + 2i$ **29.** $-2.3444 - 5.5230i$

31. $9.8163 + 1.9081i$ **33.** Answers vary.

6.3   Problem Set, pages 246–247

1. 6 cis 210°   **3.** $\dfrac{5}{2}$ cis 267°   **5.** 3 cis 230°   **7.** 81 cis 240°

9. -64 (64 cis 180°)   **11.** $-128 + 128i\sqrt{3}$ (256 cis 120°)

13. 2 cis 80°, 2 cis 200°, 2 cis 320°

15. 2 cis 40°, 2 cis 112°, 2 cis 184°, 2 cis 256°, 2 cis 328°

17. 2 cis 32°, 2 cis 104°, 2 cis 176°, 2 cis 248°, 2 cis 320°

19. 3 cis 0°, 3 cis 120°, 3 cis 240°

21. $\sqrt[8]{2}$ cis 56.25°, $\sqrt[8]{2}$ cis 146.25°, $\sqrt[8]{2}$ cis 236.25°, $\sqrt[8]{2}$ cis 326.25°

23. 2 cis 15°, 2 cis 75°, 2 cis 135°, 2 cis 195°, 2 cis 255°, 2 cis 315°

25. $2^{1/18}$ cis 15°, $2^{1/18}$ cis 55°, $2^{1/18}$ cis 95°, $2^{1/18}$ cis 135°, $2^{1/18}$ cis 175°, $2^{1/18}$ cis 215°, $2^{1/18}$ cis 255°, $2^{1/18}$ cis 295°, $2^{1/18}$ cis 335°

27. cis 0°, cis 36°, cis 72°, cis 108°, cis 144°, cis 180°, cis 216°, cis 252°, cis 288°, cis 324°

Problems 29–33 should also be shown graphically.

29. 2, 2i, −2, −2i

31. −.6840 + 1.8794i, −1.2856 − 1.5321i, 1.9696 − .3473i

33. 1.9696 + .3473i, −.3473 + 1.9696i, −1.9696 − .3473i, .3473 − 1.9696i

35. 1, .3090 ± .9511i, −.8090 ± .5878i

37. $\pm 1, \frac{1}{2} \pm \frac{\sqrt{3}}{2}i, -\frac{1}{2} \pm \frac{\sqrt{3}}{2}i$ **39.** ±2, ±2i

41. $-1, \frac{1}{2} \pm \frac{\sqrt{3}}{2}i, -\frac{1}{2} \pm \frac{\sqrt{3}}{2}i$

43. 8 cis 0°, 8 cis 72°, 8 cis 144°, 8 cis 216°, 8 cis 288°

45. Answers vary.

Chapter 6 Summary and Review

Review of Objectives, page 248

1. a. 2 + 5i   **b.** −1 − 3i   **2. a.** 8   **b.** −21 − 20i

3. a. $-\frac{1}{2} + \frac{5}{2}i$ **b.** 4i

4.   **a.** $2\sqrt{5}$ **b.** $\sqrt{26}$ **c.** $\sqrt{17}$ **d.** $2\sqrt{5}$

5. a. $7\sqrt{2}$ cis 315°   **b.** 3 cis 270°   **6. a.** $-\sqrt{3} + i$
   b. $2\sqrt{2} - 2\sqrt{2}i$ **7.** cis 154°   **8.** $-128 - 128\sqrt{3}i$

9. a. $\sqrt{7}$ cis 165° ≈ −2.5556 + .6848i, $\sqrt{7}$ cis 345° ≈ 2.5556 − .6848i
   b. cis 0° = 1; cis 90° = i; cis 180° = −1; cis 270° = −i.

Also show these roots graphically.

10. $3, -\frac{3}{2} + \frac{3}{2}\sqrt{3}i, -\frac{3}{2} - \frac{3}{2}\sqrt{3}i$

Miscellaneous Problems, page 249

1. $9\sqrt{2}$ cis 315°   **3.** 7 cis 150°   **5.** 9 cis 0°

7. 16 cis 90°   **9.** $-\frac{9}{2}\sqrt{2} + \frac{9}{2}\sqrt{2}i$ **11.** $-\frac{3}{2} - \frac{3\sqrt{3}}{2}i$

13. 4.5315 + 2.1131i   **15.** $-128 - 128\sqrt{3}i$

17. cis 228°   **19.** cis 30°; cis 150°; cis 270°

21. 2 cis 15°; 2 cis 75°; 2 cis 135°; 2 cis 195°; 2 cis 255°; 2 cis 315°

23. $2^{1/18}$ cis 15°; $2^{1/18}$ cis 55°; $2^{1/18}$ cis 95°; $2^{1/18}$ cis 135°; $2^{1/18}$ cis 175°; $2^{1/18}$ cis 215°; $2^{1/18}$ cis 255°; $2^{1/18}$ cis 295°; $2^{1/18}$ cis 335°

25. cis 0°; cis 36°; cis 72°; cis 108°; cis 144°; cis 180°; cis 216°; cis 252°; cis 288°; cis 324°

27. $2\sqrt[3]{4}$ cis 40°; $2\sqrt[3]{4}$ cis 160°; $2\sqrt[3]{4}$ cis 280°

29. .6848 + 2.5556i; −.6848 − 2.5556i

Application for Further Study:

Electric Generator, page 251

1. $I = 15 \sin 120\pi t; a = 15; p = \frac{2\pi}{120\pi} = \frac{1}{60}$

3. cos θ = .96; θ ≈ 16.3°. Since cos θ is in Quadrant I, if cos θ ≥ .96, then θ ≤ 16.3°. Thus, the maximum current lag (to the nearest degree) is 16°.

7.1   Problem Set, pages 258–259

Points in Problems 1–12 should also be plotted.

1. Polar: $\left(4, \frac{\pi}{4}\right), \left(-4, \frac{5\pi}{4}\right)$; rectangular: $(2\sqrt{2}, 2\sqrt{2})$

3. Polar: $\left(5, \frac{5\pi}{3}\right), \left(-5, \frac{5\pi}{3}\right)$; rectangular: $\left(-\frac{5}{2}, \frac{5\sqrt{3}}{2}\right)$

5. Polar: $\left(\frac{3}{2}, \frac{7\pi}{6}\right), \left(-\frac{3}{2}, \frac{\pi}{6}\right)$; rectangular: $\left(\frac{-3\sqrt{3}}{4}, -\frac{3}{4}\right)$

7. $\left(5\sqrt{2}, \frac{\pi}{4}\right)$ and $\left(-5\sqrt{2}, \frac{5\pi}{4}\right)$ **9.** $\left(4, \frac{5\pi}{3}\right)$ and $\left(-4, \frac{2\pi}{3}\right)$

11. $\left(3\sqrt{2}, \frac{7\pi}{4}\right)$ and $\left(-3\sqrt{2}, \frac{3\pi}{4}\right)$

13. a. United States   **b.** India   **c.** Greenland   **d.** Canada

15. yes   **17.** no   **19.** yes   **21.** yes   **23.** yes

25–39. Answers vary.

7.2   Problem Set, page 263

1.

3.

5.

7.

9.   **11.**

13.

15.

b.

c.

17.

19.

d.

e.

21.

23.

f.

37. a. spiral of Archimedes

25.

27.

b. spiral of Archimedes

c. hyperbolic spiral

29.

31.

d. hyperbolic spiral

e. logarithmic spiral

33.

35. a.

f. logarithmic spiral

33.

7.3 Problem Set, page 269

1. lemniscate   **3.** three-leaved rose   **5.** cardioid
7. none   **9.** none (circle)   **11.** none
13. none (it is a line)   **15.** cardioid
17.

19.

21.

23.

25.

27.

29.

31.

7.4 Problem Set, pages 274–275

1.

3.

5.

7.

9.

11.

13.

15.

17.

19.

21.

23.

2. a. $\left(3, \frac{4\pi}{3}\right), \left(-3, \frac{\pi}{3}\right)$ **b.** $\left(5, \frac{9\pi}{10}\right), \left(-5, \frac{19\pi}{10}\right)$
   c. $\left(6, \frac{2\pi}{3}\right), \left(-6, \frac{5\pi}{3}\right)$ **d.** $(5.3852, 1.1903), (-5.3852, 4.3319)$

3. a. $\left(-\frac{3}{2}, \frac{-3\sqrt{3}}{2}\right)$ **b.** $\left(-4.7553, 1.5451\right)$

4. a. no **b.** yes **c.** yes **d.** no

5. a.

b.

25.

27.

c.

$r = -2$

6.

29.

31. $x = a \cos \theta + a\theta \sin \theta, y = a \sin \theta - a\theta \cos \theta$
33–35. Answers vary.

7.5 Problem Set, page 281

Graphs should also be shown for Problems 1–30.

1. $(0, 0), \left(2\sqrt{2}, \frac{\pi}{4}\right)$ **3.** $\left(1, \frac{\pi}{3}\right), \left(1, \frac{5\pi}{3}\right)$ **5.** $(3, 0), (3, \pi)$

7. $(0, 0), \left(2, \frac{\pi}{2}\right), \left(2, \frac{3\pi}{2}\right)$ **9.** $(2, 0), (2, \pi)$ **11.** $(0, 0), \left(2, \frac{\pi}{4}\right)$

13. $(0, 0)$ **15.** $(0, 0), (a, 0), (a, \pi)$ **17.** $(0, 0), \left(\pi, \frac{\pi}{3}\right), \left(4\pi, \frac{4\pi}{3}\right)$

19. $(0, 0), \left(\frac{16}{5}, \text{Arccos}\left(-\frac{3}{5}\right)\right)$ [*Note:* $\text{Arccos}\left(-\frac{3}{5}\right) \approx 2.214$ or $126.87°$]

21. $(0, 0), (3.2, 2\pi - \text{Arccos} .8)$
   [*Note:* $2\pi - \text{Arccos} .8 \approx 5.640$ or $323.13°$]

23 $(0, 0), (1, 0), (.6, \text{Arcsin} .8)$ [*Note:* $\text{Arcsin} .8 \approx .927$ or $53.1°$]

25. $\left(2, \frac{\pi}{2}\right), \left(2, \frac{3\pi}{2}\right)$

27. $\left(2 + \sqrt{2}, \frac{5\pi}{4}\right), \left(2 + \sqrt{2}, \frac{3\pi}{4}\right), \left(2 - \sqrt{2}, \frac{\pi}{4}\right), \left(2 - \sqrt{2}, \frac{7\pi}{4}\right)$

29. $\left(2, \frac{\pi}{2}\right)$

Chapter 7 Summary and Review

Review of Objectives, page 283

1.

$(5, \sqrt{75})$ $\left(-5, -\frac{\pi}{10}\right)$ $\left(3, -\frac{2\pi}{3}\right)$ $(-2, 2)$

7.

8.

9. a. lemniscate **b.** cardioid **c.** five-leaved rose **d.** circle
11. $(4, \pi), (0, 0), (4, 0)$

10.

Miscellaneous Problems, pages 283–284

1. rose curve **3.** circle **5.** cardioid **7.** rose curve
9. lemniscate **11.** rose curve **13.** rose curve **15.** cardioid
17.

19.

21.

23.

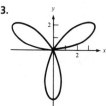

3. It is 50° away from the direction given in Problem 1. It is also a distance requiring about twice as much energy to get there.

25.

27.

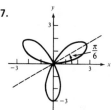

Cumulative Review for Chapters 5–7

Objective Test, page 287

1. a. Law of Cosines   **b.** Law of Cosines   **c.** Law of Cosines
(ambiguous case)   **d.** Law of Sines   **e.** no solution
f. Law of Cosines   **g.** right-triangle solution   **h.** no solution
i. Law of Cosines (ambiguous case)   **j.** Law of Sines
2. a. rose curve   **b.** lemniscate   **c.** rose curve   **d.** cardioid
e. circle   **f.** cardioid   **g.** lemniscate   **h.** line   **i.** cardioid
j. cardioid
3. $\alpha = 121°; \beta = 32°; \gamma = 27°; a = 83; b = 52; c = 43$
4. $\alpha = 57°; \beta = 68°; \beta' = 112°; \gamma = 55°; \gamma' = 11°; a = 55;$
$b = 61; c = 53; c' = 13$
5. $46°, 102°, 32°$

29.

31.

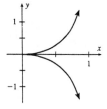

6. a. polar, $\left(3, \frac{7\pi}{4}\right), \left(-3, \frac{3\pi}{4}\right)$; rectangular, $\left(\frac{3\sqrt{2}}{2}\right), \left(\frac{-3\sqrt{2}}{2}\right)$

    b. polar, $\left(5, \frac{5\pi}{6}\right), \left(-5, \frac{11\pi}{6}\right)$; rectangular, $\left(\frac{-5\sqrt{3}}{2}, \frac{5}{2}\right)$

    c. polar, $(3, 0), (-3, 3.1416)$; rectangular, $(3, 0)$

7. $\sqrt[3]{5}$ cis 50°; $\sqrt[3]{5}$ cis 170°; $\sqrt[3]{5}$ cis 290°   **8.** $2\sqrt{2}$ cis 135°; $2\sqrt{2}$ cis 315°

9.

10.

33.

35.

37.

39.

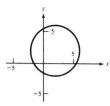

Application for Further Study:

Communication of Bees, page 286

1. The food source is 90° away from the sun.

Multiple-Choice Test, pages 288–289

1. $a^2 + b^2 - 2ab \cos \gamma; \dfrac{a^2 + c^2 - b^2}{2ac}$

2. $\dfrac{\sin \alpha}{a} = \dfrac{\sin \beta}{b} = \dfrac{\sin \gamma}{c}$ **3.** $r \cos \theta; r \sin \theta$

4. $r_1 r_2$ cis $(\theta_1 + \theta_2)$ **5.** $r = f(\theta)$; through an angle α **6.** C
7. B   **8.** A   **9.** C   **10.** A   **11.** C   **12.** A
13. E   **14.** B   **15.** D   **16.** E   **17.** B   **18.** C
19. D   **20.** C

8.1   Problem Set, pages 297–298

1.

3.

5.

7.

9.

11.

13.

15.

17.

19.

21.

23.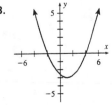

25. $y^2 = 20x$ **27.** $(x + 3)^2 = -12(y - 2)$ **29.** 2.25 m
31–33. Answers vary.

8.2   Problem Set, pages 305–306

1.

3.

5.

7.

9.

11.

13.

15.

17.

19.

21.

23. circle; center, $(-2, 3)$; radius 3; $(x + 2)^2 + (y - 3)^2 = 9$
25. horizontal ellipse; center, $(0, 0)$; $a = 10$, $c = 6$, $b^2 = 64$;
$$\frac{x^2}{100} + \frac{y^2}{64} = 1$$

27. vertical ellipse; center, $(0, 0)$; $a = 7, c = 5, b^2 = 24$;
$$\frac{x^2}{24} + \frac{y^2}{49} = 1$$

29. horizontal ellipse; center, $(0, 0)$; $a = 10, c = 2, b^2 = 96$;
$$\frac{x^2}{100} + \frac{y^2}{96} = 1$$

31. Aphelion is 94,500,000 and perihelion is 91,500,000.

33. .053   **35.** Answers vary.

8.3 Problem Set, pages 311–312

1.

3.

5.

7.

9.

11.

13.

15.

17.

$(-3, 1)$

19.

21.

$(2, -3)$

23. $\dfrac{y^2}{25} - \dfrac{x^2}{24} = 1$ **25.** $\dfrac{x^2}{1} - \dfrac{y^2}{24} = 1$ **27.** $\dfrac{x^2}{9} - \dfrac{(y+3)^2}{7} = 1$

29–31. Answers vary.

8.4 Problem Set, pages 316–317

1. line

3. parabola

5. hyperbola

7. ellipse (degenerate)

$(-3, 2)$

9. hyperbola (degenerate)

11. ellipse

13. hyperbola

15. ellipse

17. parabola

19. ellipse (circle)

$(-1, 2)$

21. $\theta = 45°$; $x = \frac{1}{\sqrt{2}}(x' - y')$; $y = \frac{1}{\sqrt{2}}(x' + y')$;

$$\frac{x'^2}{16} - \frac{y'^2}{16} = 1$$

45°

21. hyperbola

$(3, -1)$

23. parabola

$V(7, -2)$ $\bullet F$

23. $\theta = 45°$; $x = \frac{1}{\sqrt{2}}(x' - y')$; $y = \frac{1}{\sqrt{2}}(x' + y')$;

$$\frac{x'^2}{9} + \frac{y'^2}{4} = 1$$

45°

25. parabola

$V\left(-\frac{1}{3}, \frac{4}{3}\right)$

$F \bullet$

27. ellipse (circle)

$(2, -5)$

25. $\tan \theta = 2$; $\theta \approx 63.4°$; $x = \frac{1}{\sqrt{5}}(x' - 2y')$; $y = \frac{1}{\sqrt{5}}(2x' + y')$;

$$(x' + 1)^2 = 4\left(y + \frac{7}{10}\right)$$

$\tan \theta = 2$

29. ellipse

$(2, 1)$

27. $\theta = 30°$; $x = \frac{1}{2}\left(\sqrt{3}x' - y'\right)$; $y = \frac{1}{2}\left(x' + \sqrt{3}y'\right)$;

$$\frac{x'^2}{4} - \frac{y'^2}{9} = 1$$

30°

31. $(4, 14.7)$, $(8, 13.7)$, $(12, 12.0)$, $(16, 9.0)$, $(20, 0.0)$

8.5   Problem Set, pages 323–324

1–3. Answers vary.   **5.** hyperbola   **7.** ellipse
9. parabola   **11.** hyperbola   **13.** parabola
15. ellipse   **17.** ellipse

19. $\theta = 45°$; $x = \frac{1}{\sqrt{2}}(x' - y')$;

$y = \frac{1}{\sqrt{2}}(x' + y')$;

$$\frac{y'^2}{2} - \frac{x'^2}{2} = 1$$

45°

29. $\theta = 30°$; $x = \frac{1}{2}\left(\sqrt{3}x' - y'\right)$; $y = \frac{1}{2}\left(x' + \sqrt{3}y'\right)$;

$$x'^2 = -\frac{1}{16}(y' - 4)$$

$\theta = 30°$

31. $\theta = 45°$; $x = \frac{1}{\sqrt{2}}(x' - y')$; $y = \frac{1}{\sqrt{2}}(x' + y')$;

$$\frac{y'^2}{4} - \frac{x'^2}{16} = 1$$

33. $\tan \theta = \frac{4}{3}$; $\theta \approx 53.1°$; $x = \frac{1}{5}(3x' + 4y')$; $y = \frac{1}{5}(4x' + 3y')$;

$$\frac{x'^2}{\frac{9}{26}} + \frac{y'^2}{9} = 1$$

21.

23.

25.

27.

29.

31.

33. Answers vary.

8.6   Problem Set, page 329

1. $\epsilon = 1$; parabola   **3.** $r = \dfrac{\frac{5}{3}}{1 + \frac{5}{3} \cos \theta}$; $\epsilon = \frac{5}{3}$; hyperbola

5. $r = \dfrac{1}{1 - \frac{4}{5} \cos \theta}$; $\epsilon = \frac{4}{5}$; ellipse

7. $r = \dfrac{4}{1 - 2 \cos \theta}$; $\epsilon = 2$; hyperbola

9. $r = \dfrac{4}{1 - \cos \theta}$; $\epsilon = 1$; parabola   **11.** $\epsilon = 1$; parabola

13. $r = \dfrac{\frac{5}{4}}{1 - \frac{1}{2} \cos \left(\theta - \frac{\pi}{6} \right)}$; $\epsilon = \frac{1}{2}$; ellipse   **15.** $\epsilon = 1$; parabola

17.

19.

Chapter 8 Summary and Review

Review of Objectives, pages 331–332

1. a.

b.

2. a.

b.

3. a. $(y - 3)^2 = 20(x - 6)$ **b.** $(x + 3)^2 = -10\left(y - \frac{1}{2}\right)$
c. $(x + 3)^2 = -24(y - 5)$

4. a

b.

5. a.

b.

6. a.

b.
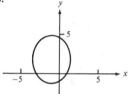

7. a. $\dfrac{(x - 4)^2}{4} + \dfrac{(y - 1)^2}{3} = 1$ **b.** $\dfrac{(x + 5)^2}{36} + \dfrac{(y - 4)^2}{32} = 1$
c. $(x + 1)^2 + (y + 2)^2 = 64$

8. a.

b.

9. a.

b.
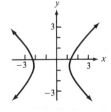

10. a. $\dfrac{x^2}{9} - \dfrac{y^2}{16} = 1$ **b.** $\dfrac{y^2}{9} - \dfrac{x^2}{16} = 1$
c. $\dfrac{(x + 4)^2}{1} - \dfrac{(y - 1)^2}{5} = 1$

11. a. degenerate circle; a point at $(2, -3)$
b. degenerate hyperbola; two lines passing through the origin with slopes $\frac{3}{2}$ and $-\frac{3}{2}$.

12. a. $\dfrac{(x - h)^2}{a^2} + \dfrac{(y - k)^2}{b^2} = 1$ **b.** $\dfrac{(y - k)^2}{a^2} - \dfrac{(x - h)^2}{b^2} = 1$
c. $(y - k)^2 = 4c(x - h)$ **d.** $(x - h)^2 + (y - k)^2 = r^2$

13. a. parabola   **b.** line   **c.** ellipse   **d.** ellipse (circle)
e. hyperbola   **f.** ellipse   **g.** hyperbola   **h.** parabola
i. parabola   **j.** hyperbola

14. a.

b.

c.

d.

15. a. $45°$ **b.** $26.6°$

16.

17. parabola

18. ellipse

Miscellaneous Problems, page 332

1.

3.

5.

7.

9.

11.

13.

15.

17.

19.

21.

23.

25.

27.

29.

Application for Further Study:

Planetary Orbits, page 334

1. $1.5 \cdot 10^8$; $1.3 \cdot 10^8$
3. $3.7 \cdot 10^9$; $3.6 \cdot 10^9$; almost a circular orbit   **5.** 5,500 mi
7. $1.2x^2 + 1.3y^2 = 1.6 \cdot 10^{15}$
9. $8.802x^2 + 8.805y^2 = 7.751 \cdot 10^{16}$

B Problem Set, page 345

1. $4914 \approx 4{,}900$ **3.** $5517840 \approx 5{,}520{,}000$
5. $5937 \approx 5{,}940$ **7.** $378193771 \approx 380{,}000{,}000$
9. $40 \approx 40.0$ **11.** $7714.00791 \approx 7{,}714.01$ **13.** 12.2
15. .264   **17.** .008353   **19.** 2.62   **21.** .310   **23.** 92.3
25. .24   **27.** 1.0   **29.** $-.379$

INDEX

A

AAA (triangle), 189, 192
  summary, 200
AAS (triangle), 189, 197–200
  summary, 200
Absolute value of a complex
    number, 237
Abû'l-Wefâ, 92
Accuracy (calculations), 6, 11, 181,
    342–344
  agreement in this book, 344
  angles, 344
  sides, 344
Acute angle, 5, 31
Acute triangle, 347
Adams, John Couch, 333
Addition laws, 112–115
Addition of ordinates, 74–76
  graphing by, 76
  radio modulation and, 77
Addition of vectors, 211
Adjacent angles, 211, 347
Adjacent side, 180
Alexandria, Egypt, 17
Algebraic logic, 339
Algebraic representation of a vector, 217
Algebraic solutions, 139, 189, 291, 315
Alpha (α), 4
Alternate interior angles, 347
Alternating current, 250
Altitude, 347
AM radio waves, 77
Ambiguous case (SSA), 194–195
  summary, 200
Amplitude
  computer use with, 82
  graphing arbitrary, 62–63
  modulation, 77
Angle(s), 3–17. *See also* Angles
  (standard position)

Angles (*continued*)
  accuracy in measuring, 6, 11, 181
  acute, 5, 31
  adjacent, 211, 347
  alternate interior, 347
  between vectors, 221
  central, 5
  classification of, 346
  complementary, 346
  conversion to decimal degrees, 6
  corresponding, 347
  coterminal, 7
  definition of, 4, 5
  degree measure of, 4–8
  of depression/elevation, 184–185
  exact measurement, 5
  exterior, 347
  geometric classification, 346–348
  grad measurement, 32
  of inclination, 189
  initial side of, 5
  interior, 347
  measurement agreement in this book, 8
  negative, 5, 7
  obtuse, 5
  positive, 5, 7
  primary, 137
  quadrantal, 7
  radian measure of, 8–17
  real-number measure, 8
  reference, 29
  right, 5, 347
  sides of, 4–5
  standard position, 7
  straight, 5, 10
  supplementary, 346
  symbols for, 4
  theta (θ), 18
  trigonometric functions of, 27
  vertex of, 4
  vertical, 346

"Angles of Elevation of the Pyramids..."
    (Smith), 187, 209
Angular velocity, 42–47
  of bicycle wheel, 42
  of pulley, 44
Answers, 350
  rounding of, 182, 343
Aphelion, 306, 333
Apogee, 306
Apollonius, 290
Applications for Further Study
  angular velocity, 42–47
  communication of bees, 285–286
  computer graphing of trigonometric
    curves, 82–88
  electric generator, 250–251
  linear interpolation, 132
  linear velocity, 42–47
  parking-lot design, 228–229
  planetary orbits, 333–335
  waves, harmonic motion and
    resonance of, 167–173
Arc, 5, 349
Arc length, 13–15, 132–133
Arccosecant function, 153
Arccosecant relation, 136
Arccosine function, 153
Arccosine relation, 136
Arccotangent function, 153
Arccotangent relation, 136
Archimedes, spiral of 263
Arcsecant function, 153
Arcsecant relation, 136
Arcsine function, 153
Arcsine relation, 136
Arctangent function, 153
Arctangent relation, 136
Area, 205–207
  of a right triangle, 189
  of a sector, 206–207
  of a triangle, 205–207

Area (*continued*)
ASA, 206
   SAS, 205
   SSS, 206
Argument, 237
Arithmetic logic, 339
Arlington, VA, 188
Aryabhata, 48
ASA (triangle), 189, 197–200, 206, 348
  summary, 200
Associative law, 236
Astronaut apparatus, 216
Astronomy, 2, 92
  problems, 17, 133, 188, 306, 335
Asymptote(s), 54, 69, 71
  cosecant, 69
  cotangent, 71
  secant, 70
  slant, of hyperbola, 308
  tangent, 54, 66, 71
Aviation problems, 122–123, 196, 214, 215, 226
Axis (axes)
  of cone, 292
  conjugate (of hyperbola), 307
  major (of ellipse), 298–299, 302
  minor (of ellipse), 298–299, 302
  of parabola, 293
  real and imaginary, 236
  rotation of, 317–321
  shift of, 59
  translation of, 59–60
  transverse (of hyperbola), 307

B

Babylonians, 2, 8, 27
  Plimpton-322 tablet, 27
Ballistics problem, 196
Banneker, Benjamin, 178
Base of a triangle, 347
Baseball stadium problem, 187
Bearing, 185
Bees, communication of, 285
Benard, John, 197
Bernoulli, Jacob, 261
Beta (β), 48
Bifolium, 269
Billard, Jules B., 209

Black Mathematicians and Their Works
  (Newell, et al), 178
Bow waves, 123
Brady, Joseph, 333
Brahmagupta's formula, 209
Brittany, France, 186
Bryn Mawr College, 252
Buffalo, NY, 196
Butler, N., 252
Business problems, 78

C

Calculator(s), 13, 25, 134, 338–341
  evaluation of trigonometric functions, 24–25
  for exponents, 340
  graphics, 65, 79, 81, 85, 95, 97, 142
  graphing parametric equations, 271, 327
  graphing polar-form curves, 266,
  for inverse functions, 156
  logic, 339
  for mixed operations, 341
  rectangular-polar conversion, 239
  solving equations, 138
  solving triangles, 179, 191, 194
  types of, 338–339
Calculus, (Finney/Thomas), 135
Calculus, Fourth Edition (Grossman), 253
Calculus, Second Edition (Stewart), 231
Canon mathematicus (Viète), 191
Cardioid, 261–262, 268
  standard position, 261
Carpenter, Edna M. , 333
Carson, Johnny, 202
Cartesian coordinate system, 253, 260
Center
  circle, 349
  ellipse, 298
  hyperbola, 307
Central angle, 5
Central rectangle, 308
Chattanooga, TN, 133
Cheops (pyramid), 187
Chicago, IL, 16
Chord, 106–107, 349
  length of, 106

Circle(s), 8, 17, 303–304, 349
  arc, 5, 349
  chord, 106, 349
  circumference, 5, 349
  circumscribed, 349
  degree measure, 8
  diameter, 349
  eccentricity, 303
  equation, 303
  geometric definition, 349
  identification, 313
  inscribed, 349
  involute, 275
  parametric form, 270–271
  polar form, 260, 268
  radius, 8, 303, 349
  secant, 349
  standard-form equation, 303
  tangent, 349
  unit, 17
Circular functions. *See* Trigonometric functions
Circumference, 5, 349
Circumscribed circle, 349
Circumscribed polygon, 349
Clayton, NM, 47
Close Encounters of the Third Kind (movie), 188
Cofunction identities, 107
Cohen, Martin, 197
Columbia, SC, 16
Communication of bees, 285–286
Commutative law, 236
Compass directions, 185
Complementary angles, 346
Completing the square, 315
Complex analysis, 231
Complex conjugate, 234
Complex numbers, 231–246, 251, 253
  absolute value (modulus), 237
  associative law, 236
  commutative law, 236
  definition, 232
  distributive law, 236
  De Moivre's formula, 241–246
  equality of, 233
  operations with, 232–2333
  plotting, 236–237
  power of, 243–244

Complex numbers (*continued*)
   product in rectangular form, 233
   product in trigonometric form, 242
   quotient in trigonometric form, 242
   rectangular to trigonometric form , 239
   simplified, 232
   square roots of, 244–245
   trigonometric form, 238
   trigonometric form to rectangular, 240
Complex plane, 236
Components, 211
Component vectors, 211
   rectangular, 211
Computer applications for plotting
     cycloids, 273
Computer graphics, 58, 81, 82–87, 273.
    See also Calculators
   Converge program (JEMware), 58, 81,
     266, 271, 327
   of trigonometric curves, 82–87
Computer science problems, 58, 197
Cone, 346
   generator, 292
   volume of, 207
Conic sections (ellipse, hyperbola,
    parabola), 291–332
   alternate definition, 324
   alternate equations for polar-form, 328
   classifications, 323
   degenerate, 312
   graphing, 294, 301, 303, 308, 323
   graphing rotated, 321–323
   planetary orbits and, 333–335
   polar form, 324–329
   recognition of, 313–314
   rotations of, 317–324
   rotated polar-form, 328–329
   summary of standard-position, 314
   translating by completing square,
    315–316
Conjugate, 234
Conjugate axis, 307
   length of, 308
Construction problems, 209
Consumer problems, 196
Converge program (JEMware), 58, 81,
    266, 271, 327
Coordinate systems, 290
   Cartesian, 50, 253, 260

Coordinate systems (*continued*)
   complex plane, 236
   Gaussian plane, 236
   polar, 253, 256–257
   relationship of polar to rectangular,
    256–257
   rectangular, 253
Copernicus, Nicholas, 2
Corresponding angles, 347
cos *x*. *See* Cosine function
Cosecant function, 18, 72
   domain, 70, 73
   evaluation, 22, 25
   exact values, 35
   general form, 72
   generalized definition, 27
   graphing, 69–70, 72
   period, 70
   principle values of, 157
   range, 70, 73
   real-number definition, 32
   right-triangle definition, 180
   unit-circle definition, 18
   values not defined, 18
Cosine function, 18, 53, 179
   domain, 54, 73
   evaluation, 22, 25
   exact values, 35
   general form, 64
   generalized definition, 27
   graphing, 50, 53–54, 59, 61, 64–66
   introduction, 18
   period, 54
   range, 54
   real-number definition, 32
   right-triangle definition, 180
   series expansion, 37
   unit-circle definition, 18
   values not defined, 18
Cosines, Law of, 190, 200
   alternate forms, 190
   historical origin, 197
cot *x*. *See* Cotangent function
Cotangent function, 18
   domain, 71, 73
   evaluation, 22, 25
   exact values, 35
   general form, 72
   generalized definition, 27

Cotangent function (*continued*)
   graphing, 71
   range, 71, 73
   real-number definition, 32
   right-triangle definition, 180
   unit-circle definition, 18
   values not defined, 18
Coterminal angles, 7
Cotes, Roger, 230
Counterexamples, 102
Critical values for graphing, 64
csc *x*. *See* Cosecant function
ctn *x*. *See* Cotangent function
Cumulative Review
    Chapters 1–2, 89
    Chapters 3–4, 175
    Chapters 5–7, 287
Current, 68, 79, 88, 250
   alternating, 251
Cunitz, Marie, 2
Cycloid, 272–273
   epicycloid, 273
   hypocycloid, 273
Cylinder, 346

D

Dallas-Fort Worth, TX, 15
Damped harmonic motion, 170
Dance Language and Orientation of Bees
    (von Frisch), 285
Data points, graphing, 67
Davis, R. F., 204
De revolutionibus orbium coelestium
    (Copernicus), 2
De Moivre, Abraham, 230
   Stirling's formula and, 230
De Moivre's formula, 243
   proof of, 247
De Morgan, Augustus, 48
Death Star, 202
Decagon, 346
Degenerate conic, 312
Degree
   definition, 5
   minutes and seconds of, 6
   symbol (°), 5
Degree measure
   of angles, 4–8

Degree measurement (*continued*)
  conversion table, 13
  decimal, conversion to, 6, 8
  protractor, 5
  radian, conversion to, 11–13
Deimos (satellite), 335
Delta (δ), 4
Denver, CO, 196
Depression, angle of, 184
Descartes, René, 290
Destructive waves, 78
Devil's Tower, WY, 188
Diagonals of a parallelogram, 211
Diameter, 349
Direction, 211
Directrix, 314
  of ellipse, 314
  of hyperbola, 314
  of parabola, 292
Discriminant (conics), 313
Disneyland, 16
Disproving identities, 102
Distance formula, 105–106
Distance problems, 15–17, 199,
    226
Distributive law, 236
Dodecagon, 346
Domain of trigonometric functions,
    55–56, 65, 70, 73
Dot product, 220
Double-angle identities, 116–118
Dreyfuss, Richard, 188

E

Earth, orbit of, 131, 334, 335
Eccentricity
  of conic sections, 303, 314
  of planetary orbits, 334
Edwards, I. E., 209
Egyptians, 2, 178, 187
Eiffel Tower, 203
Electric generator, 250
Electrical problems, 88, 152, 251
Electricity, 251
  alternating currents/voltage, 251
Electromotive force, 152
Elevation, angle of, 185
Eliminating a parameter, 272

Ellipse, 291, 292, 298–306
  axes, 298–299, 302
  center, 298
  compared with other conic sections,
    314
  covertices, 300
  definition, 298, 314
  degenerate, 312
  directrix, 314, 324
  eccentricity, 303, 304, 314
  equations, 301
  finding graph when given equation,
    301–302
  finding equation when given graph,
    304
  foci, 298, 304
  graph paper for definition, 298–299
  graphing, 301–302, 312
  identification, 313
  polar form, 324, 328
  recognition, 313
  rotated, 321–322
  standard form, 299
  standard-form equation
    with center (0, 0), 301
    with center (*h, k*), 302
    for translated, 302
  translated, 302
  vertices, 300, 304
  x-intercepts, 300
  y-intercepts, 300
Ellis, Robert, 93
Empire State Building, 188
Engineering problems, 47, 68, 202, 229,
    297, 298
  electrical, 68, 79
Entebbe, Uganda, 17
Entomology problems, 286
Epicycloid, 273
Equality of complex numbers, 233
Equations. *See also* Trigonometric
    equations
  first-degree (linear), 292
  procedure for solving trigonometric,
    140
  second-degree (quadratic), 292
  trigonometric, 136
Equilateral triangles, 347, 348
Eratosthenes, 17, 178

Euclid, 191
Euler, Leonard, 134
Euler's formula, 231
Euler's number, (*e*) 241
Evaluation of inverse trigonometric
    functions, 154–159
Evaluation of inverse trigonometric
    relations, 136
Evaluation of trigonometric functions,
    22, 28
  by addition identities, 159–160
  by calculator, 22, 24–25
  by definition, 22–24
  by trigonometric table, 22, 92
Eves, Howard, 8, 27, 48, 92, 134, 197,
    230
Exact measurement, 9
Exact values, 6, 11, 33–36
  solving equations with, 137
  table of, 35
Exterior angles, 347

F

Fakhry, Ahmed, 209
Faraday, Michael, 250
Ferndale, CA, 89, 225
Fine, H. B., 230
First-degree equation, 292
FM radio waves, 77
Focal chord, 294
  length of, 306
Foci (plural for focus)
  of ellipse, 298
  of hyperbola, 306
Focus of parabola, 292
Folium of Descartes, 269
Force, 209
  inclined plane, 213
Fortuna, CA, 225
45°–45°–90° triangle, 348
Fourier, Joseph, 79, 86
Fourier series, 79, 86
  graphing, 86–87
Framing curves, 49–57
  height and length of frames, 51–55,
    64, 72
Frequency modulation, 77
Frisch, Karl, 285

Frischlinus, N., 252
Fundamental identities, 19–21, 94. *See also* Identities

G

Galactic Empire, 202
Gamma (γ), 4
Gauss, Karl, 92, 230, 236
Gaussian plane, 236
General form of trigonometric curves, 64
    computer graphics, 85
Geography problems, 16
Geometry problems, 196, 197, 215
Geometry, review of, 346–349
Goodman, Terry, 197
Gordon, Florence, 273
Gordon, Sheldon, 273
Grad measurement, 32
Graph of an equation, 257
Graphic calculators, 142, 339
Graphing. *See also* Trigonometric
        functions
    by adding ordinates, 74–76
    with arbitrary amplitude, 62–63
    with arbitrary period, 63
    by calculator, 65, 78, 82–87
    circle, using a parameter, 270–271
    complex numbers, 236
    by computer, 82, 87
    conics, 294, 301, 303, 308
        by completing the square, 315–316
    with Converge (program), 58, 81
    cosecant, 72
    cosine, 61
    cotangent, 71
    cycloid, using a parameter, 274
    from data points, 67
    double-angle identity, 117
    Fourier series, 86–87
    framing curves, 49–57
    generalized reciprocal functions, 69,
        71–73
    generalized trigonometric curves, 64
    inverse trigonometric functions, 158
    inverse trigonometric relations, 158
    *n*th roots of a complex equation, 344
    with a negative angle or number, 111
    opposites, 111

Graphing (*continued*)
    polar-form curves, 259–279
    products of trigonometric curves,
        76–79
    reduction identity, 163
    secant, 70
    sine, 50
    sums of sines and cosines, 163
    sum of squares, 86
    sum of trigonometric functions, 74–77,
        163
    tangent, 54
    translations and phase shifts, 59–67
*Great Moments in Mathematics Before
    1650* (Eves), 92
Great Pyramid, 187
Greek letters, 4
Greenwich meridian, 15
Gulick, Denny, 93
Gunter, Edmund, 48

H

Half-angle identities, 119–123
Harmonic motion, 170–172
    damped, 170–171
    simple, 170, 171
Harmonic waves, 170–172
Harpsichord problem, 203
Heading, 210
Height of frame, 51–55, 64, 72
Height of a triangle, 347
Heptagon, 346
Hero's formula. *See* Heron's formula
Herodotus, 178
Heron's formula, 206, 209
Hexagon, 346
Hipparchus of Nicaea, 2, 48, 92
Historical questions
    Babylonians, 8
    Egyptian pyramids, 187
    grad measurement, 32
    law of cosines, 197
    measuring earth's radius, 17
    Plimpton table, 27
    Pythagorean triplets, 27
Huff, Darrell, 204
Hyperbola, 291, 292, 306–312
    center, 307

Hyperbola (*continued*)
    central rectangle, 308
    compared with other conic sections,
        314
    conjugate axis, 307
    definition, 306, 310, 314
    degenerate, 312, 313
    directrix, 314, 324
    eccentricity, 314
    equation, 307–308
    finding graph when given equation,
        308
    finding equation when given graph,
        309
    foci, 307
    graph paper for definition, 307
    graphing, 308, 309, 313
    polar-form, 324, 325–326
    psuedovertex, 308
    recognition, 313
    rotated, 321
    slant asymptotes, 308
    standard form, 307
    standard-form equation
        with center (0, 0), 307
        with center (*h, k*), 308
    translated, 308–309
    transverse axis, 307
    vertices, 307, 309
Hyperbolic spiral, 263
Hypocyloid, 273
Hypotenuse, 180

I

i (imaginary unit), 231–236
i vector, 217–219
Identities, 93–130
    addition laws, 112
    cofunction, 107–108
    disproving, 102–104
    double-angle, 116–118, 121–122
    fundamental, 19–21, 94
    half-angle, 119–123
    hints for proving, 101
    inverse trigonometric, 155
    opposite-angle, 108–111
    procedure for disproving, 102
    procedure for proving, 95, 99

Identities (*continued*)
  product-to-sum, 124
  proving, 93–101, 122
  Pythagorean, 20–21, 94
  ratio, 19–20, 94
  reciprocal, 19, 94
  reduction, 161–164
  sum-to-product, 125
  solving equations using, 149
Imaginary axis, 236
Imaginary unit (*i*), 231–236
Inclination, angle of, 189
Inclined plane, 213
Index of refraction, 146
Initial angle, 5
Initial side, 5, 7
Inner product, 220
Inscribed circle, 349
Inscribed polygon, 349
Intercept an arc, 349
Interior angles, 347
Internal-combustion engine, 44–46
Interpolation, 131
Intersection of polar curves
  analytic solution, 279–281
  graphical solution 275–279
  procedure for finding, 277
An Introduction to Mathematics
    (Whitehead), 290
Introductory Oceanography
    (Thurman), 171
Inverse cosecant function, 153
  domain, 158
  range 158
Inverse cosecant relation, 136
Inverse cosine function, 153
  domain, 157
  range, 157
Inverse cosine relation, 136
Inverse cotangent function, 153
  domain, 158
  range, 158
Inverse cotangent relation, 136
Inverse secant function, 153
  domain, 158
  range, 158
Inverse secant relation, 136
Inverse sine function, 153
  domain, 158

Inverse sine function (*continued*)
  range, 158
Inverse sine relation, 136
Inverse tangent function, 153
  domain, 158
  range, 158
Inverse tangent relation, 136
Inverse trigonometric functions, 152–160
  definition, 153
  evaluating, 154
  graphs of, 157–158
  vs. inverse relations, 152
  principal values of, 152–153
Inverse trigonometric identities, 155–156
Inverse trigonometric relations, 136
  definition, 136
  evaluating, 136
  graphs of, 158
Involute of a circle, 275
Irrational number(s), 6
Isosceles triangle, 347, 348

J

j vector, 217–219
Juneau, AL, 133

K

Kepler's third law, 334, 335
Khufu (pyramid), 187
Kondratieff, Nikolai, 49

L

Lambda (λ), 4
Lang, Les, 209
Langley Research Center
    (NASA), 216
Latitude, 15, 132–133
Law of Cosines, 170–171, 189–195, 200
  alternate forms, 190
  historical origin, 197
  summary, 200
Law of Sines, 179, 195–204
  summary, 200
Law of Tangents, 171, 179, 203–204
Lawrence Livermore Laboratory, 333
Leaning Tower of Pisa, 202
Leaves of a rose curve, 268
Le Havre, France, 203

Lemniscate, 264–269
Leverrier, Urbain, 333
Lightthill, Michael J., 3
Limaçon, 262–263
Line, identification of, 313
Linear equation, 292
Linear interpolation, 131–133
Linear velocity, 42–47
  of bicycle wheel, 43
  problems for study, 47
Logarithmic spiral, 263
London, England, 259
Longitude, 15
Longitudinal wave, 167
  graph of, 168
Los Angeles, CA, 226, 259

M

Mach, Ernst, 134
Mach number, 122–123
Magnitude, 211, 218
  definition, 219
Major axis, 298
Malibu, CA, 202
Malone, Jim, 66
Mars, 333–335
  Deimos (satellite), 335
  Phobos (satellite), 335
In Mathematical Circles (Eves), 8, 27,
    48, 134
Mathematical Syntaxis (Ptolemy), 2
"Mathematics Discovery...Hypocycloids
    and Epicycloids" (Gordon), 273
McMahon, Francis H., 333
Measurement(s), 3, 13
  accuracy of, 6, 11, 342–344
  accuracy of sides and angles, 181
  angles, agreement in this book, 9
  calculators and, 13
  in degrees, 4–8, 9
  degree-radian relationship, 9, 13
  on radians, 8–17
Mendelssohn, Kurt 187, 209
Menelaus, 2, 92
Mensuration formulas (geometry), 346
Mercury, orbit of, 334, 335
Mesopotamians, 2
Mexico City, Mexico, 259
Miami, FL, 259

Minor axis, 298
Minutes of degree, 6
Modulation, 77
Modulus, 237
"Moire Fringes and the Conic Sections"
    (Cullen), 291
Moire patterns, 291
Mollweide's formula, 204
Moon curves, 78
Mu (μ), 4
Music problems, 88

N

NASA, 216
Navigation problems, 123, 186, 187,
    202, 203, 215
Negative angle, 5, 7
Neptune, orbit of, 333, 335
New degree, 32
New Orleans, LA, 196
New York City, 16
Newton's formula, 204
Newton's method, 139
Nonagon, 346
nth Root, principal, 244
nth Root theorem, 244
The Number System of Algebra
    (Fine), 230
Number(s)
    complex, 231–246, 251, 253
    irrational, 6
    Mach, 122–123
    real, 8–9, 32–37, 253

O

Objectives
    Chapter 1, 38
    Chapter 2, 79
    Chapter 3, 127
    Chapter 4, 164
    Chapter 5, 223
    Chapter 6, 247
    Chapter 7, 282
    Chapter 8, 330
Oblique triangle, 189–204
    Law of Cosines, 189–197
    Law of Sines, 197–204

Oblique triangle (*continued*)
    summary for solving, 200
Obtuse angle, 5
Obtuse triangle, 348
Oceanography problems, 78
Octagon, 346
Omaha, NE, 17
Omega, 4
One revolution, 9
Opposite-angle identities, 108–111
Opposite side, 180
Opus palatinum de triangulis
    (Rheticus), 2
Ordinates
    addition of, 74–77
    graphing by, 76
    multiplying, 76
Orthogonal vectors, 221–222

P

Page, Warren, 291
"Pappus's Extension of the Pythagorean
    Theorem" (Eves), 197
Parabola, 291–298
    axis, 293, 296
    compared with other conic sections, 314
    definition, 314
    degenerate, 312
    directrix, 292–294, 324
    eccentricity, 314
    equation, 296
    finding graph when given equation,
        294
    finding equation when given graph,
        295
    graph paper for definition, 293
    focal chord, 293, 294
    focus, 292
    geometric definition, 292
    graphing, 293–297
    graph paper, 293
    identification, 313
    intersection with cone, 292
    polar-form, 324–325, 328
    recognition, 313
    rotated, 322
    standard-form, 293
    standard-form equations

Parabola (*continued*)
    with vertex at (0, 0), 294
    with vertex at (h, k), 296
    translated, 296–297
    vertex, 293
Parabolic archway, 317
Parabolic reflector, 298
Parallel lines (geometry), 347
Parallelogram, 211–212, 349
Parameter(s), 270
    eliminating, 272
Parametric equations, 269–274
    graphing, 269–274
Paris Observatory, 259
Parking-lot design, 228
Parts of a triangle, 180
Pentagon, 188, 346
Perihelion, 306, 333
Period(s), 51–56, 62. *See also* particular
    curves
    computer use for, 83–84
Perpendicular lines, 221
Phase displacement (shift), 60. *See also*
    Translations
    computer use and, 84–85
Phi (ϕ or φ), 4
Phobos (satellite), 335
Physics problems, 16, 47, 78–79,
    173–174, 215, 216
Pi (π), 4, 10
Pisa, Tower of, 203
Piston, position of, 45–47
Plane, inclined, 213
Planet X, 333, 335
Planetary orbits, 306, 333–335
    aphelion, 306, 333–334
    eccentricity, 334
    Kepler's third law and, 334
    perihelion, 306, 333–334
Plotting points in the polar system, 254
Plus/minus symbol ($\pm$) in trigonometry,
    119
Pluto, orbit of, 333, 335
Poincaré, Henri, 230
Polar axis, 254
Polar coordinates, 253–257, 261
    communication of bees, 285–286
Polar-form curve(s), 253–286
    analytic solution of, 279–281

Polar-form curve(s) (*continued*)
  cardioid, 261–263
  conic sections, 324, 328
  conversion to/from rectangular form, 256
  graphing, 259–281
  intersection of, 275–281
  lemniscate, 264–269
  limaçon, 262–263
  rose curve, 263–266, 269
  rotation of, 261–263
  spiral, 263
Polar-form points, 254–258, 260
  plotting, 254–255
  primary representations of, 255–256
Polar graph paper, 260
Pole, 254
Polygon, circumscribed and inscribed, 349
Popular Science puzzle, 204
Positive angle, 5, 7
Precision, 181
Primary angle, 137
Primary representation of a point, 255
Primitive quadruple, 197
Principal *n*th root, 245
Principal values, 154–153
  definition, 153
  of sine and cosine functions, 157–158
Product identities, 124–125
Product of ordinates, 76
Proportion, 349
Protractor
  for degree measure, 5
  for radian measure, 9–10
Proving identities
  changing sines to cosines, 97
  combining fractions, 97
  comparing methods for, 96–97
  by fractions to sums, 98
  multiplying by conjugate, 99
Pseudovertex, 308
Ptolemy, Claudius, 2, 92
Ptolemy's theorem, 92
Puget Sound, WA, 173
Pulley, finding angular velocity of, 44
Pure imaginary number, 232
Pyramids, angles of elevation, 187
Pythagorean identities, 20–21, 94

Pythagorean theorem, 20, 179, 190, 348
Pythagorean triplets, 27, 197

Q

Quadrant, 7
Quadrantal angle, 7
Quadratic equations, 292
Quadratic formula, 142–143, 201
Quadrilateral, 346

R

Radar antenna, 298
Radian, 9
Radian measure, 9
  conversion to degree, 11–13
  protractor, 9–10
  real numbers and, 8–9
Radio waves, 77, 167
Radius, 349
Rainbow Bridge, UT, 225
Range of trigonometric functions, 73. *See also* particular trigonometric functions
Rappaport, Karen, 252
Rate of travel/time problems, 199. *See also* Distance problems
Ratio identities, 94
Ratio(s), trigonometric, 180–181, 349
Real axis, 236
Real-number definitions of trigonometric functions, 32
Real numbers, 8–9, 32–37, 253
  radian measure and, 8–9
  trigonometric functions and, 32–37
Reciprocal identities, 19, 94
Rectangle, 346
Rectangular components, 211
Rectangular form (complex number), 237
  conversion to polar form, 256
Reduction identity, 161–164
Reduction principle, 28–29
  exact values using, 36
  statement of, 30
Reference angle(s), 28–29
Refraction, index of, 146
Refraction problems, 146
Regiomontanus, 92
Resolving a vector, 211

Resonance, 172–173
Resultant vector, 211
Revolution(s), 5, 8. *See also* Circle
Rheticus, 2, 92
The Riddle of the Pyramids (Mendelssohn), 187, 209
Right angle, 5, 347
Right triangle, 180, 347–348
Right triangles, solving, 180–187
Right-triangle definition of trigonometric functions, 180
Right-circular cone, 346
Right-circular cylinder, 346
Roots, complex, 244–246
  algebraic representation, 244
  graphical representation, 246
Rose curve, 263–266
  leaves of, 268
Rotation
  axes, 318–321
  conic sections in polar form, 328
  formula for conic sections, 318
  formula for angle of, 319
  of polar-form graphs, 262
  procedure for sketching, 321
Rounding procedure for calculations, 182, 342–343
RPN logic, 339

S

Samala River, Guatemala, 215
San Francisco, CA, 202, 226
Santa Rosa, CA, 131, 133, 226
SAS (triangle), 189, 192, 205, 348
  summary, 200
Satellite orbit, 152
Satellite position, 68
Satisfying a polar-form equation, 258
Scalar multiplication, 216
Scalar product, 220
Scalene triangle, 348
Scott, Charlotte Angas, 252
sec *x*. *See* Secant function
Secant
  to a circle, 349
  origin of word, 48
Secant function, 18, 48
  domain, 70

Secant function (*continued*)
   evaluation, 22, 25
   exact values, 35
   general form, 72
   generalized definition, 27
   graphing, 70, 72
   period, 70
   range, 70, 73
   real-number definition, 32
   right-triangle definition, 180
   unit-circle definition, 18
   values not defined, 18
Seconds (of a degree), 6
Second-degree equation, 292
Sector, 206
   area of, 206–207
Semielliptic arch, 317
Semimajor axis, 302
Semiminor axis, 302
Sesostris, 178
Seward, AL, 133
Shock waves, 3–4
Sides of angles, 4
   initial, 5
   standard-position, 7
   terminal, 5, 7, 28
Sides of triangles, 180, 347
   accuracy in measuring, 181
   adjacent, 180
   hypotenuse, 180
   opposite, 180
Significant digits, 181, 342–344
   rules for, 343
Signs of trigonometric functions, 23–24
Similar triangles, 27, 192, 349
Simple harmonic motion, 170
sin *x*. *See* Sine function
Sine, origin of word, 48
Sine function, 18, 48
   domain, 51
   evaluation, 22, 25
   exact values, 35
   framing, 51
   general form, 64
   generalized definition, 27
   graph of, 50
   introduction, 18
   period, 51, 63
   principle values of, 158

Sine function (*continued*)
   real-number definition, 32
   right-triangle definition, 180
   series expansion, 37
   unit-circle definition, 18
Sines, Law of, 109, 200
Skywalker, Luke, 202
Slope of a line, 189
Smith, Arthur F., 187, 209
Solving triangles, 179–180, 185–187, 200
   ambiguous case, 194
   calculator for, 179
   right, 180
   summary, 200
Solzhenitzyn, Alexsandr, 49
Sound waves, 74, 152
   graphing, 86–87
Space science problems, 47, 68, 152, 216
Sphere, 346
Spiral, hyperbolic, 263
Spiral, logarithmic, 263
Spiral of Archimedes, 263, 263
Square, 346
SSA (triangle), 189, 192–195, 200
   summary, 200
"SSA and the Law of Cosines" (Cohen, Goodman, Benard), 197
SSS (triangle), 189, 190–191, 206, 348
   summary, 200
Standard position angle, 7
Stanley Falls, Zaire, 17
Star Wars (movie), 202
Stirling's formula, 230
Straight angle, 5, 10
Strophoid, 269
Study hints, 39
Subtended angle, 5
Subtraction, vector, 216
Sum identities, 124
Sum of ordinates, 74
Sum of vectors, 211–212
Summaries,
   Chapter 1, 38
   Chapter 2, 79
   Chapter 3, 127
   Chapter 4, 164
   Chapter 5, 223
   Chapter 6, 247

Summaries (*continued*)
   Chapter 7, 282
   Chapter 8, 330
Sun curves, 78
Sunrise and sunset, 131
   table of times, 132
Supplementary angles, 346
Surveying problems, 17, 178, 186–188, 196, 202, 203, 226
Surveying, use of transit/vernier in, 185
Syene, Egypt, 17
Syntaxis mathematica (Ptolemy), 92

T

Tacoma Narrows Bridge, collapse of, 173
Tampa, FL, 133
tan *x*. *See* Tangent function
Tangent
   to a circle, 349
   origin of word, 48
Tangent function, 18, 48, 54–56
   domain, 55
   evaluation, 22, 25
   exact values, 35
   framing, 55
   general form, 64
   generalized definition, 27
   graph of, 55
   period, 55, 62
   range, 55
   real-number definition, 32
   right-triangle definition, 180
   unit-circle definition, 18
   values not defined, 18
Tangents, Law of, 191, 203
Terminal side of an angle, 5, 7
Terms
   Chapter 1, 39
   Chapter 2, 80
   Chapter 3, 128
   Chapter 4, 165
   Chapter 5, 224
   Chapter 6, 248
   Chapter 7, 282
   Chapter 8, 331
The Teaching of Mathematics (Young), 2
Theta (θ), 4

"Thirty-Seven-Year-Old Puzzle," 204
30°–60°–90° triangle, 348
Tide curves, 78, 171
Tide-generating force, 171
Tower of Pisa, 202
Traffic flow, 3–4
Transit, 185
Translations, 60–62
  conic section, 315
  parabola, 296
  using a computer, 85
Transversal, 347
Triangle(s), 27, 179–180, 346–349
  acute, 347
  altitude (height), 347
  area of, 205
  base, 348
  classification by angles, 347–348
  classification by sides, 347–348
  congruent, 348
  correctly labeled, 180
  equilateral, 347, 348
  inequality, 241
  isosceles, 347
  oblique, 179, 191
  obtuse, 347
  properties of, 348
  right, 180–187, 347, 348
  scalene, 347
  sides of, 180, 347–348
  similar, 27, 192, 349
  six parts of, 180
  solving, 179–180, 185–187, 200
  vector, 209–216
  vertex, 347
Trigonometric equations, 135–146
  algebraic solution, 139
  calculator solution, 138
  exact value solution, 137
  extraneous roots, 144
  factoring and, 141, 143
  general form, 141
  identities and, 135, 143, 149
  multiple angles and, 139, 147–152
  Newton's method for solving, 135
  procedure for solving, 140
  quadratic formula and, 142–143

Trigonometric equations (*continued*)
  solving, 140
Trigonometric functions, 3–4, 18, 50. *See*
    also particular trigonometric
    functions
  domains of, 73
  evaluation of, 22–25, 28
  generalized definition of, 27
  graphing, 49–58, 69–75
  inverse, 152–168
  ranges of, 73
  real number definition, 32
  right-triangle definition, 180
  signs of, 23
  unit-circle definition, 18
Trigonometric identities. *See* Fundamen-
    tal identities, Identities
Trigonometric ratios. 180–181. *See also*
    Trigonometric functions
  right triangle definition of, 180–181
Trigonometric tables, 22, 92, 314
Trigonometry, origin of word, 2,
Trigonometry and Double Algebra
    (De Morgan), 48
True course, 210
"Two American Women Mathe-
    maticians" (Rappaport), 253
Two Arms Are Better Than One
    (Sinnott), 227

U

Ulfborg, Denmark, 47
Undecagon, 346
Unit circle, 17
  chord length on, 106–107
Unit circle definition of trigonometric
    functions, 18
Urania Propitia (Cunitz), 2
Uranus, orbit of, 333, 335

V

Vector(s)
  addition of, 211
  algebraic representation, 217–218
  angle between, 220–221

Vectors (*continued*)
  components, 211
  definition of, 210–211
  direction, 211
  i and **j**, 217–219
  magnitude of, 210–211, 218–219
  operations with, 216–222
  orthogonal, 221–222
  product, 220
  resolving, 211, 212–213
  resultant, 211
  scalar multiplication of, 216
  sum of, 211–212
  triangles, 209–216
Velocity, angular/linear, 42–47
Venus, orbit of, 334, 335
Vernier, 185

Vertex, 4–5
  of ellipse, 300
  of hyperbola, 307
  of parabola, 293
  of triangle, 347
Vibrating string, 87
Viète, François, 191
The VNR Concise Ency. of Mathematics
    (Gellert et al.), 179
Voltage, 251
Volume, 205, 207–209
  of a cone, 207–208
  slice cut from a cylinder, 209
Von Peurback, Georg, 92

W

Walker, Jearl, 3, 204
WARNINGS
  agreement about accuracy, 181, 344
  agreement on rounding, 24, 343, 182
  center of frame for tangent curve, 56
  dividing by sine, 141
  factoring, 66
  framing a curve, 52, 61
  frequently used angles, 3
  inverse distinguished form reciprocal,
    136
  period of standard cosine curve, 53

WARNINGS (*continued*)
  period of standard sine curve, 51
  period of standard tangent curve, 55
  plus/minus symbol ($\pm$), 119
  proving identities, 95
  simplifying polar-form representations, 255
  single value for inverse functions, 154
  use of cos, sin, and tan, 18
  use of degree and radian measure, 9, 138
Washington, D.C., 178

Wave equations, 168–173
Wave motion
  bow, 123
  destructive, 78
  electromagnetic, 167
  equations for, 168–173

  harmonic, 170–172
  longitudinal, 167
  radio, 77, 167
  sound, 79, 87, 167
  tidal, 78

Wave motion (*continued*)
  transverse, 167–168
  water, 123, 167, 168
Whitehead, Alfred North, 252, 290
Whitham, Gerald B., 3
Wind generator, 47
Windmill, 47
Wichita, KS, 17
Winnipeg, Canada, 133

Z

Zunil , Guatemala, 215

SOLVING TRIANGLES

RIGHT TRIANGLES

Pythagorean Theorem:

$$c^2 = a^2 + b^2$$

Angle Theorem:

The sum of the measures of the angles in any triangle is 180°.

Solve by using the trigonometric ratios in a right triangle.

Given	Conditions on the given information	Method of solution
1. AAA		No solution
	More than one side given:	**Law of Cosines**
2. SSS	**a.** The sum of the lengths of the two smaller sides is less than or equal to the length of the larger side.	No solution
	b. The sum of the lengths of the two smaller sides is greater than the length of the larger side.	Law of Cosines
3. SAS	**a.** The given angle is greater than or equal to 180°.	No solution
	b. The given angle is less than 180°.	Law of Cosines
4. SSA	**a.** The given angle is greater than or equal to 180°.	No solution
	b. There are no solutions, one solution, or two solutions as determined by the quadratic formula.	Law of Cosines
	More than one angle given:	**Law of Sines**
5. ASA or AAS	**a.** The sum of the given angles is greater than or equal to 180°.	No solution
	b. The sum of the given angles is less than 180°.	Law of Sines

LAW OF COSINES

$$a^2 = b^2 + c^2 - 2bc \cos \alpha \qquad b^2 = a^2 + c^2 - 2ac \cos \beta \qquad c^2 = a^2 + b^2 - 2ab \cos \gamma$$

$$\cos \alpha = \frac{b^2 + c^2 - a^2}{2bc} \qquad \cos \beta = \frac{a^2 + c^2 - b^2}{2ac} \qquad \cos \gamma = \frac{a^2 + b^2 - c^2}{2ab}$$

LAW OF SINES

$$\frac{\sin \alpha}{a} = \frac{\sin \beta}{b} = \frac{\sin \gamma}{c}$$

TRIGONOMETRIC GRAPHS